ELECTRONIC DESIGN DATA BOOK

Rudolf F. Graf

Senior Member of the Institute of Electrical
and Electronics Engineers

VNR Van Nostrand Reinhold Company
New York / Cincinnati / Toronto / London / Melbourne

To My Mother and Father

Van Nostrand Reinhold Company Regional Offices:
New York Cincinnati Chicago Millbrae Dallas

Van Nostrand Reinhold Company International Offices:
London Toronto Melbourne

Copyright © 1971 by Litton Educational Publishing Inc.

Library of Congress Catalog Card Number: 74-122669

Manufactured in the United States of America

Published by Van Nostrand Reinhold Company
450 West 33rd Street, New York, N.Y. 10001

Published simultaneously in Canada by
Van Nostrand Reinhold Ltd.

15 14 13 12 11 10 9 8 7 6 5 4 3 2 1

PREFACE

The word *knowledge* brings to mind the staggering body of facts and data accumulated by mankind since his descent from the trees. Once, thousands of years ago, it was possible for a man to know all that his kind had discovered. But, time has added so greatly to our reservoir of wisdom, that *knowledge*, today, has assumed another meaning: knowing where to find the information needed.

This book humbly admits to being the author's attempt at simplifying the task of the busy engineer, technician, amateur, and student in locating the data he needs in the shortest possible time.

Gathered here, in one single volume, is a wealth of information in the form of timely and practical nomograms, tables, charts, and formulas.

Some of the material was available elsewhere, at some time or other, but never has all of it been gathered together under one cover. New and heretofore unpublished charts and nomograms are added because of what seemed to the author an obvious need for such material.

The book is arranged in a most readily usable format. It contains only clear-cut, theory-free data and examples that are concise, accurate, and to the point. The user of this book will be looking for answers and he will find them, without having to fight his way through lengthy derivations and proofs.

The preparation of a reference book such as this is not possible without the cooperation and assistance of numerous industry sources who have so generously made their material available. The author gratefully acknowledges, with special thanks, the contributions and critical efforts of Messrs. George J. Whalen, Arthur E. Fury, René Colen, and B. William Dudley, Jr.

If this book saves many hours of tedious computations and search for information, it will indeed have served its intended purpose.

New Rochelle, New York   *Rudolf F. Graf*

The author and publisher invite your comments and suggestions regarding any such other material as might have been included here, so that it may be considered for any subsequent edition or revision.

iii

HOW TO USE THIS BOOK

In order to assist the reader in finding the data he seeks, the book has been divided into six functional sections. That organization, together with a comprehensive index, quickly leads to the specific information needed.

1 Frequency Data:

This section first presents the entire electromagnetic spectrum. Then it selects portions of this spectrum which are of particular interest to the electrical and electronic engineer and describes them in greater detail.

2 Communication:

This section presents information useful in all segments of communication, starting with propagation characteristics, modes, standards, and transmission data. Antenna, transmission line, and waveguide characteristics and performance data are presented. Modulation and international telecommunications standards, signals, signal reporting codes, radio amateur data, and emission information are also given. Information on microphones is included to complete this section.

3 Passive Components and Circuits:

Covered in this section are resistors, attenuators, filters, inductors, transformers, and capacitors. Their characteristics and applications are treated in depth.

4 Active Components and Circuits:

Vacuum tubes, semiconductors, and integrated circuits are covered. Circuit configurations are given in which these components are employed. Definitions of integrated circuit terms and a tabulation that shows the characteristics of integrated circuit logic families currently in use, conclude this section.

5 Mathematical Data, Formulas, Symbols:

This section covers reliability, mathematical tables, charts and formulas, prefixes, geometric curves, solids, spherical as well as plane geometry, and trigonometry. Frequency, phase angle, and time relationships for recurrent wave forms are given. Power and voltage level determinations in signals circuits are explained. Letter symbols for all quantities encountered in the electronics, electrical field are defined. This section concludes with a comprehensive selection of conversion factors.

6 Physical Data:

This section covers the most often needed physical data and includes, among other items, laser radiation, radioactivity, optical data, sound, incandescent lamps, cathode ray tubes, crystals, color codes, military nomenclature, atmospheric and space data, chemical data, plastics, temperature and humidity tables, wire data, hardware, shock and vibration, cooling data, and characteristics of materials.

The book maintains uniform terminology and format which assures that data found in one section can be easily and accurately related to those in the rest of the book.

ACKNOWLEDGEMENTS

Acknowledgment is made to the following organizations and publications who have permitted use of material originally published by them. The author appreciates their cooperation during the preparation of this book.

Alpha Metals, Inc.: page 304.

The American Radio Relay League: pages 48-51, 54, 61 (all from "The Radio Amateur's Operating Manual," © 1969).

Automatic Electric Company: pages 188-189, 190, 292 (all from "Tables and Formulae").

Centralab Division of Globe-Union, Inc.: page 89.

Clairex Electronics, Inc. (and J. R. Rabinowitz): page 249.

Computers & Data Processing News: page 181.

Conrad, Inc.: page 284.

Design News: pages 194-195 (Nov. 1963); 198 (March 1967); 200 (Feb. 1959); 233 (March 1958); 286 (Sept. 1959); 299 (June 1967).

EDN: pages 23 (Nov. 1968); 40 (Sept. 1963); 43 (June 1964); 44 (Nov. 1968); 63 (Nov. 1968); 71 (Nov. 1968); 90 (May 1967); 92 (Sept. 1966); 97 (Nov. 1965); 98 (Apr. 1959); 103 (Jan. 1962); 106, 107, 108, 109, 110, 111 (Dec. 1961); 115, 116, 117, 118, 119, 120 (June 1958); 126 (Oct. 1966); 129 (Nov. 1961); 153 (Sept. 1962); 180 (Nov. 1963); 186 (July 1959); 204 (May 1968); 206 (Dec. 1966); 231, 232 (Oct. 1960); 252 (Nov. 1962); 253 (Nov. 1962); 290 (May 1963).

Electric Hotpack Company, Inc.: page 285.

Electronic Design: pages 79 (July 1956); 94, 95 (March 1959); 245 (Sept. 1966); 263, 264, 265, 266 (May 1966).

The Electronic Engineer: pages 26 (Nov. 1956); 27, 28, 29 (Jan. 1968); 30 (Nov. 1967); 31 (Oct. 1963); 39 (June 1961); 99, 100, 101 (Jan. 1948).

Electronic Equipment Engineering: pages 20 (July 1958); 124 (Aug. 1963).

Electronic Industries: page 128 (May 1966). Electronic Industries Association: pages 136-142 (from "EIA-NEMA Standards," © 1966 by Electronic Industries Association and the National Electrical Manufacturers Association).

Electronic Tube and Instrument Division of General Atronics Corporation: page 262.

Electronics World: pages 35 (June 1965); 77 (Dec. 1964); 91 (Dec. 1964); 171 (1959); 173 (Sept. 1963); 202-203 (1962); 212, 213 (July 1961); 275 (Nov. 1969).

The Garrett Corporation: pages 272, 273.

General Electric Company: pages 134, 135.

General Radio Company: pages 37, 38.

Hudson Lamp Company: page 259 (from Hudson Lamp Company Catalog).

Industrial Research, Inc.: pages 125 (March 1959); 182-183 (Apr. 1960) (all from *Electro-Technology*).

Instrument Systems Corporation: page 295.

ACKNOWLEDGEMENTS

Kepco, Inc.: page 291.

Lenkurt Electric Co., Inc.: page 22 (from Lenkurt Demodulator "Carrier and Microwave Dictionary").

Machine Design: page 247 (July 1970).

P. R. Mallory & Co., Inc.: pages 74, 250 ("Minimum Detail that the Human Eye Can Resolve") (both from "MYE Technical Manual," © 1942).

Martin Marietta Corporation: page 244.

Measurements and Data: page 283.

Microwave Journal: page 175 (Oct. 1964).

Parker Seal Company: page 199.

Popular Electronics: pages 8, 9 (from "Communications Handbook," © 1969); page 143.

PRD Electronics, Inc.: page 41.

Radio Engineering Laboratories, Inc.: page 17.

Reynolds Metals Company: pages 196–197, 201 (all from "Facts and Formulas," © 1961).

Howard W. Sams & Co., Inc.: page 104 (from "Audio Cyclopedia," © 1969); pages 145, 158–159, 294 (from "Reference Data for Radio Engineers," 5th Edition, © 1968).

Smithsonian Institution Press: page 257 (from "Smithsonian Physical Tables").

Sprague Electric Company: page 84.

Sylvania Electronic Products, Inc.: pages 160–165.

Testing Machines Inc.: pages 234–241.

Vibrac Corporation: page 297.

CONTENTS

ELECTRONIC
DESIGN
DATA
BOOK

SECTION 1

FREQUENCY DATA

THE ELECTROMAGNETIC SPECTRUM

This chart presents an overview of the complete electromagnetic radiation spectrum, extending from infrasonics to cosmic rays. The wavelength, the amount of energy required to radiate one photon, a general description, the band designation, and, the normal occurrence or use are given. Some specific bands are described in more detail on the following pages.

FREQUENCY (hertz)	WAVELENGTH (meters) / (other units)	QUANTUM OF ENERGY (ev) / (other units)	GENERAL DESCRIPTION	BAND DESIGNATION AND NUMBER	GENERAL OCCURRENCE AND APPLICATION
10	10^8	10^{-14}	ELECTRIC WAVES	ELF — 1	Infrasonics; Power Transmission
10^2	10^7	10^{-13}		SLF — 2	Sonics
10^3 (1kHz)	10^6 (1km)	10^{-12}		ULF — 3	
10^4	10^5	10^{-11}		VLF — 4	Ultrasonics
10^5	10^4	10^{-10}		LF — 5	
10^6 (1MHz)	10^3	10^{-9}	RADIO WAVES	MF — 6	AM Radio Broadcasting
10^7	10^2	10^{-8}		HF — 7	TV, FM
10^8	10	10^{-7}		VHF — 8	Radar
10^9 (1GHz)	1	10^{-6}		UHF — 9	Microwaves
10^{10}	10^{-1} (1dm)	10^{-5}		SHF — 10	
10^{11}	10^{-2} (1cm)	10^{-4}		EHF — 11	
10^{12} (1THz)	10^{-3} (1mm)	10^{-3}	LIGHT WAVES	Infrared — 12	
10^{13}	10^{-4}	10^{-2}		— 13	
10^{14}	10^{-5}	10^{-1}		— 14	
10^{15}	10^{-6} (1u)	1		Visible — 15	Optics
10^{16}	10^{-7}	10		Ultra Violet — 16	
10^{17}	10^{-8} (100Å)	10^2	X-RAY WAVES	— 17	
10^{18}	10^{-9}	10^3 (1kev)		(Soft) — 18	Gamma Rays; Laser
10^{19}	10^{-10} (1Å)	10^4		(Hard) — 19	
10^{20}	10^{-11}	10^5		— 20	
10^{21}	10^{-12}	10^6 (1Mev)	COSMIC RAYS	— 21	Cosmic Rays; Particle Accelerators
10^{22}	10^{-13} (1xu)	10^7		— 22	
10^{23}	10^{-14}	10^8		— 23	
	10^{-15}	10^9 (1Gev)			

WAVELENGTH BANDS AND FREQUENCY USED IN RADIO-COMMUNICATION

Nomenclature of the frequency and wavelength bands used in radiocommunication in accordance with Article 2, No. 12 of the "Radio Regulations," Geneva, 1959.

Band Number	Frequency Range (lower limit exclusive, upper limit inclusive)	Corresponding Metric Subdivision	Adjectival Band Designation
4	3- 30 kc/s (kHz)	Myriametric waves	VLF
5	30- 300 kc/s (kHz)	Kilometric waves	LF
6	300-3000 kc/s (kHz)	Hectometric waves	MF
7	3- 30 Mc/s (MHz)	Decametric waves	HF
8	30- 300 Mc/s (MHz)	Metric waves	VHF
9	300-3000 Mc/s (MHz)	Decimetric waves	UHF
10	3- 30 Gc/s (GHz)	Centimetric waves	SHF
11	30- 300 Gc/s (GHz)	Millimetric waves	EHF
12	300-3000 Gc/s (GHz) or 3 Tc/s (THz)	Decimillimetric waves	

BROADCASTING FREQUENCY ASSIGNMENTS

This table shows the frequency range, number of available channels, and channel width for AM, FM, and TV service in the United States.

Type of Service	Frequency Range	Number of Available Channels	Width of Each Channel
AM radio	535–1605 kHz	107	10 kHz
FM radio	88– 108 MHz	100	200 kHz
VHF television	54– 88 MHz 174– 216 MHz	12	6 MHz
UHF television	470– 890 MHz	70	6 MHz

WORLD AIR ROUTE AREA	FREQUENCY ALLOCATION (kHz)						
Alaska	2945	3411.5	4668.5	5611.5	6567		11,328
Hawaii		3453.5		5559	6649.5		
West Indies	2861		4689.5					
Central East Pacific		3432.5 3446.5 3467.5 3481.5		5551.5 5604	6612 6679.5	8879.5 8930.5	10,048 10,084 11,299.5 11,318.5	13,304.5 13,334.5 17,926.5
Central West Pacific	2966			5506.5 5536.5		8862.5		13,354.5 17,906.5
North Pacific	2987			5521.5		8939		13,274.5 17,906.5
South Pacific	2945			5641.5		8845.5		13,344.5 17,946.5
North Atlantic	2868 2931 2945 2987			5611.5 5626.5 5641.5 5671.5		8862.5 8888 8913.5 8947.5		13,264.5 13,284.5 13,324.5 13,354.5 17,966.5
Europe	2889 2910	3467.5 3481.5	4654.5 4689.5	5551.5	6552 6582	8871 8930.5	11,299.5	17,906.5
North-South America	2889 2910 2966	3404.5	4696.5	5566.5 5581.5	6567 6664.5	8820 8845.5 8871	11,290 11,337.5	13,314.5 13,344.5 17,916.5
Far East	2868 2987			5611.5 5671.5		8871 8879.5 8930.5		13,284.5 13,324.5 17,966.5
South Atlantic	2875	3432.5			6597 6612 6679.5	8879.5 8939	10,048	13,274.5 17,946.5
Middle East		3404.5 3446.5		5604	6627	8845.5	10,021	13,334.5 17,926.5
North-South Africa	2966	3411.5		5506.5 5521.5		8820 8956		13,304.5 13,334.5 17,926.5 17,946.5
Caribbean	2875 2952 2966			5499 5566.5 5619	6537	8837 8871	10,021	13,294.5 13,344.5 17,936.5
Canada	2973			5499		8871	11,356.5	

FREQUENCIES USED BY SHIP
AND SHORE STATIONS

Band (MHz)	SHIP STATIONS		SHORE STATIONS
	Calling Frequencies (kHz)	Working Frequencies (kHz)	(Approximate Limits)
2	2065 − 2107	Same as calling	2000 − 2065
4	4178 − 4186	4161 − 4176 4188 − 4236	4240 − 4400
6	6267 − 6279	6241 − 6264 6282 − 6355	6362 − 6523
8	8356 − 8372	8322 − 8352 8376 − 8473	8478 − 8742
12	12,534 − 12,558	12,474 − 12,528 12,564 − 12,709	12,714 − 13,128
16	16,712 − 16,744	16,626 − 16,704 16,752 − 16,946	16,950 − 17,285
22	22,225 − 22,265	22,151 − 22,217 22,272 − 22,395	22,400 − 22,670

COMMONLY USED LETTER-CODE DESIGNATIONS FOR MICROWAVE FREQUENCY BANDS

Band	Frequency	Wavelength	Typical Use
P	225– 390 MHz	133.3– 76.9 cm	Long range (over 200 miles) to very long range (beyond 1000 miles) surface-to-air search.
L	390–1550 MHz	76.9– 19.3 cm	Very long through medium range surface-to-air missile and aircraft detection, tracking and air traffic control, IFF transponders, beacon systems.
S	1.55– 5.2 GHz	19.3– 5.77 cm	Medium and long range surface-to-air surveillance, surface-based weather radar, altimetry, missile-borne guidance, airborne bomb-navigation systems.
C	3.9– 6.2 MHz	7.69– 4.84 cm	Airborne fire control, missile-borne beacons, recon, airborne weather avoidance, aircraft and missile target tracking.
X	5.2– 10.9 MHz	5.77– 2.75 cm	Doppler navigation, airborne fire control, airborne and surface-based weather detection, bomb-navigation systems, missile-borne guidance, precision landing approach.
K	10.9– 36 GHz	2.75–0.834 cm	Doppler navigation, automatic landing systems, airborne fire control, radar fuzing, recon, missile-borne guidance.
Q	36– 46 GHz	0.834–0.652 cm	Recon, airport surface detection.
V	46– 56 GHz	0.652–0.536 cm	High-resolution experimental short-range systems.

NBS STANDARD FREQUENCY AND TIME BROADCAST SCHEDULES

The diagrams presented here, with explanatory notes, summarize the standard frequency and time services provided by the National Bureau of Standards radio stations WWV, WWVH, WWVB, and WWVL.

Hourly Schedules of WWVB and WWVL

WWVB (Carrier: 60 kHz): Station is identified by 45° advance in carrier phase starting at 10 min after each hour and returning to normal at 15 min after each hour. Time is disseminated continuously by amplitude modulation of carrier level with once-per-second pulses. Modulation consists of 10-dB drop in carrier level during a pulse. Width of pulses carries time information in special "WWVB" format. Each time frame, which includes UT2 corrections, lasts 1 min. Note: WWVB carrier frequency is now maintained without offset with respect to United States Frequency Standard.

WWVL (Carrier: 20 kHz): Program is experimental and subject to change. At present, WWVL broadcasts continuously and is identified in Morse code on 1st, 21st, and 41st minute of each hour by station call letters, repeated 3 times, followed by frequency offset. Code is produced by "on-off" carrier keying. WWVL carrier has -150 parts in 10^{10} offset with respect to United States Frequency Standard.

Time Pulse Adjustments

The National Bureau of Standards has announced that the clock which controls the 1-sec time pulses from standard broadcast station WWVB (60 kHz) was retarded by 200 msec on Oct. 1, 1965. This change is in keeping with the previously announced policy of maintaining the WWVB time pulses within 100 msec of the UT2 time scale (the period between the WWVB time pulses is 150 parts of 10^{10} shorter than the period between UT2 sec, a difference of 0.0013 sec/day). The WWVB time pulses have already been retarded 200 msec on each of three occasions during 1965, on Jan. 1, April 1, and July 1.

There were no changes in the phases of time pulses of NBS high-frequency stations WWV and WWVH on Oct. 1, 1965. Retardations of 100 msec each were made on Jan. 1, Mar. 1, July 1, and Sept. 1, 1965.

Hourly Broadcast Schedules of WWV and WWVH

1. Call letters and Universal time in Morse code; call letters and Eastern Standard time (WWV) or Hawaiian Standard time (WWVH) by voice.

2. Frequency offset: amount in parts in 10^{10} by which transmitted frequencies are offset from U. S. Frequency Standard; broadcast in fast Morse code immediately following voice announcement on the hour only. Symbols now transmitted are M150, meaning Minus 150 parts in 10^{10}.

3. Propagation forecast (WWV only): condition of ionosphere in North Atlantic area, at time of last issue, and radio quality expected in subsequent 6-hr period (forecasts issued at 0500, 1200—1100 in summer—1700 and 2300 UT). Letter portion identifies radio quality at time of forecast and numbered portion is forecast of expected quality on typical North Atlantic path during 6 hr period following forecast, according to following scale:

Disturbed grade (W)	Unsettled grade (U)	Normal grades (N)
1. useless	5. fair	6. fair-to-good
2. very poor		7. good
3. poor		8. very good
4. poor-to-fair		9. excellent

4. Seconds pulses are transmitted continuously except for 59th second of each minute and during silent periods. Each minute starts with two pulses spaced by 0.1 sec.

5. 100 pps timing code: binary-coded 1000-Hz tone gives Universal Time in seconds, minutes, hours, and day of year for use as unambiguous time base during data recording in scientific experiments. Format used is NASA 36-bit Time Code with 2-msec pulse for "0" and 6 msec pulse for "1." Each frame lasts 1 sec, repeated once per second during 1 min broadcast interval.

NBS STANDARD FREQUENCY AND TIME BROADCAST SCHEDULES (Continued)

6. Geoalerts: slow Morse code to identify days on which outstanding solar or geophysical events are expected or have occurred during previous 24 hr. Geoalert is identified by letters "GEO" followed by one of following letters repeated five times:

M—Magnetic storm   C—Cosmic ray event
N—Magnetic quiet   W—Stratospheric
S—Solar activity   warming
Q—Solar quiet   E—No geoalert
                        issued

7. UT2 time corrections: corrections to be applied to time signals for obtaining actual UT2 time within ±3 msec. Corrections, in Morse code, begin with "UT2" followed by either "AD" (for add) or "SU" (for subtract) in turn followed by number of milliseconds to be added or subtracted from time as broadcast.

HOURLY SCHEDULES OF WWVB AND WWVL

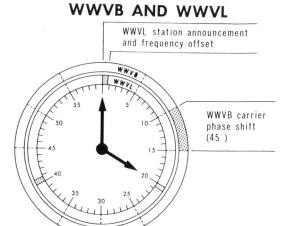

HOURLY BROADCAST SCHEDULES OF WWV AND WWVH

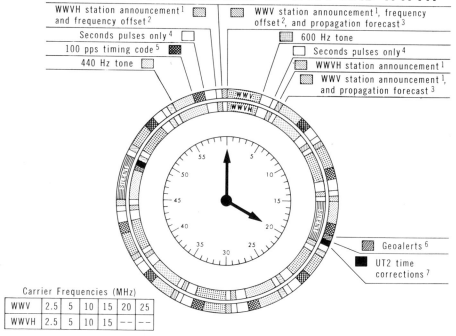

Carrier Frequencies (MHz)						
WWV	2.5	5	10	15	20	25
WWVH	2.5	5	10	15	--	--

WAVELENGTH-FREQUENCY CONVERSION SCALE

This scale is based on the formula

$$\lambda m = \frac{300}{f_{MHz}}$$

It shows the relationship between free space wavelength λ and frequency f and covers a frequency range extending from 300 Hz to 300 GHz, corresponding to wavelengths of 1000 m (1 km) to 1 mm.

FOR EXAMPLE:

A 60-MHz signal has a wavelength of 5 m. A signal whose wavelength is 3 mm has a frequency of 100 GHz.

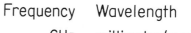

Frequency   Wavelength

GHz — millimeter (mm)

MHz — meter (m)

kHz — kilometer (km)

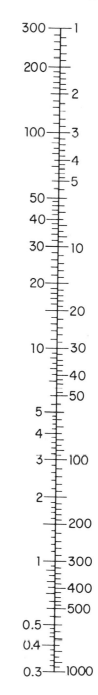

SECTION 2

COMMUNICATION

PROPAGATION CHARACTERISTICS
OF ELECTROMAGNETIC WAVES

Band	Frequency (Wavelength)	Characteristics	Applications
Very-low-frequency (VLF)	20–30 kHz (20,000–10,000 m)	Very stable; low attenuation at all times. Influenced by magnetic storms. Ground wave extends over long distances. (No fading out long-time variations occur.)	Continuously operating long-distance station-to-station communication service.
Low-frequency (LF)	30–300 kHz (10,000–1000 m)	Seasonal and daily variations greater than that of VLF; daytime absorption also greater, increasing with frequency. At night similar to VLF although slightly less reliable.	Long-distance station-to-station service (marine, navigational aids).
Medium-frequency (MF)	300–3000 kHz (1000–100 m)	Less reliable over longer distances than lower frequencies. Attenuation: low at night, high in daytime; greater in summer than in winter. Low attenuation at night is due to sky-wave reflection. Ground-wave attenuation is relatively high over land and low over salt water.	Commercial broadcasting, police, marine and airplane navigation.
High-frequency (HF)	3–30 MHz (100–10 m)	Dependent on ionospheric conditions, leading to considerable variation from day to night and from season to season. Attenuation low under favorable conditions, and high under unfavorable conditions, at medium to very long distances.	Medium and long-distance communication service of all types.
Very-high-frequency (VHF)	30–300 MHz (10–1 m)	30–60 MHz sometimes affected by ionosphere. Quasi-optical transmission (similar to light, but subject to diffraction by surface of the earth).	Television, FM commercial broadcasting, radar, airplane navigation, short-distance communications.
Ultra-high frequency (UHF)	300–3000 MHz (100–10 cm)	Substantially same as above; slightly less diffraction. Under abnormal conditions, can be refracted by troposphere similar to sky-wave refraction. This often results temporarily in abnormally long ranges of transmission.	Television, radar, microwave relay, short-distance communications.
Super-high-frequency (SHF)	3000–30,000 MHz (10–1 cm)	Same as above. 1-cm range has broad water-vapor absorption band (slight O_2 absorption).	Radar, microwave relay, short-distance communications.

COMMUNICATION MODES

Principal ground-to-ground communication modes, utilizing the microwave (70 MHz to 20 GHz) region of the spectrum. Characteristically wide-band (100 kHz to 20 MHz) service.

Mode	Range	Power/Antennas	Notes
LINE OF SIGHT (LOS)	0 to 35 miles, depending on (h).	0.1 to 10W, two to 10-ft antennas	Low-cost, high-performance wide-band system; replaces costly right-of-way maintenance of coaxial or multiple cable or overhead wiring.
LOS Space Communications	up to 1/2 circumference of earth depending on satellite orbit and (Θ)	1 to 15 kW, 30 to 85-ft antennas	Only practical system of global coverage using three active synchronous satellites (22,000 miles from earth) or a number of orbiting satellites (dependent on distance covered and altitude) in conjunction with multiple earth earth stations.
DIFFRACTION (Plane Surface)	30 to 70 miles, depending on (h) and N_S)	0.1 to 100W, six to 28-ft antennas	Diffraction mode is very specialized form of UHF used only rarely where rugged terrain prevents use of direct LOS and permits longer path with obstacle gain.
DIFFRACTION (Knife Edge)	30 to 120 miles, depending on (h), (N_S) and (G_O)	0.1 to 100W, six to 28-ft antennas	
DIFFRACTION (Rough Surface)	30 to 120 miles, depending on (h), (N_S), (G_O), and (A_O)	0.1 to 100W, six to 28-ft antennas	Great attention is being given to refining propagational computation in the diffraction region because of need for utilization in tropo path predictions.
TROPO Scatter Region	70 to 600 miles, depending on many factors	1 to 100 kW, 10 to 120-ft antennas, refined modulation and receiver techniques	Only practical wide-band, reliable ground-based method of achieving 70 to 600 mile hop where unsuitable intervening territory prevents use of LOS or diffraction modes.

(h) = height of antenna center
(N_S) = refractive index
(G_O) = obstacle gain

(A_O) = obstacle absorption
(d) = distance between stations
(Θ) = scatter angle or angle of elevation

INTERNATIONAL TELEVISION STANDARDS

This table outlines pertinent characteristics* of the current TV standards used throughout the world. The video frequency-channel arrangements are also shown. The systems have been designated by letter and are in use or proposed for use in the countries listed.

Country	Standard Used[c]
Argentina	N
Australia	B
Austria	B
Belgium	C, F
Bulgaria	D
Canada	M
Czechoslovakia	D
Denmark	B
Finland	B
France	E, L
Hungary	D
India	B
Iran	M
Ireland	A
Italy	B, G
Japan	M
Korea	M
Luxembourg	F
Mexico	M
Monaco	E
Morocco	B
Netherlands Antilles	M
New Zealand	B
Nigeria	B
Norway	B
Pakistan	B
Panama	M
Poland	D
Portugal	B
Rhodesia	B
Romania	D
Saudi Arabia	M
Spain	B
Sweden	B
Switzerland	B
The Netherlands	B
United Kingdom	A
United States of America	M
Union of Soviet Socialist Republics	D, K
West Germany	B, G

[c] Letter designations correspond to those in table on p. 19.

*CCIR Report 308, Geneva, 1963.

	A	M	N	B	C	G	H	I	D, K	L	F	E
Lines/frame	405	525	625	625	625	625	625	625	625	625	819	819
Fields/sec	50	60	50	50	50	50	50	50	50	50	50	50
Interlace	2/1	2/1	2/1	2/1	2/1	2/1	2/1	2/1	2/1	2/1	2/1	2/1
Frames/sec	25	30	—	25	25	25	25	25	25	25	25	25
Lines/sec	10 125	15 750	—	15 625	15 625	15 625	15 625	15 625	15 625	15 625	20 475	20 475
Aspect ratio[1]	4/3	4/3	—	4/3	4/3	4/3	4/3	4/3	4/3	4/3	4/3	4/3
Video band (MHz)	3	4.2	4.2	5	5	5	5	5.5	6	6	5	10
RF band (MHz)	5	6	6	7	7	8	8	8	8	8	7	14
Visual polarity[2]	+	−	−	−	+	−	−	−	−	+	+	+
Sound modulation	A3	F3	—	F3	A3	F3	F3	F3	F3	F3	A3	A3
Pre-emphasis in microseconds	—	75	—	50	50	50	50	50	50	—	50	—
Deviation (kHz)	—	25	—	50	—	50	50	50	50	—	—	—
Gamma of picture signal	0.45	0.45	—	0.5	0.5	0.5	0.5	0.5	0.5	0.5	0.5	0.6

Notes:

[1] In all systems the scanning sequence is from left to right and top to bottom.

[2] All visual carriers are amplitude modulated. Positive polarity indicates that an increase in light intensity causes an increase in radiated power. Negative polarity (as used in the US—Standard *M*) means that a decrease in light intensity causes an increase in radiated power.

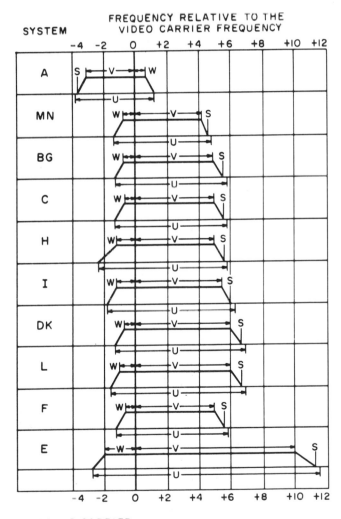

S = SOUND CARRIER
U = LIMITS OF RADIO-FREQUENCY CHANNEL
V = NOMINAL WIDTH OF MAIN SIDEBAND
W = NOMINAL WIDTH OF VESTIGIAL SIDEBAND

FREE SPACE TRANSMISSION NOMOGRAM

This nomogram relates receiver–transmitter distance, wavelength and free space attenuation. It can also be used to convert between nautical and statute miles and between frequency and wavelength.

FOR EXAMPLE:
A signal from a 200-MHz transmitter will be attenuated 125 dB before it reaches a receiver located 100 nautical miles away.

At a distance of 200 nautical miles, and a system gain of 130 dB, the highest usable frequency is 180 MHz.

The maximum distance between a transmitter-receiver-antenna system with a total gain of 125 dB operating at 500 MHz is 45 statute miles.

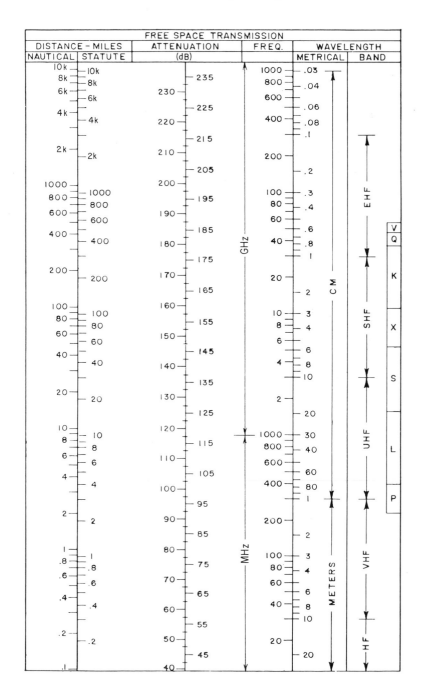

This nomogram is used to compute signal-strength input at the receiver based on a formula that converts field intensity at the receiving antenna to receiver input voltage.

If field intensity ϵ, in microvolts per meter, of a given signal f, in MHz, is known, the signal strength E_r, in microvolts, is determined for an input impedance of 50 ohms (E_r in μV for $R = 50$) and may be adjusted for any value of input impedance between 30 and 5000 ohms (E_r in μV for $30 \leqslant R \leqslant 5,000$). An isotropic antenna, no-loss transmission line is assumed.

Signal strength for receiving antennas of gain $>$ 1 (0 dB) are solved first by finding from the chart the voltage input for a system with an isotropic antenna and then adjusting the answer using the relation: $G = 20 \log (E_r'/E_r)$ where G is the gain of the antenna referred to isotropic; E_r' is the voltage input to be found; and E_r is the voltage input.

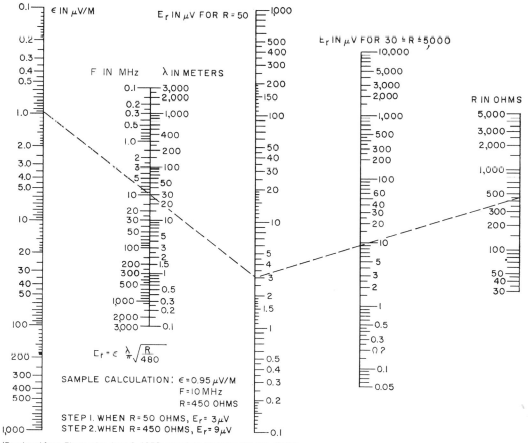

$$E_r = \epsilon \frac{\lambda}{\pi} \sqrt{\frac{R}{480}}$$

SAMPLE CALCULATION: $\epsilon = 0.95\,\mu$V/M
$F = 10\,$MHz
$R = 450\,$OHMS

STEP 1. WHEN $R = 50$ OHMS, $E_r = 3\,\mu$V
STEP 2. WHEN $R = 450$ OHMS, $E_r = 9\,\mu$V

(Reprinted from *Electronics*, June 6, 1958; copyright McGraw-Hill, Inc., 1958).

NOMOGRAM RELATING TRANSMITTER OUTPUT, TRANSMISSION LOSS, AND RECEIVER INPUT

This monogram shows the available input voltage (microvolts into 50 ohms), if transmitter output in watts and transmission loss in decibels are known. It can also show the maximum permissible transmission loss if transmitter power and receiver requirements are given, or it can be used to determine the required transmitter output for a given transmission loss and receiver input voltage. Microvolts (into 50 ohms) may be directly converted to dBm

on the left scale and watts may be converted to dBm on the center scale.

FOR EXAMPLE:
1. For a transmitter output of 5W and a transmission loss of 90 dB, the receiver input will be 500 μV.
2. For a minimum of 50 μV at the receiver, and a transmitter output of 5W, the transmission loss may not exceed 110 dB.

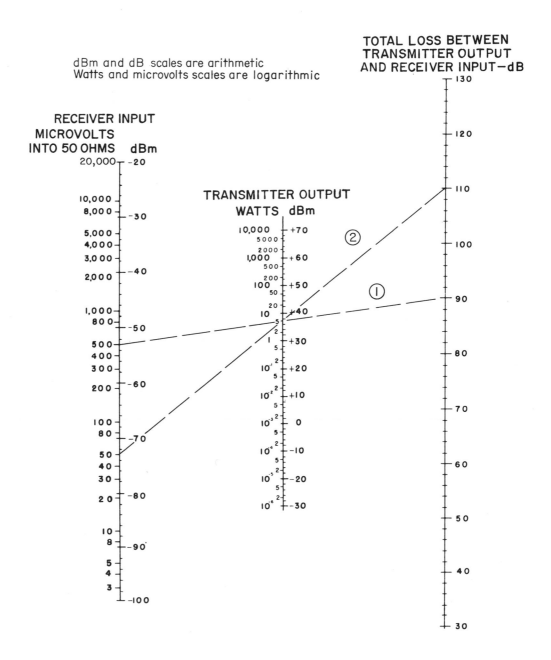

RECEIVER BANDWIDTH-SENSITIVITY-NOISE FIGURE NOMOGRAM

This nomogram is based on the noise figure of a receiver as given by the equation:

$$NF = \frac{(mE_o\sqrt{P_n/P_s})^2}{2R(4KT\Delta f)}$$

where NF = noise figure; m = modulation index; P_n = noise power; P_s = signal power; K = Boltzmann's constant or 1.38×10^{-23} joules/°K; R = antenna resistance; T = degrees Kelvin; Δf = 6-dB audio bandwidth, and E_o = signal generator output in μV.

Nominal antenna impedance is 52 ohms and the temperature can be approximated at 300°K.

To find the noise figure of a receiver, it is only necessary to place a straightedge across the sensitivity and audio bandwidth points, extending it to intersect the noise figure line.

FOR EXAMPLE:

Sensitivity of 10 μV and bandwidth of 6 kHz gives a noise figure of 100, or 20 dB.

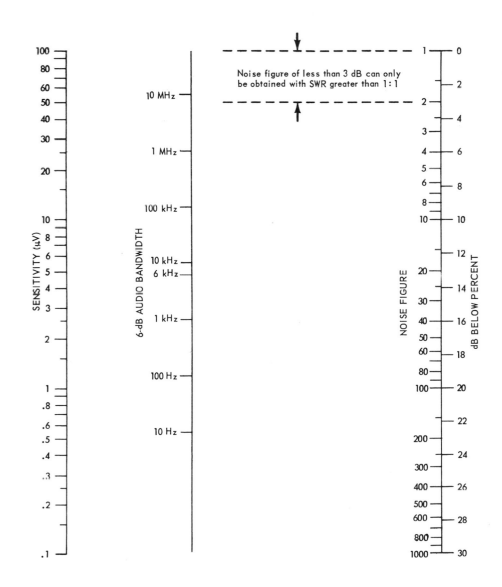

LINE-OF-SIGHT TRANSMISSION RANGE NOMOGRAM SHOWING THE APPROXIMATE TRANSMISSION RANGE OF SIGNALS IN THE VHF BAND

The theoretical maximum distance that can be covered is equal to the geometrical or "optical" horizon distance of each antenna, and is defined by the formula $D = 1.23\sqrt{H_r} + 1.23\sqrt{H_t}$, where D is in miles and H_r and H_t are the height in feet, above effective ground level, of the receiving and transmitting antennas. Atmospheric diffraction increases the distance by a factor of $2/\sqrt{3}$ which defines the "radio" path under normal or standard diffraction, by the formula $D = 1.41\sqrt{H_r} + 1.41\sqrt{H_t}$.

FOR EXAMPLE:

With a receiving antenna height of 30 ft and a transmitting antenna height of 100 ft, the "optical" horizon is 19 miles and the "radio" horizon is 21.5 miles.

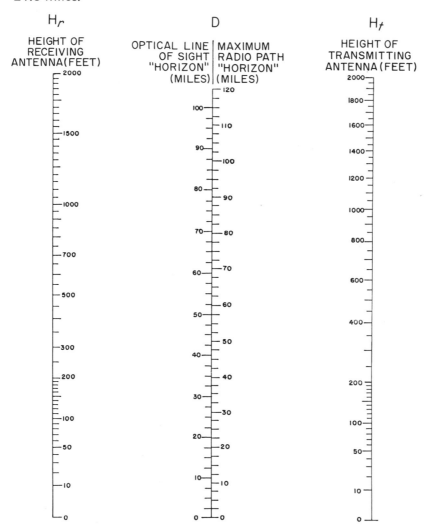

TYPES OF RADAR INDICATORS

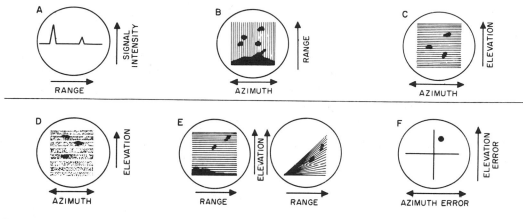

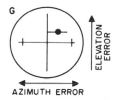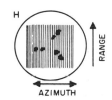

Coarse range information is provided by position of signal in broad azimuthal trace.

Single signal only. In the absence of a signal, the spot may be made to expand into a circle

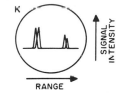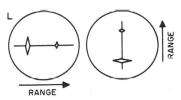

Single signal only. Signal appears as "wingspot," position giving azimuth and elevation errors. Length of wings inversely proportional to range.

Signal appears as two dots. Left dot gives range and azimuth of target. Relative position of right dot gives rough indication of elevation.

Antenna scan is conical. Signal is a circle, the radius proportional to range. Brightest part indicates direction from axis of cone to target.

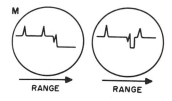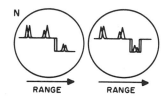

Same as type A, except time base is circular, and signals appear as radial pips.

Type A with lobe-switching antenna. Spread voltage splits signals from two lobes. When pips are of equal size, antenna is on target.

Same as type K, but signals from two lobes are placed back to back.

Type A with range step or range notch. When pip is aligned with step or notch, range can be read from a dial or counter.

A combination of type K and type M.

Range is measured radially from the center.

RADAR POWER-ENERGY
NOMOGRAM

The energy available from a radar transmitter is often the limiting factor in determining the maximum free space range. This nomogram relates the four interdependent radar equations involving peak power, average power, energy, duty cycle, pulse width, pulse repetition rate and pulse interval based on the following equations:

$$\frac{P_{AV}}{P_P} = d = \tau f_r \quad \text{and} \quad P_P \tau = E = P_{AV} t$$

where P_P = peak power in watts
P_{AV} = average power
E = energy in joules
d = duty cycle

τ = pulse width in microseconds
f_r = pulse repetition rate in pulses/sec
t = pulse interval in microseconds

FOR EXAMPLE:

A pulse repetition rate of 1000 pulses/sec with a pulse width of 5 μsec will give a duty cycle of 0.005. For a peak power of 100 kW, join this value on the P_P scale with 0.005 on the duty-cycle scale and read an average power of 500 W. Joining the 100 kW point with the pulse width of 5 μsec shows the energy as 0.5 J. (To crosscheck, connect the average power of 500 W with 1000 pps rep rate, which also yields 0.5 J.)

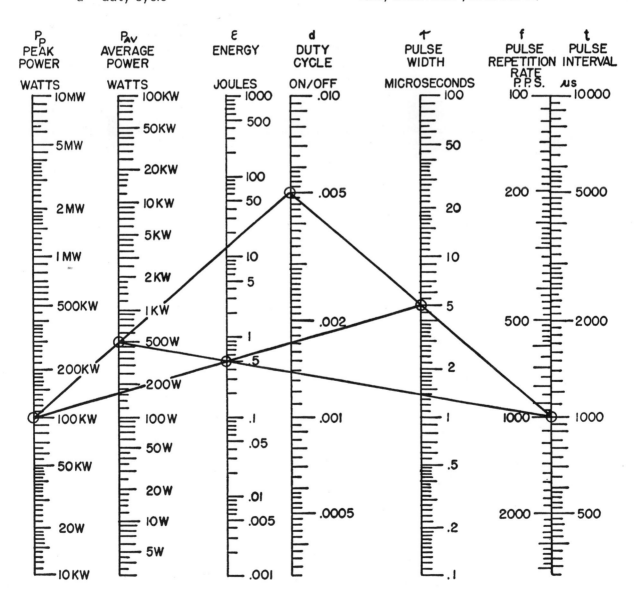

Antennas may be classified as linear radiators or elements, apertures arrays, and traveling wave types. Basic information on a few types of antennas is tabulated. For each type the following is given: the antenna name, physical size in wavelengths, a line drawing superimposed on coordinate axis, the impedance R in ohms at the resonant frequency f_r, the half-power (3dB) bandwidth in percent, the gain in dB above an isotropic radiator, as well as the conventional half-wavelength dipole, the polarization for the given configuration, and a set of Fraunhofer Zone field strength patterns for each of the three orthogonal planes of the axis system shown.

An isotropic radiator is given, even though such an antenna for electromagnetic waves does not exist. It is a convenient and frequent reference, however, for gain and pattern measurements.

The antennas tabulated may be vertically or horizontally polarized radiators. The configuration shown in the chart is the one most frequently used in practice. The antennas listed may be fed by balanced transmission lines, by coaxial lines and a balun (balanced-to-unbalanced transformer) when necessary, or in some cases by waveguides. Aperture antennas, such as parabolic dishes and horns, are usually fed by waveguides and, for such feed systems, impedance is not too meaningful.

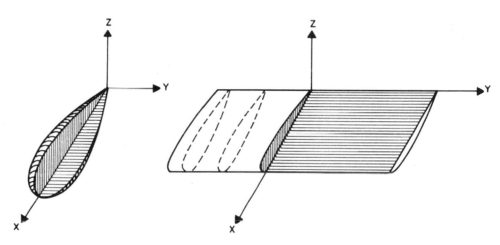

Pencil beam radiation pattern

Fan beam radiation pattern

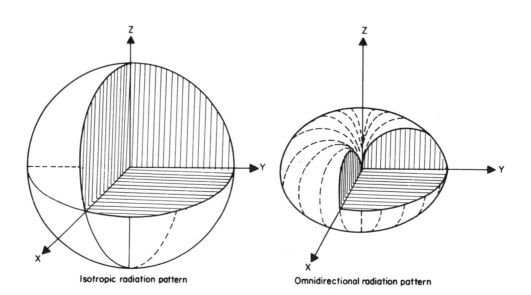

Isotropic radiation pattern

Omnidirectional radiation pattern

ANTENNA REFERENCE CHART
(Continued)

Upper chart

TYPE	IMPEDANCE Resistive at f_r, R, ohms	−3dB BANDWIDTH%	GAIN dB above Isotrope	GAIN dB above Dipole	POLARIZATION	PATTERN #
Dipole over small ground plane ($L = \lambda/4$, $L/D = 53$, $l = 2\lambda$)	28	40	2.14	0	V	E
Folded Unipole over small ground plane ($L = \lambda/4$, $L/D = 53$, $l = 2\lambda$, $l = 13$)	150	45	2.14	0	V	E
Coaxial Dipole ($L/D = 40$)	50	16	2.14	0	V	E
Biconical Coaxial Dipole ($L = \lambda/2$, $d = \lambda/8$, $L = 3\lambda/8$)	72	200	2.14	0	V	E
Disc-Cone or Rod Disc-Cone ($L = \lambda/4$, $l = \lambda$)	50	300	2.14	0	V	E
Biconical Horn ($L = 9\lambda/2$, $l = 14\lambda$)	20	25	14.14	12	V	E
Slot in Large Ground Plane ($L = \lambda/2$, $l = 29$)	350	70	2.14	0	H	F
Vertical Full Wave Loop ($D = \lambda/\pi$)	45	13	3.14	1	H	B
Helical over reflector screen, tube 6λ long coiled into 6 turns, λ/4 apart ($D/d = 36$)	130	200	10.14	8	Circ.	G
Rhombic ($L = 9\lambda$, $l = 9\lambda/2$)	600	100	16.74	14.5	H	H
Parabolic with folded dipole feed ($\lambda/2$) ($l = 3\lambda$)	300	30	14.74	12.5	H	H
Horn, coaxial feed ($l = 3\lambda$, $L = 3\lambda$)	50	35	15.14	13	H	H

PATTERN TYPES: A, B, C, D, E, F, G, H

(Pattern F — "NO SIGNAL RESPONSE IN THIS PLANE"; Pattern G — "NO SIGNAL RESPONSE IN THIS PLANE")

Lower chart

TYPE	PATTERN #	POLARIZATION	GAIN dB above Dipole	GAIN dB above Isotrope	−3dB BANDWIDTH%	IMPEDANCE Resistive at f_r, R, ohms
Isotropic Radiator (theoretical)	A	none	−2.14	0	—	—
Small Dipole ($L < \lambda/2$)	B	H	−0.4	1.74	very small	very high
Thin Dipole ($L = \lambda/2$, $L/D = 276$)	B	H	0	2.14	34	60
Thick Dipole ($L = \lambda/2$, $L/D = 51$)	B	H	0	2.14	55	49
Cylindrical Dipole ($L = \lambda/2$, $L/D = 10$)	B	H	0	2.14	100	37
Folded Dipole ($L = \lambda/2$, $L/d = 13$)	B	H	−0.5	1.64	5	6000
Folded Dipole	B	H	0	2.14	45	300
Cylindrical Dipole ($L = \lambda/2$, $L/d = 9.6$)	B	H	1.5	3.64	130	150
Biconical ($L = \lambda/2$)	B	H	0	2.14	100	72
Biconical ($L = \lambda$)	B	H	0	2.14	200	350
Turnstile ($L = \lambda/2$, $L/d = 25.5$)	C	H	−3	−0.86	50	150
Folded Dipole over reflecting sheet ($L = \lambda/2$, $L/d = 25.5$, $\lambda/4$ above sheet)	D	H	5	7.14	20	150

28

Broadside Array
$$L = \lambda/2$$
polarization: vertical

Theoretical Gain of Broadside ½ λ elements at different spacings "a".

Spacing in wavelengths "a"	Gain, dB above Dipole
5/8	4.8
3/4	4.6
1/2	4.0
3/8	2.4
1/4	1.0
1/8	0.3

Theoretical Gain of Broadside ½ λ elements for different numbers of elements.

Number of elements	Gain, dB above Dipole
2	4.0
3	5.5
4	7.0
5	8.0
6	9.0

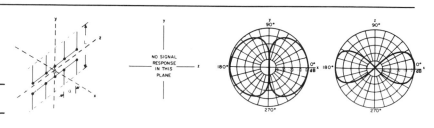

End Fire Array
$$L = \lambda/2$$
polarization: vertical

Theoretical Gain of Two End Fire ½ λ Elements for Various Spacings "a"

a	Gain, dB above Dipole
5/8	1.7
1/2	2.2
3/8	3.0
1/4	3.8
1/20	4.1
1/8	4.3

Parasitic Array
$$L = \lambda/2$$
polarization: horizontal

Number of Elements	Gain, dB above Dipole	Front to Back Ratio, dB
2	4 to 5	10 to 15
3	6 to 7	15 to 25
4	7 to 9	20 to 30
5	9	—

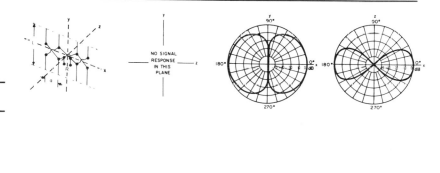

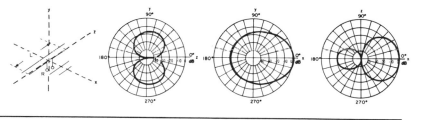

Collinear Array
$$L = \lambda/2$$
$$b = \lambda/4$$

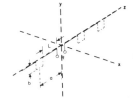

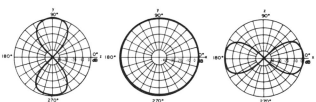

Spacing "a" between centers of adjacent ½ λ elements	Number of ½ λ elements in array versus gain in dB above a reference Dipole				
	2	3	4	5	6
a = ½ λ	1.8	3.3	4.5	5.3	6.2
a = ¾ λ	3.2	4.8	6.0	7.0	7.8

Shown here is the relationship between circular antenna aperture size, frequency, and gain. Also listed are the antenna performance requirements for various system applications. Practical factors, such as whether the antenna is solid or perforated, the type of aperture illumination, accuracy of construction, and shadowing from the feed system will tend to reduce the gain somewhat.

FOR EXAMPLE:

1. To achieve a gain of 40 dB at 10 GHz requires an antenna with a diameter of 10 m.
2. An antenna with a diameter of 100 m has a gain of 100 dB at 100 GHz.

Antenna Performance Requirements

APPLICATION	PATTERN	POLARI-ZATION	GAIN g, (dB) above isotropic rad.	BEAMWIDTH (H) degrees	POINTING ACCURACY, to degrees	TYPICAL TYPES
1. SATELLITE Link or Probe	Pencil Beam	any	10 to 40 dB or more	60 to 2 or less	6 to .2 or better	Horn, Phased array, Parabola, Cassegrain
2. POINT TO POINT RELAY a. On Earth b. Earth to Satellite to Earth c. Satellite to Satellite	Pencil Beam	any	a. 50 to 120 b. 50 to 120 c. 50 to 160	5.8×10^{-1} to 1.8×10^{-4} 5.8×10^{-1} to 1.8×10^{-6}	5.8×10^{-2} to 1.8×10^{-7}	Horn, Parabola, Cassegrain
3. BROADCAST a. Earth Trans. b. Sat. Trans.	omnidir. wide or fan beam	any	a. 3 to 40 b. 1 to 10	100 to 1.8 180 to 60	10 to .18	a. Vertical radiator b. Cylindrical parabola
4. NAVIGATION	omnidir. or fan beam	any	3 to 50	100 to .58	10 to .058	Vertical radiator, Horn, or Parabola
5. RADAR a. Search b. Track	csc^2 Pencil Beam	any	40 to 120	1.8 to 1.8×10^{-4}	.18 to 1.8×10^{-5}	Horn, Parabola, Cassegrain, Phased array
6. RADIO ASTRONOMY a. Passive b. Active	Pencil Beam	any	50 to 160 or greater	.58 to 1.8×10^{-6}	.057 to 1.8×10^{-7}	Parabola, Cassegrain, Phased array
7. RADIOMETRY Industrial	any	any	unknown	unknown	unknown	Any

Antenna Gain and Size vs Frequency for Uniformly Illuminated Circular Aperture

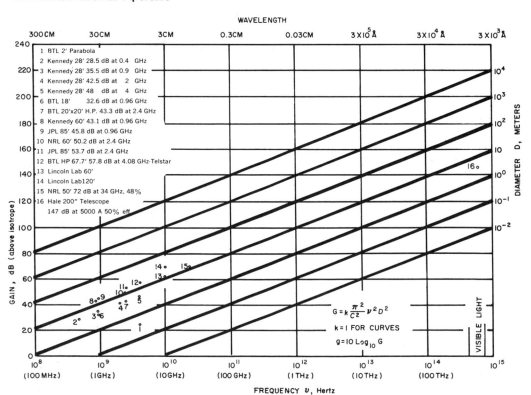

ANTENNA EFFECTIVENESS
NOMOGRAM

Antennas are judged on the basis of radiation efficiency or their VSWR. Radiation efficiency is the ratio of the radiated power to the total power fed into the antenna terminals. Total power is the sum of the radiated power and the power lost in ohmic losses in the form of heat. The power going into the antenna terminals is the power which a transmitter can put out less the power reflected due to antenna mismatch. Antenna effectiveness is the ratio of the radiated power to the power which a transmitter can put into a matched load, i.e., the forward or incident power.

$$\text{Effectiveness} = \frac{4\ \text{VSWR}}{(\text{VSWR} + 1)^2} \times \text{efficiency}$$

FOR EXAMPLE:
A 60% efficient antenna with a 2.5:1 VSWR has an effectiveness of 48% compared to a perfectly matched 100% efficient antenna.

NOTE:
In some cases an antenna can be made more effective by lessening its efficiency if this will produce a sufficient reduction in the VSWR.

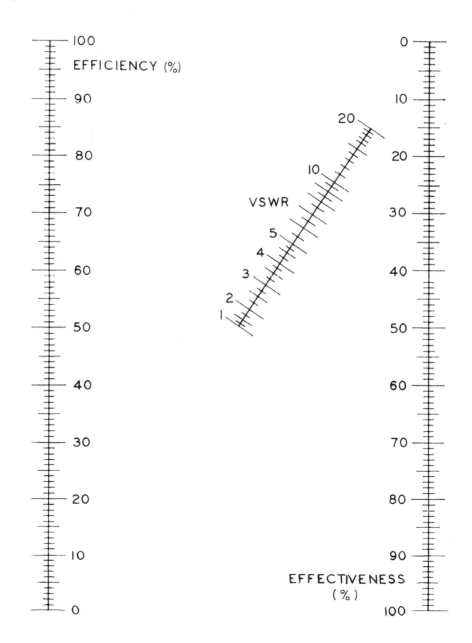

TRANSMISSION LINE CHARACTERISTICS

Characteristcs of Various Types of Transmission Lines Erected Parallel to a Perfectly Conducting Earth.

LOGARITHMS TO THE BASE 10			I_1 = GENERATOR CURRENT
LINE CONFIGURATION		CHARACTERISTIC IMPEDANCE	NET GROUND–RETURN CURRENT
Single wire		$Z_0 = 138 \, \mathrm{Log} \, \dfrac{2h}{r}$	$I_{Gnd} = I_1$
2-Wire balanced		$Z_0 = 276 \, \mathrm{Log} \, \dfrac{s}{r}$	$I_{Gnd} = 0$
2-Wire 1 wire grounded		$Z_0 \approx 276 \, \dfrac{\mathrm{Log} \frac{s}{r} \, \mathrm{Log} \left[\rho^2 \frac{s}{r} \right]}{\mathrm{Log} \left[\rho^2 \left(\frac{s}{r} \right)^2 \right]}$ $\rho = \dfrac{2h}{s}$	$I_{Gnd} \approx I_1 \, \dfrac{\mathrm{Log} \frac{s}{r}}{\mathrm{Log} \frac{2h}{r}}$
3-Wire 2 wires grounded		$Z_0 \approx 69 \left[\mathrm{Log} \, \dfrac{s^3}{2r^3} - \dfrac{\left(\mathrm{Log} \frac{s}{r} \right)^2}{\mathrm{Log} \frac{2h^2}{rs}} \right]$	$I_{Gnd} \approx I_1 \, \dfrac{\mathrm{Log} \frac{s}{2r}}{\mathrm{Log} \frac{s\rho^2}{2r}}$ $\rho = \dfrac{2h}{s}$
4-Wire balanced		$Z_0 = 138 \left(\mathrm{Log} \, \dfrac{s}{r} \right) - 21$	$I_{Gnd} = 0$
4-Wire 2-wires grounded		$Z_0 \approx 138 \left[\dfrac{\mathrm{Log} \frac{s}{r\sqrt{2}} \, \mathrm{Log} \left[\rho^4 \frac{s}{r\sqrt{2}} \right]}{\mathrm{Log} \left[\rho^4 \left(\frac{s}{r\sqrt{2}} \right)^2 \right]} \right]$ $\rho = \dfrac{2h}{s}$	$I_{Gnd} \approx I_1 \, \dfrac{\mathrm{Log} \frac{s}{r\sqrt{2}}}{\mathrm{Log} \frac{\rho^2 s}{r\sqrt{2}}}$
5-Wire 4 wires grounded		$Z_0 \approx 138 \left[\mathrm{Log} \, \dfrac{2h}{r} - \dfrac{\left[\mathrm{Log} \, 2\rho^2 \right]^2}{\mathrm{Log} \left[\rho^3 \frac{h\sqrt{2}}{r} \right]} \right]$ $\rho = \dfrac{2h}{s}$	$I_{Gnd} \approx I_1 \, \dfrac{\mathrm{Log} \frac{s}{r4\sqrt{2}}}{\mathrm{Log} \frac{s\rho^4}{r\sqrt{2}}}$
Concentric (coaxial)		$Z_0 = 138 \, \dfrac{\mathrm{Log} \frac{c}{b}}{\sqrt{1 + \frac{(\epsilon - 1) \omega}{s}}}$ ϵ = Dielectric constant of insulating material	
Double coaxial balanced		$Z_0 = 276 \, \dfrac{\mathrm{Log} \frac{c}{b}}{\sqrt{1 + \frac{(\epsilon - 1) \omega}{s}}}$	
Shielded pair balanced		$Z_0 = \dfrac{120}{\sqrt{\epsilon}} \left[2.303 \, \mathrm{Log} \left(2v \, \dfrac{1 - \sigma^2}{1 + \sigma^2} \right) - \dfrac{1 + 4v^2}{16 \, v^4} \left(1 - 4\sigma^2 \right) \right]$ ϵ = Dieletric constant of medium ϵ = Unity for gaseous medium $v = \dfrac{h}{b}; \, \sigma = \dfrac{b}{c}$	

(From "Radio Engineers' Handbook" by Frederick E. Terman. Copyright 1943 by McGraw-Hill Book Company. Used with permission of McGraw-Hill Book Company.)

CHARACTERISTIC IMPEDANCE OF
BALANCED TWO-WIRE LINES

This nomogram determines the theoretical exact impedance of air-dielectric parallel lines in air or in a vacuum, and remote from any conducting plane. It covers conductors having diameters from 0.01 to 5 in., spaced from 0.01 to 100 in. center-to-center.

$$Z_0 = 276 \log_{10} \frac{2D}{d}$$

$$D > 2d$$

FOR EXAMPLE:

1. The impedance of a line using #12 wire spaced $1\frac{1}{2}$ in. is 430 ohms.
2. What is the wire diameter required for a 300-ohm line spaced $1\frac{1}{4}$ in.? Answer: 0.20 in.

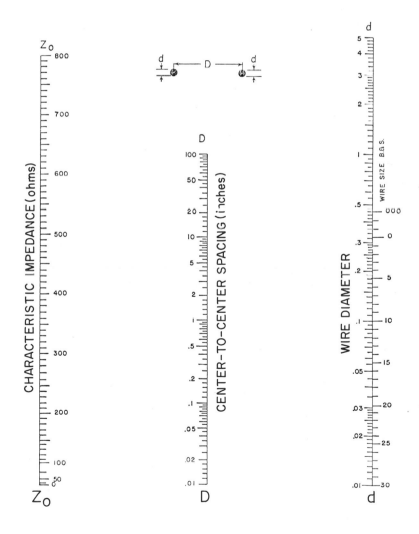

(Reprinted with permission from International Telephone and Telegraph Corporation.)

CHARACTERISTICS OF COAXIAL CABLES

▶ START HERE TO SELECT BY TYPE NUMBER

▶ START HERE TO SELECT BY CHARACTERISTIC IMPEDANCE

RG CABLE TYPE	INNER COND. MAT.	STRAND	INNER COND. O.D.	PE DIELECTRIC O.D.	SHIELD INNER	SHIELD OUTER	SHIELD O.D.	JACKET MAT.	JACKET O.D.	ARMOR (.0126 ALUM. WIRE)	OPER. VOLTS RMS	LBS. PER M FT.	NOMINAL IMPEDANCE OHMS	CAP. pF/FT.	10	50	100	200	400	600	1000	3000	VP %
5B/U	SC	1	.051	.185	SC	SC	.260	NCV	.335	—	3000	83	50 △	29.5	.65	1.6	2.4	3.6	5.2	6.6	8.8	16.7	65.9
6A/U	CW	1	.0285	.189	SC	C	.264	NCV	.336	—	2700	74	75 □	20	.70	1.8	2.9	4.3	6.5	8.3	11.2	22	65.9
8/U	C	7	.086	.295	C	—	.340	V	.415	—	4000	99	52 △	29.5	.56	1.35	2.1	3.1	5.0	6.5	8.8	17.5	65.9
8A/U	C	7	.086	.295	C	SC	.340	NCV	.415	—	4000	99	52 △	29.5	.56	1.35	2.1	3.1	5.0	6.5	8.8	17.5	65.9
9A/U	SC	7	.086	.285	SC	SC	.355	NCV	.430	—	4000	126	52 △	29.5	.45	1.26	2.3	3.4	5.2	6.5	9.0	17	65.9
9B/U	SC	7	.086	.285	SC	SC	.355	NCV	.430	—	4000	126	50 △	30	.45	1.26	2.3	3.4	5.2	6.5	9.0	17	65.9
10A/U	C	7	.048	.295	C	—	.340	NCV	.415	.475	4000	121	52 △	29.5	.56	1.35	2.1	3.1	5.0	6.5	8.8	17.5	65.9
11/U	TC	7	.048	.295	C	—	.340	V	.415	—	4000	89	75 □	20.5	.65	1.5	2.15	3.2	4.7	6.0	8.2	18	65.9
11A/U	TC	7	.048	.295	C	—	.340	NCV	.412	—	4000	89	75 □	20.5	.65	1.5	2.15	3.2	4.7	6.0	8.2	18	65.9
12A/U	TC	7	.048	.292	C	—	.340	NCV	.412	.475	4000	113	75 □	20.5	.65	1.5	2.15	3.2	4.7	6.0	8.2	18	65.9
13A/U	TC	7	.102	.290	C	—	.355	NCV	.430	—	4000	114	74 □	20.5	.65	1.5	2.15	3.2	4.7	6.0	8.2	18	65.9
14A/U	C	1	.102	.383	C	C	.463	NCV	.558	—	5500	201	52 △	29.5	.28	.85	1.5	2.3	3.5	4.4	6.0	11.7	65.9
17A/U	C	1	.188	.695	C	—	.760	NCV	.885	—	11000	446	52 △	29.5	.23	.60	.95	1.5	2.4	3.2	4.5	9.5	65.9
18A/U	C	1	.188	.695	C	C	.760	NCV	.885	.945	11000	496	52 △	29.5	.23	.60	.95	1.5	2.4	3.2	4.5	9.5	65.9
19A/U	C	1	.250	.925	C	—	.990	NCV	1.135	—	14000	720	52 △	29.5	.14	.42	.69	1.1	1.8	2.45	3.5	7.7	65.9
20A/U	C	1	.250	.925	C	C	.990	NCV	1.135	1.195	14000	786	52 △	29.5	.14	.42	.69	1.1	1.8	2.45	3.5	7.7	65.9
34B/U	C	7	.075	.470	C	—	.535	NCV	.640	—	5200	195	75 □	20	.29	.85	1.3	2.1	3.3	4.5	6.0	12.5	65.9
35B/U	C	1	.1045	.690	C	TC	.760	NCV	.880	.945	10000	425	75 □	20.5	.23	.61	.85	1.25	1.95	2.47	3.5	8.6	65.9
55/U	C	1	.032	.121	TC	TC	.176	PE	.206	—	1900	31	53.5 △	28.5	1.3	3.2	4.8	7.0	10.5	13.0	17	32	65.9
55A/U	SC	1	.035	.121	SC	SC	.176	NCV	.216	—	1900	36	50 △	29.5	1.3	3.2	4.8	7.0	10.5	13.0	17	32	65.9
55B/U	SC	1	.032	.121	SC	TC	.176	PE	.206	—	1900	32	53.5 △	28.5	1.3	3.2	4.8	7.0	10.5	13.0	17	32	65.9
58/U	C	1	.032	.121	C	—	.150	V	.200	—	1900	24	53.5 △	28.5	1.4	3.5	5.3	8.3	11.5	17.8	20	40	65.9
58A/U	TC	19	.0375	.120	TC	TC	.150	V	.199	—	1900	25	50 △	29.5	1.6	4.1	6.2	9.2	14.0	17.5	23.5	45	65.9
58C/U	TC	19	.0375	.120	TC	—	.150	V	.199	—	1900	25	50 △	29.5	1.6	4.1	6.2	9.2	14.0	17.5	23.5	45	65.9
59/U	CW	1	.0253	.150	C	—	.191	V	.250	—	2300	36	73 □	21	1.1	2.7	4.0	5.7	8.5	10.8	14.0	26	65.9
59B/U	CW	1	.023	.150	C	—	.191	NCV	.246	—	2300	36	75 □	20.5	1.1	2.7	4.0	5.7	8.5	10.8	14.0	26	65.9
62/U	CW	1	.025	.151	C	—	.191	V	.250	—	750	34	93 ▲	13.5	.82	1.9	2.7	3.9	5.8	7.0	9.0	17	84
62A/U	CW	1	.025	.151	C	—	.191	NCV	.249	—	750	34	93 ▲	13.5	.82	1.9	2.7	3.9	5.8	7.0	9.0	17	84
63B/U	CW	1	.0253	.295	C	—	.340	NCV	.415	—	1000	78	125 ●	10	.60	1.4	2.0	2.9	4.1	5.1	6.5	11.3	84
71/U	CW	1	.025	.151	C	TC	.198	PE	.259	—	750	42	93 ▲	13.5	.82	1.9	2.7	3.9	5.8	7.0	9.0	17	84
71A/U	TC	1	.025	.151	TC	—	.198	V	.245	—	750	42	93 ▲	13.5	.82	1.9	2.7	3.9	5.8	7.0	9.0	17	84
71B/U	TC	1	.025	.151	TC	—	.208	PE	.250	—	750	42	93 ▲	13.5	.82	1.9	2.7	3.9	5.8	7.0	9.0	17	84
74A/U	C	1	.102	.383	C	C	.564	PE	.558	.615	5500	230	52 △	29.5	.28	.85	1.5	2.3	3.5	4.4	6.0	11.7	65.9
79B/U	CW	1	.025	.295	C	—	.340	NCV	.415	.475	1000	122	125 ●	10	.60	1.4	2.0	2.9	4.1	5.1	6.5	11.3	84
164/U	C	1	.1045	.690	C	—	.760	NCV	.890	—	10000	392	75 □	20.5	.23	.61	.85	1.25	1.95	2.47	3.5	8.6	65.9
174/U	CW	7	.019	.060	TC	—	.069	V	.105	—	—	—	50 △	30	.23	.60	.95	1.5	2.0	2.7	4.5	8.6	65.9
177/U	C	1	.195	.690	SC	SC	.760	V	.910	—	14000	465	50 △	30	.23	.60	.85	1.25	1.95	2.47	3.5	8.6	65.9
212/U	SC	7	.056	.189	SC	SC	.265	NCV	.336	—	3000	85	50 △	29.5	.65	1.6	2.4	3.6	5.2	6.6	8.8	16.7	65.9
213/U	C	7	.090	.292	C	—	.340	NCV	.412	—	4000	100	50 △	30.5	.56	1.35	2.1	3.1	5.0	6.5	8.8	17.5	65.9
214/U	SC	7	.090	.292	SC	SC	.360	NCV	.432	—	4000	122	50 △	30.5	.45	1.26	2.3	3.4	5.2	6.5	9.0	17	65.9
215/U	C	7	.090	.292	C	—	.340	NCV	.412	—	4000	122	50 △	30.5	.56	1.35	2.1	3.1	5.0	6.5	8.8	16.7	65.9
216/U	TC	7	.048	.292	TC	C	.360	NCV	.432	—	4000	115	75 □	20.5	.65	1.5	2.15	3.2	4.7	6.0	8.2	18	65.9
217/U	C	1	.106	.380	C	—	.463	NCV	.555	—	5500	202	50 △	30	.28	.85	1.5	2.3	3.5	4.4	6.0	11.7	65.9
218/U	C	1	.195	.690	C	—	.760	NCV	.880	—	11000	457	50 △	30	.225	.60	.95	1.5	2.4	3.2	4.5	9.5	65.9
219/U	SC	1	.195	.690	SC	SC	.760	NCV	.880	.945	11000	507	50 △	30	.225	.60	.95	1.5	2.4	3.2	4.5	9.5	65.9
220/U	C	1	.260	.910	C	—	.990	NCV	1.120	—	14000	725	50 △	29.5	.17	—	—	1.12	1.85	—	3.6	7.7	65.9
221/U	C	1	.260	.910	C	—	.990	NCV	1.120	1.195	14000	790	50 △	30	.17	—	.69	1.12	1.85	—	3.6	7.7	—
223/U	SC	7	.036	.120	SC	SC	.176	NCV	.216	—	1900	36	50 △	30	1.3	3.2	4.8	7.0	10.5	13.0	17.0	32	65.9
224/U	C	1	.106	.380	C	C	.463	NCV	.555	.615	5500	232	50 △	30	.28	.85	1.5	2.3	3.5	4.4	6.0	11.7	65.9

ATTENUATION (dB/100 ft.) FREQUENCY IN MEGAHERTZ

Ohms Code: △ Through to 55   □ 56 Through 80   ▲ 81 Through 100   ● 101 Through 200

SC—silver plated copper, C—bare copper, PE—polyethylene, NCV—non-contaminating vinyl, V—polyvinylchloride, TC—tinned copper, CW—copperweld

ULTRA-HIGH FREQUENCY
HALF-WAVE SHORTING-
STUB NOMOGRAM

This nomogram is used to determine the length in inches of shorting stubs required to eliminate interference in the UHF television range.

FOR EXAMPLE:

To eliminate an interfering signal at 575 MHz (channel 31) requires an $8\frac{1}{2}$ in. long half-wave shorting stub, if 300-ohm twin lead is used. If 75-ohm twin lead is used, the stub has to be $7\frac{1}{4}$ in. for the same frequency.

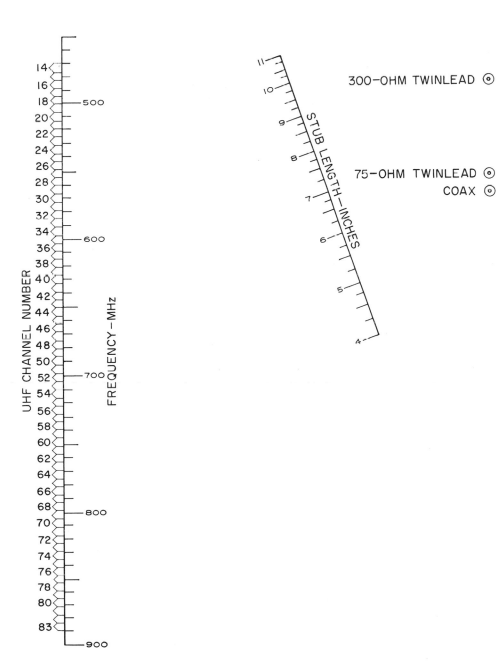

TRANSMISSION LINE NOMOGRAM

This nomogram gives the actual length of line in centimeters and inches when given the length in electrical degrees and the frequency provided that the velocity of propagation on the transmission line is equal to that in free space. The length is equal to that in free space and is given on the L scale intersection by a line between λ on ℓ°.

FOR EXAMPLE:

$$f = 600\ \text{MHz} \qquad \ell = 30^\circ$$
$$\text{Length}\ L = 1.64''\ \text{or}\ 4.2\ \text{cm}$$

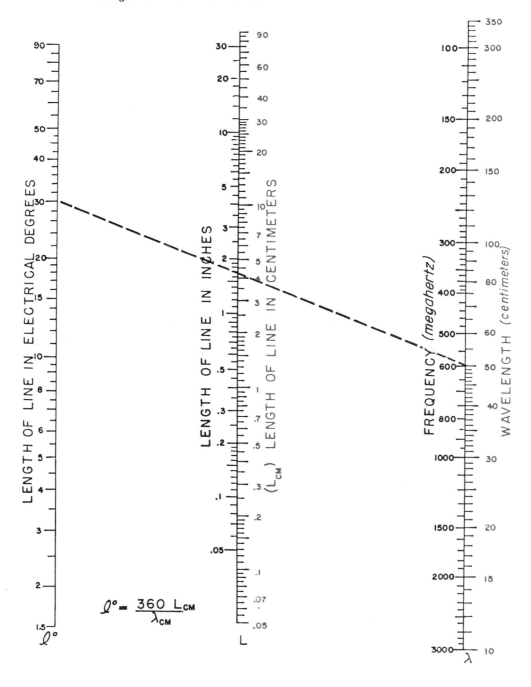

$$\ell^\circ = \frac{360\ L_{CM}}{\lambda_{CM}}$$

SLOTTED-LINE WIDTH-OF-MINIMUM VSWR NOMOGRAM

This nomogram is used to determine the VSWR and the magnitude of the reflection coefficient by the use of width-of-minimum measurement technique. This technique relies on the fact that there are two comparatively easy-to-find 3-dB points straddling any minimum, as illustrated.

FOR EXAMPLE:

A slotted-line width-of-minimum measurement of 0.18 cm, with a 1-GHz source, indicates a VSWR of 53 or a reflection coefficient magnitude of 0.963.

NOTE:

The signal-to-noise ratio at the bottom of the minimum must be at least 10 dB for accurate results.

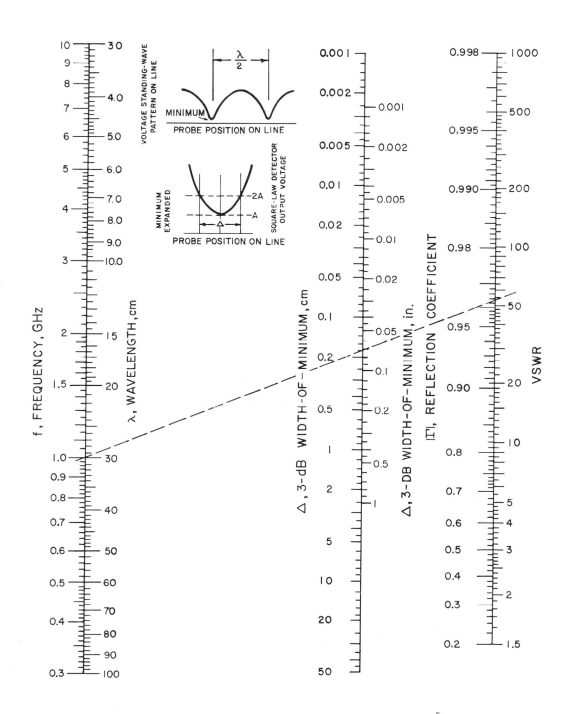

SLOTTED LINE WIDTH-OF-MINIMUM ATTENUATION CALCULATION NOMOGRAM

This nomogram is used to determine the total attenuation between the probe position and the reference plane based on width-of-minimum measurements.

FOR EXAMPLE:
With a short circuit termination at the reference plane, if the width-of-the-minimum measured 30 cm from the reference plane is 0.014 cm at 3.5 GHz, then the attenuation is 0.045 dB.

NOTE:
The signal-to-noise ratio at the bottom of the minimum should be at least 10 dB for accurate results.

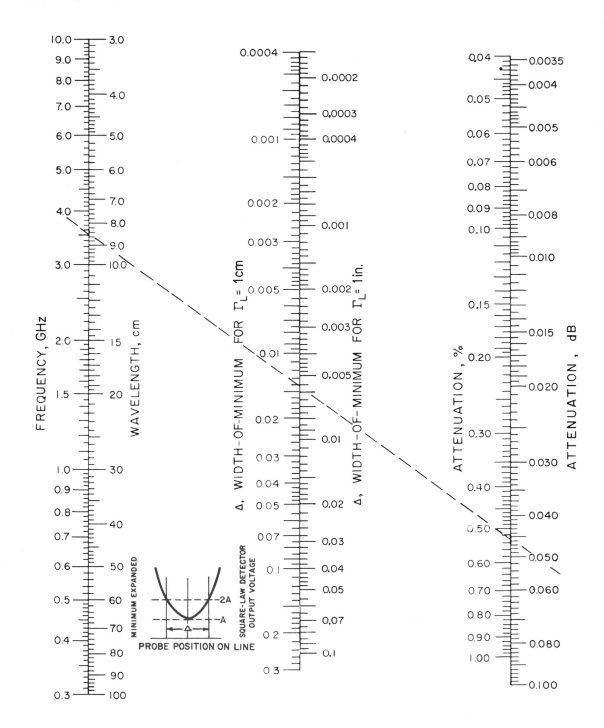

WAVEGUIDE NOMOGRAM

This nomogram relates three significant waveguide characteristics:

  waveguide wavelength (λ_g)
  free space wavelength (λ_o) or frequency (f)

and

  cutoff wavelength (λ_c)

The vertical scale gives waveguide wavelength in centimeters. The horizontal scale is for the cutoff wavelength, and the points corresponding to the cutoff wavelength in the TE_{10} mode of three common waveguides are indicated. The sloping center scale is calibrated in free space wavelength and frequency.

FOR EXAMPLE:

1. The waveguide wavelength at 6 GHz (5 cm free space wavelength) in an RG-50 waveguide is 7.17 cm.
2. Measurement on an RG-51 waveguide shows the waveguide wavelength to be 6.5 cm. The frequency is 7 GHz, which corresponds to a free space wavelength of 4.27 cm.

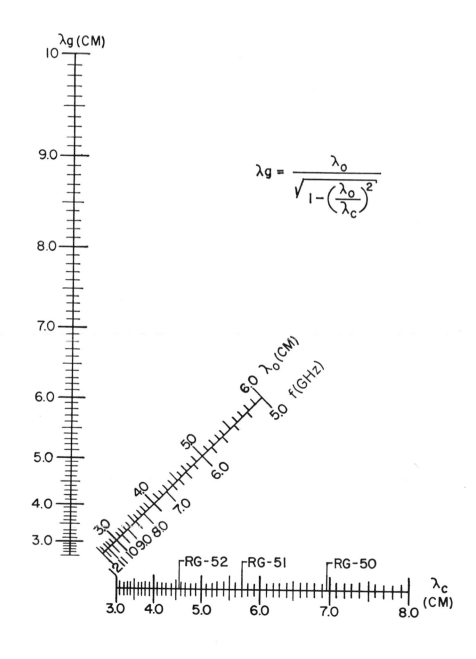

$$\lambda g = \frac{\lambda_o}{\sqrt{1 - \left(\frac{\lambda_o}{\lambda_c}\right)^2}}$$

VSWR NOMOGRAM

If a transmission line is not terminated in its characteristic impedance, then some of the energy sent along the line will be reflected back, and standing waves form on the line. The ratio of the maximum to the minimum voltage of the standing waves is the VSWR (voltage standing wave ratio) and indicates the effectiveness of the match between line and load. For a perfectly matched line, the VSWR is 1. The VSWR can be given in a number of ways:

$$VSWR = \frac{Z_L}{Z_o} = \frac{E_{max}}{E_{min}} = \frac{1 + \sqrt{\dfrac{\text{Reflected power}}{\text{Forward power}}}}{1 - \sqrt{\dfrac{\text{Reflected power}}{\text{Forward power}}}}$$

This nomogram is based on the last expression and solves for VSWR from measurements of reflected power and forward power.

FOR EXAMPLE:
For a forward power of 180 W and a reflected power of 2.7 W, the VSWR is 1.27.

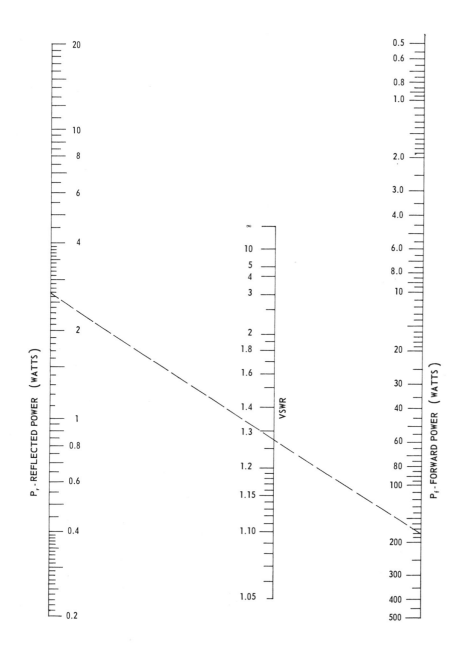

VSWR REDUCTION AS A RESULT OF ATTENUATION

This nomogram relates load VSWR, input VSWR, and attenuation. It can be used to find the resultant VSWR with a given amount of attenuation, or to determine the attenuation required for a given VSWR.

FOR EXAMPLE:
1. A 5-dB attenuator will reduce input VSWR to 1.23 if the load VSWR is 2.0.
2. The required attenuation to reduce a load VSWR of 1.8 to an input VSWR of 1.06 is 10.0 dB.

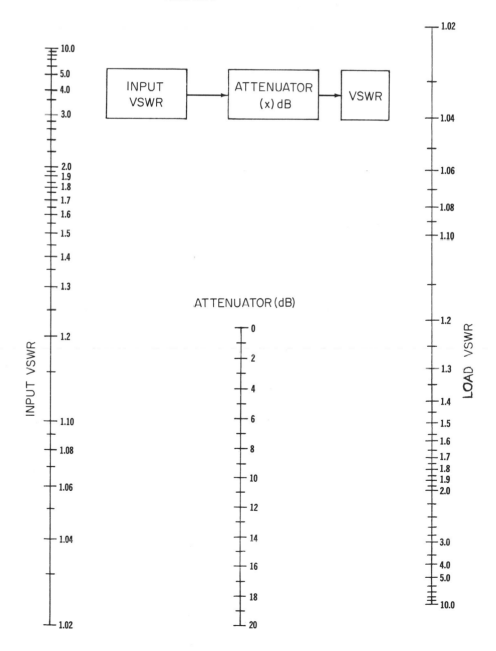

This nomogram solves for the Doppler frequency, which is produced as a result of relative motion between a transmitter and its receiver or target. The Doppler frequency is a function of transmitted frequency and velocity of motion. The angle to the velocity vector determines the actual relative velocity. For a navigation system (Fig. A) in an airplane, the earth is the target, and the angle A is the acute angle between the aircraft heading and the radar beam. In this case the Doppler shift is downward. A forward-looking radar will produce an upward Doppler shift. For surveillance-type radars (Fig. B), the angle A is the acute angle between the radar beam and target velocity. (Note that the nomogram is based on the Doppler equation for radar and that the Doppler shift for a passive listening device will be half the frequency indicated.)

FOR EXAMPLE:

A helicopter navigation system transmits at 10 GHz at an angle of 70°. What is the audio bandwidth required for aircraft velocities of 10 through 200 mph? On the left scales, connect 10 GHz and 10 mph to the turning scale. From that point on, the turning scale connecting through 70° gives 100 Hz as the lowest frequency. Repeating the steps using 200 mph in place of 10 mph shows the highest frequency to be 2 kHz. Thus the required bandwidth is 100 to 2000 Hz. The nomogram is based on the formula

$$f_d = 89.4 \frac{V}{\lambda}$$

where

f_d = Doppler frequency (Hz)

V = velocity in miles per hour

λ = transmitted wavelength in centimeters

Angle-to-velocity vector depends on type of target.

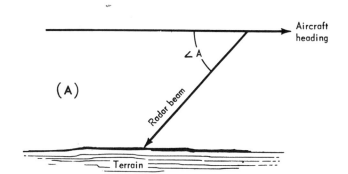

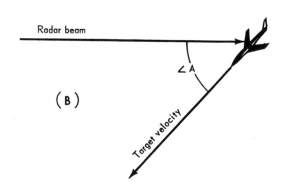

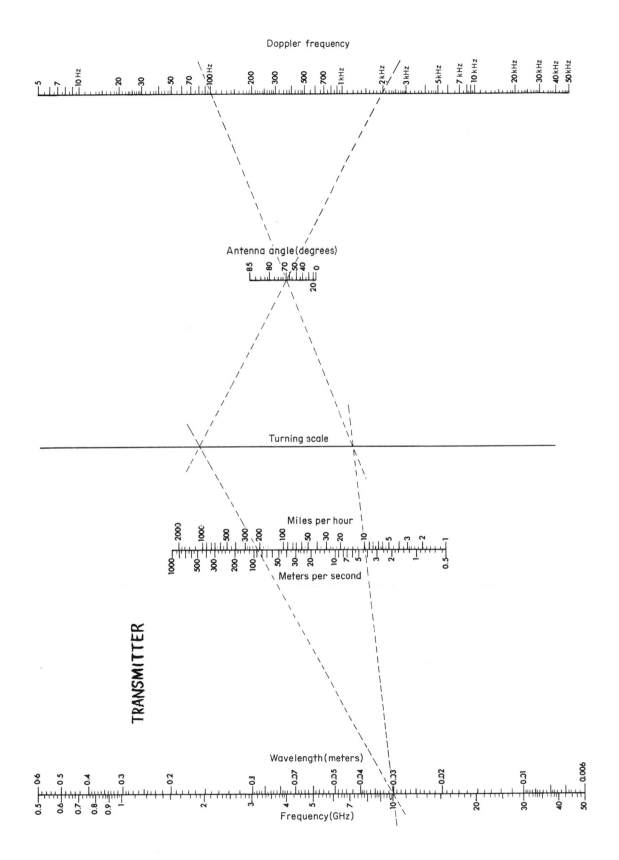

DOPPLER TO SPEED CONVERSION NOMOGRAM

Radar or sonar frequency may be converted to hundreds of miles per hour or knots per hour by using this chart. The base sonar frequency in kHz is given on the left scale and the base radar frequency in GHz is given on the right. Doppler frequency, in Hz for sonar and hundreds of Hz for radar, is shown at the bottom. The diagonals represent target rate of change of range, which is the velocity speed vector in the source's direction.

The basic formula for Doppler speed is:

$$\text{Doppler frequency} = \frac{\text{base } f. \times \text{target range rate}}{\text{signal velocity in medium}}.$$

The signal velocity in medium is 5000 ft/sec for sonar and 186,000 mi/sec for radar.

FOR EXAMPLE:

1. The base frequency of a sonar system is 40 kHz and its Doppler frequency is 55 Hz. The speed vector is found by the intersection of these two lines on the chart to be approximately 4.1 knots.

2. The base frequency of a radar system is 11 GHz, and the Doppler frequency is 8000 Hz. The speed vector of the aircraft in miles per hour is found (from the intersection of these two lines) to be approximately 480 mph.

By moving the decimal points on the scales, other values not shown on the graph may be computed.

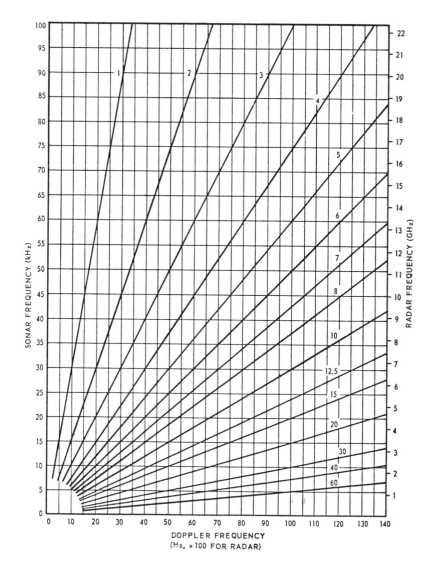

SONAR FREQUENCY (kHz)

RADAR FREQUENCY (GHz)

DOPPLER FREQUENCY
(Hz, ×100 FOR RADAR)

GRAPH FOR ADDING TWO IN-PHASE SIGNALS

This graph determines the combined signal level and shows the number of dB that must be added to the larger signal.

FOR EXAMPLE:
Two in-phase signals are -25 dB and -27 dB respectively. The difference is 2 dB and, from the graph, 2.2 dB must be added to the larger signal. Thus, the combined signal power level is -25 dB plus 2.2 dB or -22.8 dB.

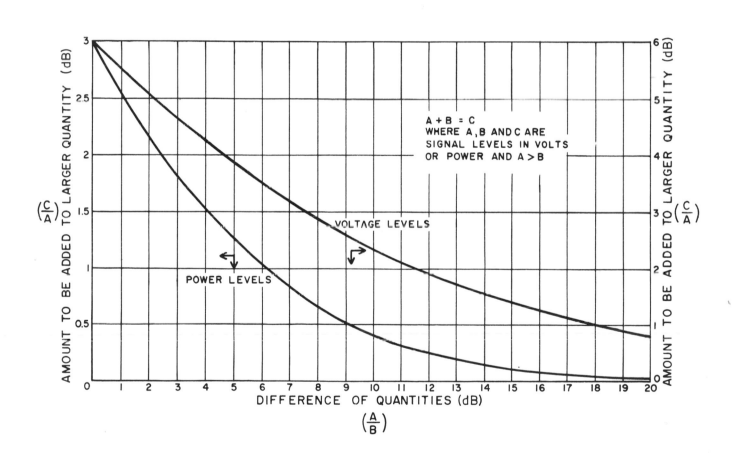

GRAPH FOR SEPARATING SIGNAL POWER FROM NOISE POWER

When making transmission loss or crosstalk measurements, the presence of noise is a potential source of error. If the total voltage measured across the load resistance when a signal is being transmitted is 15 dB or more greater than the noise voltage alone, the error in the received voltage measurement will be negligible. If, however, the dB difference between the combined signal and noise voltage and the noise voltage alone is less than 15 dB, a correction must be made. To do so, two voltage measurements must be made. Namely, (1) the noise power in dBm, and (2) the combined noise and signal power in dBm. On the horizontal axis locate the point equal to the difference between the two powers and read on the vertical axis the number of dB to be subtracted from the noise plus signal power and obtain the power of the signal alone.

FOR EXAMPLE:
The difference between the measurements of combined noise and crosstalk and noise alone is 5 dB. Thus, 1.7 dB must be subtracted from the combined signal and noise level to obtain the level of the signal alone.

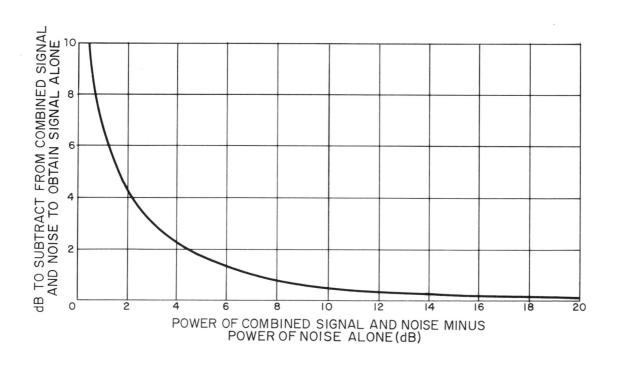

FIELD POWER CONVERSION CHART

Power density is related to field strength by the equation

$$P = \frac{E^2}{120\pi}$$

where

P = the power density

E = the field strength

and

120π = the resistance of free space

This chart converts between field strength and power density.

FOR EXAMPLE:

A field strength of 3000 μV/m corresponds to a power density of 0.024 μW/m² and is 70.5 dB above 1μV/m.

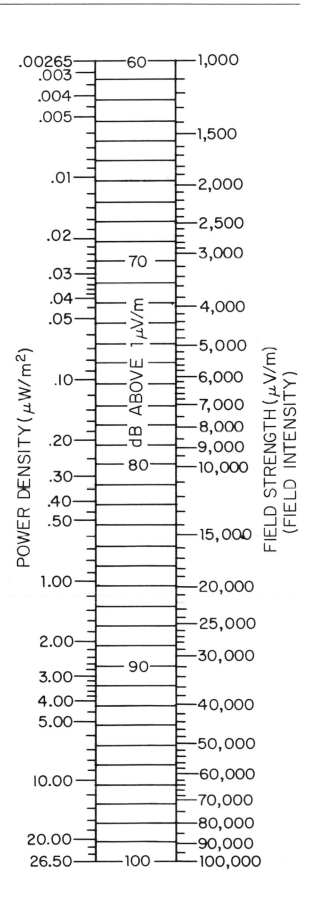

Q SIGNALS (MNEMONIC CODE)

The Q code was first adopted in 1912 by international treaty agreement to overcome the language barriers faced by ship operators of all nations as they tried to communicate with shore stations all over the world. Many of the original list of 50 signals are still in use with their definitions unchanged. Many more have been added from time to time, and the official meanings of some signals have been changed. In addition, many signals have been informally adopted for use by amateurs in situations not covered by the official lists.

The list below includes virtually every Q signal which could, even remotely, be thought to have an application in amateur radio communication. To simplify the task of finding the definition of an unfamiliar signal, we have combined all the signals into a single alphabetical list, mixing "official" and unofficial signals. The definitions listed are, in most cases, the official ones, taken verbatim from the treaty. In other cases, where definitions are not the official ones, they are as amateurs universally understand them, for purposes of amateur communications. The QN signals, adopted by ARRL for traffic net use, have official definitions which refer to aeronautical situations.

QAM   What is the latest available meterological observation for (place)?
The observation made at (time) was . . . .

QAP   Shall I listen for you (or for . . .) on . . . kHz?
Listen for me (or for . . .) on . . . kHz.

QAR   May I stop listening on the watch frequency for . . . minutes?
You may stop listening on the watch frequency for . . . minutes?

QBF   Have we worked before in this contest?
We have worked before in this contest.

QHM   I will tune from the high end of the band toward the middle.
(Used after a call or CQ.)

QIF   What frequency is . . . using?
He is using . . . kHz.

QJA   Is my RTTY (1—tape, 2—M/S) reversed?
It is reversed.

QJB   Shall I use (1—TTY, 2—reperf)? (For RTTY use.)
Use (1—TTY, 2—reperf).

QJC   Check your RTTY (1—TC, 2—auto head, 3—reperf, 5—Printer, 7—keyboard).

QJD   Shall I transmit (1—letters, 2—figs)? (For RTTY)
Transmit (1—letters, 2—figs).

QJE   Shall I send (1—wide, 2—narrow, 3—correct) RTTY shift?
Your RTTY shift is (1—wide, 2—narrow, 3—correct).

QJF   Does my RTTY signal check out OK?
Your RTTY signal checks out OK.

QJH   Shall I transmit (1—test tape, 2—test sentence) by RTTY?
Transmit (1—test tape, 2—test sentence) by RTTY.

QJI   Shall I transmit continuous (1—mark, 2—space) RTTY signal?
Transmit continuous (1—mark, 2—space) signal.

QJK   Are you receiving continuous (1—mark, 2—space, 3—mark bias, 4—space bias)?
I am receiving continuous (1—mark, 2—space, 3—mark bias, 4—space bias).

QKF   May I be relieved at . . . hours?
You may expect to be relieved at . . . hours by. . . .

QLM   I will tune for answers from the low end of the band toward the middle.

QMD   I will tune for answers from my frequency down.

QMH   I will tune for answers from the middle of the band toward the high end.

QML   I will tune for answers from the middle of the band toward the low end.

QMU   I will tune for answers from my frequency upward.

QMT   Will you mail the traffic?
I will accept the traffic for delivery by mail.

QNA*   Answer in prearranged order.

QNB*   Act as relay between . . . and. . . .

QNC   All net stations copy.
I have a message for all net stations.

QND*   Net is directed (controlled by net control station).

QNE*   Entire net stand by.

QNF   Net is free (not controlled).

QNG   Take over as net control station.

QNH   Your net frequency is high.

QNI   Net stations report in.*
I am reporting into the net. (Follow with list of traffic or QRU.)

QNJ   Can you copy me?
Can you copy . . . ?

QNK*   Transmit messages for . . . to . . . .

QNL   Your net frequency is low.

QNM*   You are QRMing the net. Stand by.

QNN   Net control station is . . . .*
What station has net control?

*For use only by Net Control Station.

QNO   Station is leaving the net.

QNP   Unable to copy you.
Unable to copy . . . .

QNQ*   QSY to . . . . . . .and wait for . . . to finish. Then send him traffic for . . . .

QNR*   Answer . . . and receive traffic.

QNS   Following stations are in the net.* (Follow with list.) Request list of stations in the net.

QNT   I request permission to leave the net for . . . minutes.

QNU*   The net has traffic for you. Stand by.

QNV   Establish contact with . . . on this freq. If successful QSY to . . . and send traffic for . . . .

QNW   How do I route messages for . . . ?

QNX   You are excused from the net.*
Request to be excused from the net.

QNY*   Shift to another frequency (or to . . . kHz) to clear traffic with . . . .

QNZ*   Zero beat your signal with mine.

QRA   What is the name of your station?
The name of my station is . . . .

QRB   How far approximately are you from my station?
The approximate distance between our station is . . . nautical miles (or kilometers).

QRD   Where are you bound for and where are you from?
I am bound for . . . from. . . .

QRE   What is your estimated time of arrival at . . . (or over . . .) (place)?
My estimated time of arrival at . . . (or over . . .) (place) is . . . hours.

QRF   Are you returning to . . . (place)?
I am returning to . . . (place).
or
Return to . . . (place).

QRG   Will you tell me my exact frequency (or that of . . .)?
Your exact frequency (or that of . . .) is . . . kHz (or MHz).

QRH   Does my frequency vary?
Your frequency varies.

QRI   How is the tone of my transmission?
The tone of your transmission is (1—good, 2—variable, 3—bad.

QRJ   Are you receiving me badly? Are my signals weak?
I am receiving you badly. Your signals are too weak.

QRK   What is the intelligibility of my signals (or those of . . .)?
The intelligibility of your signals (or those of . . .) is 1—bad, 2—poor, 3—fair, 4—good, 5—excellent.

*For use only by Net Control Station.

QRL   Are you busy?
I am busy (or I am busy with . . .). Please do not interfere.

QRM   Are you being interfered with?
I am being interfered with (1—nil, 2—slightly, 3—moderately, 4—severely, 5—extremely).

QRN   Are you troubled by static?
I am troubled by static (1—nil, 2—slightly, 3—moderately, 4—severely, 5—extremely).

QRO   Shall I increase transmitter power?
Increase transmitter power.

QRP   Shall I decrease transmitter power?
Decrease transmitter power.

QRQ   Shall I send faster?
Send faster (. . . words per minute).

QRR   Are you ready for automatic operation?
I am ready for automatic operation. Send at . . . words per minute.

QRRR   Distress call signal for use by amateur c.w. and RTTY stations. To be used only in situations where there is danger to human life or safety.

QRS   Shall I send more slowly?
Send more slowly.

QRT   Shall I stop sending?
Stop sending.

QRU   Have you anything for me?
I have nothing for you.

QRV   Are you ready?
I am ready.

QRW   Shall I inform . . . that you are calling him on . . . kHz?
Please inform . . . that I am calling him on . . . kHz.

QRX   When will you call me again?
I will call you again at . . . hours (on . . . kHz).

QRY   What is my turn?
(Relates to communication)
Your turn is Number . . . (or according to any other indication).
(Relates to communication)

QRZ   Who is calling me?
You are being called by . . . (on . .kHz).

QSA   What is the strength of my signals (or those of . . .)?
The strength of your signals (or those of . . .) is (1—scarcely perceptible, 2—weak, 3—fairly good, 4—good, 5—very good).

QSB   Are my signals fading?
Your signals are fading.

QSD   Is my keying defective?
Your keying is defective.

QSG   Shall I send . . . messages at a time?
Send . . . messages at a time.

QSH   Are you able to home on your D/F equipment?
I am able to home on my D/F equipment (on station . . .).

QSI   I have been unable to break in on your transmission.
or
Will you inform . . . (call sign) that I have been unable to break in on his transmission (on . . . kHz).

QSK   Can you hear me between your signals and if so can I break in on your transmission?
I can hear you between my signals; break in on my transmission.

QSL   Can you acknowledge receipt?
I am acknowledging receipt.

QSM   Shall I repeat the last telegram which I sent you (or some previous telegram)?
Repeat the last telegram which you sent me (or telegram(s) number(s) . . .).

QSN   Did you hear me [or . . . (call sign)] on . . . kHz?
I did hear you [or . . . (call sign)] on . . . kHz.

QSO   Can you communicate with . . . direct (or by relay)?
I can communicate with . . . direct (or by relay through . . .).

QSP   Will you relay to . . . free of charge?
I will relay to . . . free of charge.

QSQ   Have you a doctor on board [or is . . . (name of person) on board] ?
I have a doctor on board [or . . . (name of person) is on board] .

QSR   Shall I repeat the call on the calling frequency?
Repeat your call on the calling frequency; did not hear you (or have interference).

QSS   What working frequency will you use?
I will use the working frequency . .kHz.

QST   Calling all radio amateurs.

QSU   Shall I send or reply on this frequency (or on . . . kHz?
Send or reply on this frequency (or on . . . kHz.

QSV   Shall I send a series of V's on this frequency (or . . . kHz)?
Send a series of V's on this frequency (or . . . kHz).

QSW   Will you send on this frequency (or on . . . kHz)?
I am going to send on this frequency (or on . . . kHz).

QSX   Will you listen to . . . (call sign(s)) on . . . kHz?

I am listening to . . . (call sign(s)) on . . . kHz.

QSY   Shall I change to transmission on another frequency?
Change to transmission on another frequency (or on . . . kHz).

QSZ   Shall I send each word or group more than once?
Send each word or group twice (or . . . times).

QTA   Shall I cancel message number . . . ?
Cancel message number. . . .

QTB   Do you agree with my counting of words?
I do not agree with your counting of words; I will repeat the first letter or digit of each word or group.

QTC   How many messages have you to send?
I have . . . messages for you (or for . . .).

QTG   Will you send two dashes of ten seconds each followed by your call sign (repeated . . . times) (on . . . kHz)? or Will you request . . . to send two dashes of ten seconds followed by his call sign (repeated . . . times) on . . . kHz?
I am going to send two dashes of ten seconds each followed by my call sign (repeated . . . times) (on . . . kHz). or I have requested . . . to send two dashes of ten seconds followed by his call sign (repeated . . . times) on . . . kHz.

QTH   What is your position in latitude and longitude (or according to any other indication)?
My position is . . . latitude . . . longitude (or according to any other indication).

QTN   At what time did you depart from . . . (place)?
I departed from . . . (place) at . . . hours.

QTO   Have you left dock (or port)? or Are you airborne?
I have left dock (or port). or I am airborne.

QTP   Are you going to enter dock (or port)? or Are you going to alight (or land)?
I am going to enter dock (or port). or I am going to alight (or land).

QTQ   Can you communicate with my station by means of the International Code of Signals?
I am going to communicate with your station by means of the International Code of Signals.

QTR   What is the correct time?
The correct time is . . . hours.

QTS   Will you send your call sign for tuning

purposes or so that your frequency can be measured now (or at . . . hours) on . . . kHz?

I will send my call sign for tuning purposes or so that my frequency may be measured now (or at . . . hours) on . . . kHz.

QTU   What are the hours during which your station is open?

My station is open from . . . to . . . hours.

QTV   Shall I stand guard for you on the frequency of . . . kHz (from . . . to . . . . hours)? Stand guard for me on the frequency of . . . kHz (from . . . to hours).

QTX   Will you keep your station open for further communication with me until further notice (or until . . . hours)?

I will keep my station open for further communication with you until further notice (or until . . . hours).

QTY   Are you proceeding to the position of incident and if so when do you expect to arrive?

I am proceeding to the position of incident and expect to arrive at . . . hours on . . . (date).

QTZ   Are you continuing the search?

I am continuing the search for . . . (aircraft, ship, survival craft, survivors, or wreckage).

QUA   Have you news of . . . (call sign)?

Here is news of . . . (call sign).

QUB   Can you give me in the following order information concerning: the direction in

degrees TRUE and speed of the surface wind; visibility; present weather; and amount, type, and height of base of cloud above surface elevation at . . . (place of observation)?

Here is the information requested: . . . (The units used for speed and distances should be indicated.)

QUC   What is the number (or other indication) of the last message you received from me [or from . . . (call sign)] ?

The number (or other indication) of the last message I received from you [or from . . . (call sign)] is . . . .

QUE   Can you use telephony in . . . (language), with interpreter if necessary; if so, on what frequencies?

I can use telephony in . . . (language) on . . . kHz.

QUF   Have you received the distress signal sent by . . . (call sign of station)?

I have received the distress signal sent by . . . (call sign of station) at . . . hours.

QUH   Will you give me the present barometric pressure at sea level?

The present barometric pressure at sea level is . . . (units).

QUK   Can you tell me the condition of the sea observed at . . . (place or coordinates)?

The sea at . . . (place or coordinates) is . . . .

QUM   May I resume normal working?

Normal working may be resumed.

RADIO TELEPHONE CODE

General Station Operation

10-1 Receiving poorly.

10-2 Signals good.

10-3 Stop transmitting.

10-4 Okay—Affirmative—Acknowledged.

10-5 Relay this message.

10-6 Busy, stand by.

10-7 Leaving the air.

10-8 Back on the air and standing by.

10-9 Repeat message.

10-10 Transmission completed, standing by.

10-11 Speak slower.

10-13 Advise weather and road conditions.

10-18 Complete assignment as quickly as possible.

10-19 Return to base.

10-20 What is your location? My location is . . . .

10-21 Call . . . by telephone.

10-22 Report in person to . . . .

10-23 Stand by.

10-24 Have you finished? I have finished.

10-25 Do you have contact with . . . ?

Emergency or Unusual

10-30 Does not conform to Rules and Regulations.

10-33 Emergency traffic this station.

10-35 Confidential information.

10-36 Correct time.

10-41 Tune to channel . . . for test, operation, or emergency service.

10-42 Out of service at home.

10-45 Call . . . by phone.

10-54 Accident.

10-55 Wrecker or tow truck needed.

10-56 Ambulance needed.

Net Message Handling

10-60 What is next message number?

10-62 Unable to copy, use CW.

10-63 Net clear.

10-64 Net is clear.

10-66 Cancellation.

10-68 Repeat dispatch on message.

10-69 Have you dispatched message . . . ?

10-70 Net message.

10-71 Proceed with transmission in sequence.

Personal

10-82 Reserve room for . . . .

10-84 What is your telephone number?

10-88 Advise present phone number of . . . .

Technical

10-89 Repairman needed.

10-90 Repairman will arrive at your station . . .

10-92 Poor signal, have transmitter checked.

10-93 Frequency check.

10-94 Give a test without voice for frequency check.

10-95 Test with no modulation.

10-99 Unable to receive your signals.

INTERNATIONAL MORSE CODE

Alphabetical

A · —	J · — — —	S · · ·
B — · · ·	K — · —	T —
C — · — ·	L · — · ·	U · · —
D — · ·	M — —	V · · · —
E ·	N — ·	W · — —
F · · — ·	O — — —	X — · · —
G — — ·	P · — — ·	Y — · — —
H · · · ·	Q — — · —	Z — — · ·
I · ·	R · — ·	

By Groups

Group One	Group Two	Group Three
E ·	A · —	R · — ·
I · ·	W · — —	F · · — ·
S · · ·	J · — — —	L · — · ·
H · · · ·	N — ·	U · · —
T —	D — · ·	V · · · —
M — —	B — · · ·	
O — — —		

Group Four

K — · —	Q — — · —
X — · · —	G — — ·
C — · — ·	Z — — · ·
Y — · — —	P · — — ·

Numerals and Punctuation

1 · — — — —	6 — · · · ·
2 · · — — —	7 — — · · ·
3 · · · — —	8 — — — · ·
4 · · · · —	9 — — — — ·
5 · · · · ·	0 — — — — —

Period · — · — · —
Comma — — · · — —
Question mark · · — — · ·
Error · · · · · · · ·
Double dash — · · · —
Fraction bar — · · — ·
Wait · — · · ·
Invitation to transmit — · —
End of message (AR) · — · — ·
End of transmission · · · — · —

Special Foreign Letters

Ä (German) · — · —
Á or À (Spanish-Scandinavian) · — — · —
CH (German-Spanish) — — — —
É (French) · · — · ·
Ñ (Spanish) — — · — —
Ö (German) — — — ·
Ü (German) · · — —

SIGNAL REPORTING CODES

RST Code

The standard amateur method of giving signal strength reports. For phone operation only the first two sets of numbers are used with the words "readability" and "strength."

READABILITY (R)

1 Unreadable
2 Barely readable, occasional words distinguishable
3 Readable with considerable difficulty
4 Readable with practically no difficulty
5 Perfectly readable

SIGNAL STRENGTH (S)

1 Faint; signal barely perceptible
2 Very weak signal
3 Weak signal
4 Fair signal
5 Fairly good signal
6 Good signal
7 Moderately strong signal
8 Strong signal
9 Extremely strong signal

TONE (T)

1 Extremely rough, hissing signal
2 Very rough ac signal
3 Rough, low-pitched ac signal
4 Rather rough ac signal
5 Musically modulated signal
6 Modulated signal, slight whistle
7 Near dc signal, smooth ripple
8 Good dc signal, trace of ripple
9 Purest dc signal

If the signal has the steadiness of crystal control, add "X" after the RST report; add "C" for a chirp; and "K" for a keying click.

A typical report might be: "RST579X," meaning "Your signals are perfectly readable, moderately strong, have a perfectly clear tone, and have the stability of a crystal-controlled transmitter."

This reporting system is used on both CW and voice, leaving out the "Tone" report on voice.

SINPO Code

A reporting method used in the shortwave field. All the numbers after the letters range from one to five. Q-code equivalents for each characteristic are also shown.

For example:

A typical report for a station that is coming in loud and clear would read: SINPO 55555.

S Signal Strength (QSA)	I Interference (QRM)	N Atmospheric Noise (QRN)	P Propagation Disturbance (QSB)	O Overall Merit (QRK)
5 Excellent	5 None	5 None	5 None	5 Excellent
4 Good	4 Slight	4 Slight	4 Slight	4 Good
3 Fair	3 Moderate	3 Moderate	3 Moderate	3 Fair
2 Poor	2 Severe	2 Severe	2 Severe	2 Poor
1 Barely audible	1 Extreme	1 Extreme	1 Extreme	1 Unusable

555 Code

Another reporting code sometimes used in the shortwave field.

Signal Strength	Interference	Overall Merit
0 Inaudible	0 Total	0 Unusable
1 Poor	1 Very severe	1 Poor
2 Fair	2 Severe	2 Fair
3 Good	3 Moderate	3 Good
4 Very good	4 Slight	4 Very good
5 Excellent	5 None	5 Excellent

COMMERCIAL RADIO OPERATOR AND AMATEUR OPERATOR LICENSES REQUIREMENTS

Amateur Operator Licenses

Class	Prior Experience	Code Test	Written Examination	Privileges	Term
Novice	None	5 w.p.m.	Elementary theory and regulations	Telegraphy in 3.7–3.75, 7.15–7.2, 21.1–21.25, 145–147 MHz. Crystal control required; 75 W maximum input.	2 years, not renewable
Technician	None	5 w.p.m.	General theory and regulations	All amateur privileges in 50.25–54.0, 145–147, and above 220 MHz.	5 years, renewable
General Conditional[b]	None	13 w.p.m.	General theory and regulations	1.8–2,[a] 3.55–3.8, 3.9–4, 7.05–7.2, 7.25–7.3, 14.05–14.2, 14.275–14.35, 21.05–21.25, 21.35–21.45 MHz, and all higher amateur frequencies except 50–50.25 MHz.	5 years, renewable
Advanced	None	13 w.p.m. (Credit is given to General Class Licensees)	General theory and regulations, plus intermediate theory.	1.8–2,[a] 3.55–3.8, 3.825–4, 7.05–7.3, 14.05–14.35, 21.05–21.25, 21.275–21.45, and all higher amateur frequencies.	5 years, renewable
Amateur Extra	Two years since 1934,[c] except as Novice or Technician	20 w.p.m.[c]	General theory and regulations, intermediate theory,[c] plus special exam[c] on advanced techniques	All amateur privileges	5 years, renewable

[a] The 1.8–2 band frequency and power assignments differ from state to state. Check with nearest FCC office.
[b] Same as General Class, except examination taken by mail.
[c] Waived for persons who hold or can qualify for a General or Advanced Class license and who can show that they held amateur license prior to May 1917.

COMMERCIAL RADIO OPERATOR
AND AMATEUR OPERATOR
LICENSES REQUIREMENTS (Continued)

Commercial Radio Operator Licenses

Type of License	Age Minimum	Code Requirement	Written Test	Term of License
Restricted Radiotelephone Permit[a]	14 years	None	None; obtained by declaration (FCC Form 753)	Permanent
Third Class Radiotelephone Permit	None	None	Elements 1, 2	5 years, renewable
Second Class Radiotelephone License	None	None	Elements 1, 2, 3	5 years, renewable
First Class Radiotelephone License	None	None	Elements 1, 2, 3, 4	5 years, renewable
Third Class Radiotelegraph Permit	None	16 code groups per minute	Elements 1, 2, 5	5 years, renewable
Second Class Radiotelegraph License	None	16 code groups per minute	Elements 1, 2, 5, 6	5 years, renewable
First Class Radiotelegraph License	21 years; one year experience	20 code groups, 25 plain words per minute	Elements 1, 2, 5, 6	5 years, renewable

[a]Intended mainly for private boats, planes, etc.

Examination Elements

1. Basic law, 20 questions.
2. Basic operating practice, 50 questions.
3. Basic radiotelephone, 100 questions.
4. Advanced radiotelephone, 50 questions.
5. Radiotelegraph operating practice, 50 questions.
6. Advanced radiotelegraph, 100 questions.

CLASSIFICATION OF EMISSIONS

In accordance with Federal Communications Commission Rules and Regulations 2.201, Subpart C, the following system of designating emission, modulation, and transmission characteristics is employed.

(a) Emissions are designated according to their classification and their necessary bandwidth.

(b) Emissions are classified and symbolized according to the following characteristics.

(1) Type of modulation of main carrier.
(2) Type of transmission.
(3) Supplementary characteristics.

(c) Types of modulation of main carrier:

	Symbol
(1) Amplitude	A
(2) Frequency (or Phase)	F
(3) Pulse	P

(d) Types of transmission:

(1) Absence of any modulation intended to carry information _____ 0
(2) Telegraphy without the use of a modulating audio frequency _____ 1
(3) Telegraphy by the on-off keying of a modulating audio frequency or audio frequencies, or by the on-off keying of the modulated emission (special case: an unkeyed modulated emission) _____ 2

Symbol

(4) Telephony (including sound broadcasting) _____ 3
(5) Facsimile (with modulation of main carrier either directly or by a frequency modulated sub-carrier) _____ 4
(6) Television (visual only) _____ 5
(7) Four-frequency diplex telegraphy _____ 6
(8) Multichannel voice-frequency telegraphy _____ 7
(9) Cases not covered by the above _____ 9

(e) Supplementary characteristics:

(1) Double sideband _____ (None)
(2) Single sideband:
    (i) Reduced carrier _____ A
    (ii) Full carrier _____ H
    (iii) Suppressed carrier _____ J
(3) Two independent sidebands _____ B
(4) Vestigial sideband _____ C
(5) Pulse:
    (i) Amplitude modulated _____ D
    (ii) Width (or duration) modulated _____ E
    (iii) Phase (or position) modulated _____ F
    (iv) Code modulated _____ G

(f) The classification of typical emissions is tabulated as follows:

Type of modulation of main carrier	Type of transmission	Supplementary characteristics	Symbol
Amplitude modulation	With no modulation		A0
	Telegraphy without the use of a modulating audio frequency (by on-off keying).		A1
	Telegraphy by the on-off keying of an amplitude modulating audio frequency or audio frequencies, or by the on-off keying of the modulated emission (special case: an unkeyed emission amplitude modulated).		A2
	Telephony	Double sideband	A3
		Single sideband, reduced carrier	A3A
		Single sideband, suppressed carrier	A3J
		Two independent sidebands	A3B
	Facsimile (with modulation of main carrier either directly or by a frequency modulated subcarrier).		A4
	Facsimile	Single sideband, reduced carrier	A4A
	Television	Vestigial sideband	A5C
	Multichannel voice-frequency telegraphy	Single sideband, reduced carrier	A7A
	Cases not covered by the above, e.g., a combination of telephony and telegraphy.	Two independent sidebands	A9B
Frequency (or Phase) modulation	Telegraphy by frequency shift keying without the use of a modulating audio frequency: one of two frequencies being emitted at any instant.		F1
	Telegraphy by the on-off keying of a frequency modulating audio frequency or by the on-off keying of a frequency modulated emission (special case: an unkeyed emission, frequency modulated).		F2
	Telephony		F3
	Facsimile by direct frequency modulation of the carrier.		F4
	Television		F5
	Four-frequency diplex telegraphy		F6
	Cases not covered by the above, in which the main carrier is frequency modulated.		F9
Pulse modulation	A pulsed carrier without any modulation intended to carry information (e.g. radar).		P0
	Telegraphy by the on-off keying of a pulsed carrier without the use of a modulating audio frequency.		P1D
	Telegraphy by the on-off keying of a modulating audio frequency or audio frequencies, or by the on-off keying of a modulated pulsed carrier (special case: an unkeyed modulated pulsed carrier).	Audio frequency or audio frequencies modulating the amplitude of the pulses.	P2D
		Audio frequency or audio frequencies modulating the width (or duration) of the pulses.	P2E
		Audio frequency or audio frequencies modulating the phase (or position) of the pulses.	P2F
	Telephony	Amplitude modulated pulses	P3D
		Width (or duration) modulated pulses	P3E
		Phase (or position) modulated pulses	P3F
		Code modulated pulses (after sampling and quantization).	P3G
	Cases not covered by the above in which the main carrier is pulse modulated.		P9

CLASSIFICATION OF EMISSIONS
(Continued)

Class	Name	Code	Action of Modulating Signal
A	Pulse-time modulation	PTM	Varies some characteristic of pulse with respect to time.
	Pulse-position modulation	PPM	Varies position (phase) of pulse on time base.
	Pulse-duration modulation	PDM	Varies width of pulse (also called PWM, or Pulse-Width Modulation).
	Pulse-shape modulation		Varies shape of pulse.
	Pulse-frequency modulation	PFM	Varies pulse recurrence frequency.
B	Pulse-amplitude modulation	PAM	Varies amplitude of pulse—consists of two types: one using unipolar pulses, the other using bipolar pulses.
C	Pulse-code modulation	PCM	Varies the makeup of a series of pulses and spaces. Individual systems are classified as follows: Binary-pulse and spaces, or positive and negative pulses. Ternary-positive pulses, negative pulses, and spaces. N-ary-more complex combinations of pulses and spaces.

INTERNATIONAL PHONETIC
ALPHABET

To avoid errors or misunderstanding during voice communication, the new international phonetic alphabet has been adopted.

Letter	Name	Pronunciation
A	Alfa	AL-fah
B	Bravo	BRAH-voh
C	Charlie	CHAR-lee
		(or SHAR-lee)
D	Delta	DELL-tah
E	Echo	ECK-oh
F	Foxtrot	FOKS-trot
G	Golf	GOLF
H	Hotel	HOH-tel
I	India	IN-dee-ah
J	Juliett	JEW-lee-ett
K	Kilo	KEY-loh
L	Lima	LEE-mah
M	Mike	MIKE
N	November	No-VEM-ber
O	Oscar	OSS-cah
P	Papa	Pah-PAH
Q	Quebec	Keh-BECK
R	Romeo	ROW-me-oh
S	Sierra	See-AIR-rah
T	Tango	TANG-go
U	Uniform	YOU-nee-form
		(or OO-nee-form)
V	Victor	VIK-tah
W	Whiskey	WISS-key
X	X-ray	ECKS-ray
Y	Yankee	YANG-key
Z	Zulu	ZOO-loo

ARRL (AMERICAN RADIO RELAY LEAGUE) WORD LIST FOR VOICE COMMUNICATION

A—Adam	N—Nancy
B—Baker	O—Otto
C—Charlie	P—Peter
D—David	Q—Queen
E—Edward	R—Robert
F—Frank	S—Susan
G—George	T—Thomas
H—Henry	U—Union
I—Ida	V—Victor
J—John	W—William
K—King	X—X-Ray
L—Lewis	Y—Young
M—Mary	Z—Zebra

Example: W1AW . . . W1
ADAM WILLIAM . . .
W1AW

TRANSMISSION TRAVEL TIME

The time required for electromagnetic energy to travel interplanetary distances is significant. Shown here are some typical times and distances related to the earth's position.

Moon	(overhead)	= 23.9 X 10^4 n mi	1.27	sec one way
Venus	(nearest)	= 22.4 X 10^6 n mi	139.00	sec one way
	(farthest)	= 139.0 X 10^6 n mi	859.00	sec one way
Mars	(nearest)	= 42.4 X 10^6 n mi	262.00	sec one way
	(farthest)	= 203.9 X 10^6 n mi	1259.00	sec one way
Jupiter	(nearest)	= 339.8 X 10^6 n mi	2099.00	sec one way
	(farthest)	= 501.2 X 10^6 n mi	3096.00	sec one way

MICROPHONE OUTPUT NOMOGRAM

This nomogram determines the output voltages for various microphone ratings and relates this output to actual sound pressure levels.

Two methods of specifying microphone levels are in general use. Acoustic input and electrical output are specified so that the microphone can be considered as a generator, with sound pressure input and voltage or power output.

For low-impedance microphones, output is given in decibels referenced to 1 mW for 10 dynes/cm^2 sound pressure.

For high-impedance microphones, output is given in decibels referenced to 1 V for 1 dyne/cm^2 sound pressure.

(In both of the above, output is into a resistive load equal to the impedance of the microphone.)

This nomogram is prepared for microphone pre-amplifiers with low input impedances matched to the microphone impedance. (Open-circuit voltage is 6 dB higher than the nomogram value.) Connecting the microphone impedance and the decibel rating solves for the voltage across a matched load for the standard 10 dynes/cm^2 sound pressure field. By referring to the absolute sound pressure vs decibel scale, any other sound pressure level can be found and the decibel difference (with respect to 10 dynes/cm^2) can be determined, and adjustments can be made in the output voltage by adding or subtracting decibels.

For high-impedance microphones, the nomogram is used in the same way, except that the impedance is always considered as 40,000 ohms, and the reading will be that for a 10 dynes/cm^2 field. These microphones are usually operated into a very high impedance circuit, hence 6 dB must be added to the output voltage. (Use of this method results in an error of approximately 2 dB.)

MICROPHONE OUTPUT
NOMOGRAM (Continued)

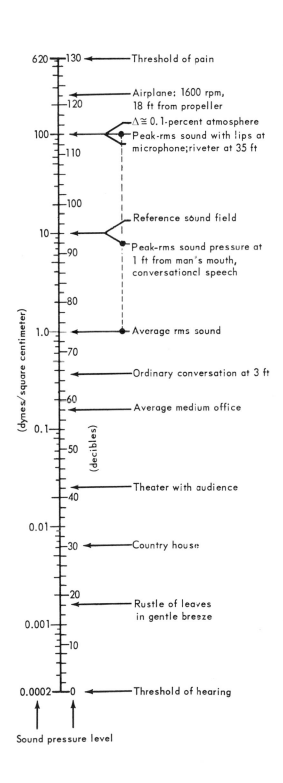

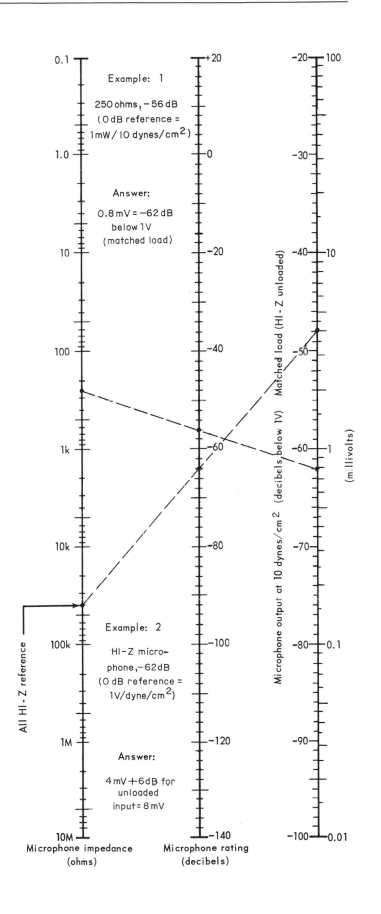

SECTION 3

PASSIVE COMPONENTS AND CIRCUITS

ATTENUATOR NOMOGRAMS

These two nomograms solve for the resistor values required for the following: T, Pi, H, O, lattice, bridged T, bridged H, L, and U-type attenuators. The nomograms are based on the equations shown. The keys next to the nomograms show which scales must be used for a particular type of attenuator.

FOR EXAMPLE:

1. Z_o is 600 ohms and the required attenuation is 20 dB. Design T, H, and Pi attenuators. From nomogram 1, for a T type, R_1 is 480 ohms and R_4 is 120 ohms. For an H type each of the four series arms would be 240 ohms. For Pi type (middle key) R_2 is 750 ohms and R_3 is 3000 ohms.

2. A lattice attenuator (key three, nomogram 1) that gives 20 dB of attenuation at 500 ohms requires R_1 to be 410 ohms and R_2 to be 610 ohms.

3. A bridged T attenuator (nomogram 2, first key) with an attenuation of 20 dB and terminal impedances of 450 ohms has R_5 as 4000 ohms and R_6 as 50 ohms.

4. Design an L-type attenuator (middle key, nomogram 2) with an attenuation of 14 dB, and an impedance of 50 ohms with the shunt arm at the output end. In this case R_5 is 200 ohms and R_8 is 62.5 ohms.

Note: In all cases the input and output impedances are the same.

$$R_1 = Z_o \left(\frac{K-1}{K+1} \right) \qquad R_3 = Z_o \left(\frac{K^2-1}{2K} \right) \qquad R_5 = Z_o (K-1) \qquad R_7 = Z_o \left(\frac{K-1}{K} \right)$$

$$R_2 = Z_o \left(\frac{K+1}{K-1} \right) \qquad R_4 = Z_o \left(\frac{2K}{K^2-1} \right) \qquad R_6 = Z_o \left(\frac{1}{K-1} \right) \qquad R_8 = Z_o \left(\frac{K}{K-1} \right)$$

where $K = \dfrac{E_{in}}{E_{out}}$

NOMOGRAM 2 FOR BRIDGED T, H, L, AND U TYPE ATTENUATORS.

NOMOGRAM 1 FOR T, Pi, H, O, AND LATTICE TYPE ATTENUATORS.

(Reprinted from *Radio-Electronics*, copyright Gernsback Publications, Inc., December 1953).

Twin-T filters with symmetrical response curves are frequently used to reject specific frequencies, or they may be included in the negative feedback loop of a frequency-selective amplifier as the tuning element. Other component combinations may be used, but the one selected here has the greatest possible selectivity. With this general configuration, any filter exhibits infinite attenuation at the notch frequency (f_o) which is specified by the values of R_1 and C_1. If it is only desired to reject f_o, then the choice of these values is arbitrary. However, if it is desired to design a filter with a symmetrical response curve so the dc gain is equal to that at high frequencies, that is accomplished when $R_1 = \sqrt{R_g R_L /2}$, and the notch

frequency is determined by the expression $f_o = 1/4\pi C_1 R_1$. The nomograms are based on these two equations. Usually R_g, R_L, and f_o are known, and the values of R_1 and C_1 are to be determined. It is also possible to use chart 2 alone and select arbitrary values for R_1 or C_1 if symmetrical response is not essential.

FOR EXAMPLE:

Design a filter with infinite attenuation at 800 Hz which is to be inserted between a 2000-ohm source impedance and a load resistance of 100,000 ohms. From nomogram 1 determine that R_1 should be 10,000 ohms, and with that value determine from nomogram 2 that C_1 must be 0.01 μF to achieve a symmetrical response curve.

Twin-T notch filter, with component values related as shown, yields maximum selectivity and symmetrical gain-frequency response.

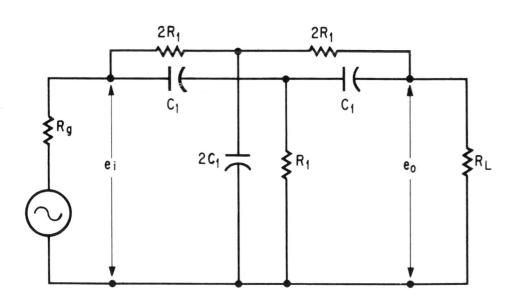

(Reprinted from *Electronics*, April 18, 1966; copyright McGraw-Hill, Inc., 1966.)

TWIN-T FILTER NOMOGRAM

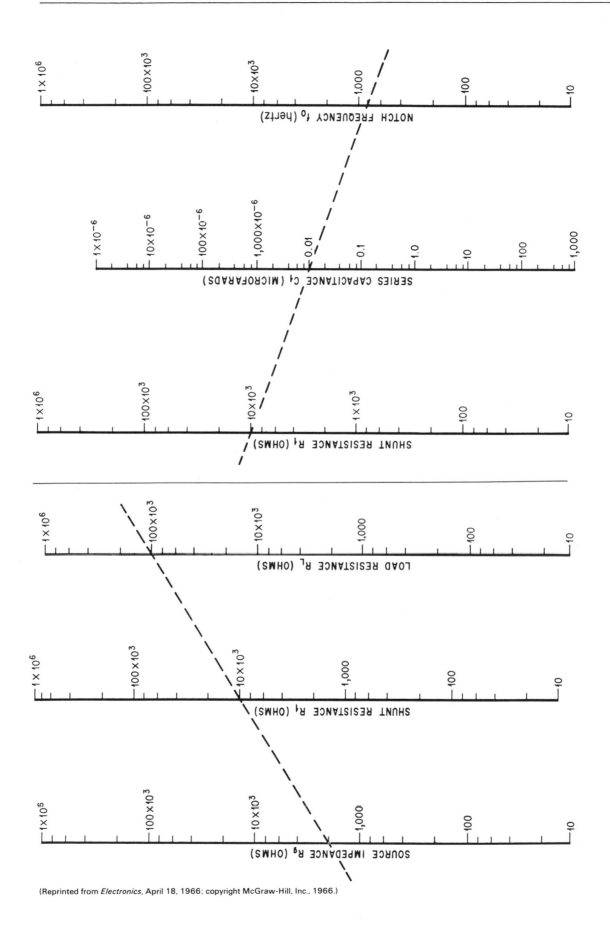

(Reprinted from *Electronics*, April 18, 1966; copyright McGraw-Hill, Inc., 1966.)

This nomogram solves for the resistance values needed for an impedance matching pad having a minimum of attenuation. Z_1 is the greater and Z_2 is the lesser terminal impedance in ohms. To use the nomogram, calculate the ratio of Z_2/Z_1 and connect that point on the center scale with Z_1 to find R_1, and with Z_2 to find R_3. Insertion loss is also read on the center scale.

FOR EXAMPLE:

If Z_2 is 400 ohms and Z_1 is 500 ohms, the value of R_1 must be 225 ohms and of R_3 890 ohms for a minimum insertion loss pad that has 4.2 dB of insertion loss.

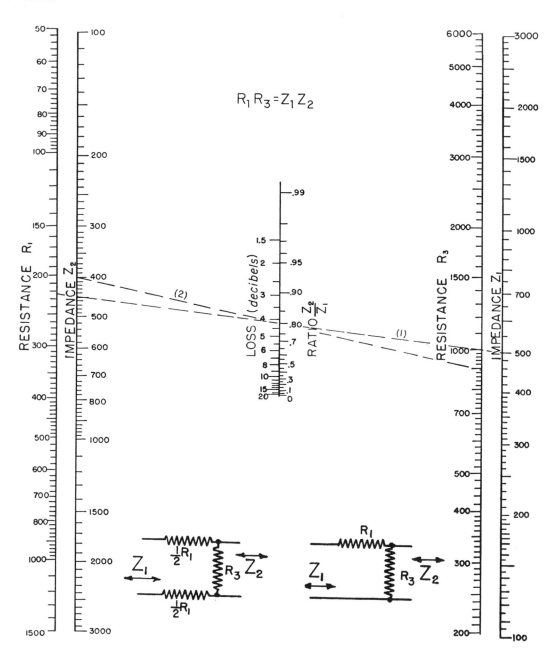

$$R_1 R_3 = Z_1 Z_2$$

THERMAL NOISE VOLTAGE
NOMOGRAM (A)

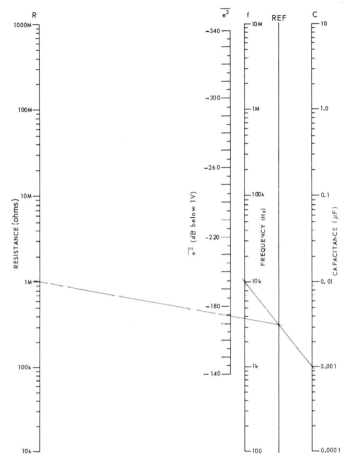

Given frequency, input C, and amplifier input Z, only two operations are required to find the equivalent thermal noise voltage.

When an amplifier is fed from a capacitive source, the spot (one frequency) noise is generated by the real part of the impedance. This nomogram reduces the calculation required to arrive at the noise value. Impedance at the amplifier input is

$$Z = \frac{R - jR^2 \omega C}{R^2 \omega^2 C^2 + 1} \qquad (1)$$

Thermal noise is generated by the real part of this expression, which is

$$(REAL\ Z) = \frac{R}{R^2 \omega^2 C^2 + 1} \approx \frac{1}{R\omega^2 C^2} \qquad (2)$$

The mean square thermal noise voltage associated with the real part of Z is given by

$$\bar{e}^2 = 4\,k\,T\,df\,(REAL\ Z) \qquad (3)$$

For this case

$$df = 1 \text{ (spot frequency)}$$
$$T = 25^\circ C$$

Combining (2) and (3)

$$\bar{e}^2 = 4\,k\,T\,df\,\frac{1}{R\omega^2 C^2} \qquad (4)$$

Equation (4) forms the basis for the nomogram.
Nomogram of equivalent spot thermal noise voltage of the parallel combination of a capacitor and an amplifier input resistance.
Using the nomogram:
1. Choose f, C, and R (in the example $f = 10$ kHz, $C = 0.001\ \mu F$, and $R = 1$ M ohm).
2. Draw a line between the chosen f and C.
3. Mark its intersection on the reference line.
4. Draw a line from the marked point on the reference scale to the chosen R.
5. The intersection of this line with the $\bar{e}^2$ scale is the desired equivalent thermal noise voltage in dB re 1V.

THERMAL NOISE VOLTAGE
NOMOGRAM (B)

Thermally produced noise voltage of any linear conductor is determined by Nyquist's equation

$$E = 2\sqrt{RkTB}$$

where E = noise voltage in rms microvolts

k = Boltzmann's constant, 1.38×10^{-23} J/$^\circ$K

R = resistance

T = absolute temperature ($^\circ$K)

B = bandwidth in hertz

This nomogram solves the above equation if any three of the four variables are given.

FOR EXAMPLE:

An amplifier has a voltage gain of 1000, an input resistance of 470,000 ohms, and a bandwidth of 2 kHz. Find the output noise level due to the input resistance if the amplifier is operated at an ambient temperature of 100°C.

Connect 100°C (T scale) with 470 K (R scale) and note intersect point on turning scale. Connect that point with 2 kHz (B scale) and read noise voltage as 4.4 μV on E scale. The amplifier has a gain of 1000; thus, the output noise of the amplifier due to the input resistance is 4.4 mV.

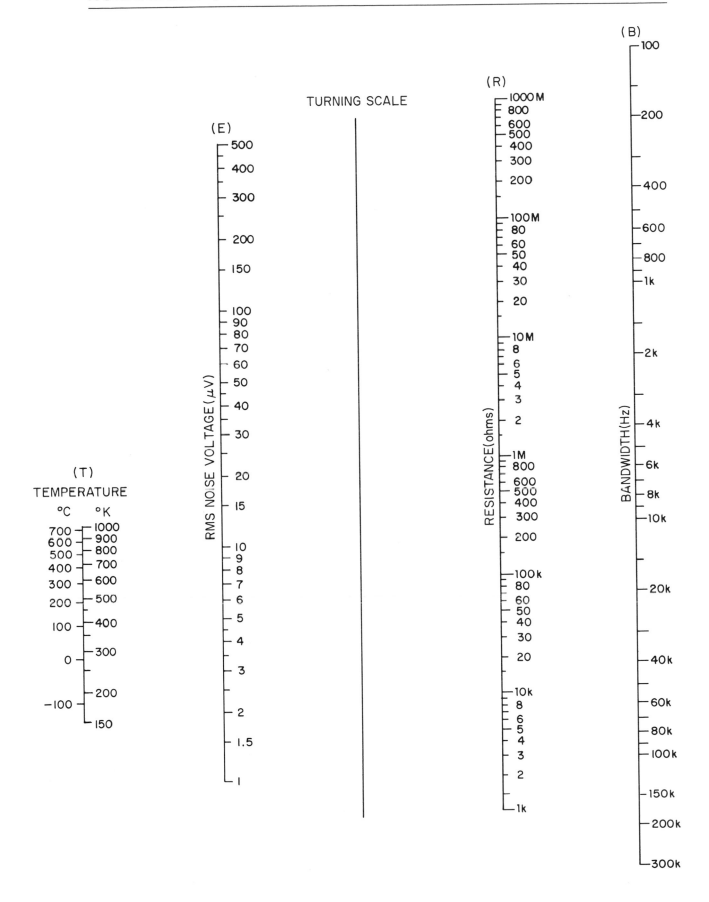

TURNING SCALE

(E)

(R)

(B)

RMS NOISE VOLTAGE (μV)

RESISTANCE (ohms)

BANDWIDTH (Hz)

(T)
TEMPERATURE

°C   °K

SINGLE-LAYER COIL DESIGN
NOMOGRAM (A)

This nomogram is based on the formula for the inductance of a single-layer coil

$$L = \frac{a^2 N^2}{9a + 10b}$$

where L = inductance in microhenries
a = coil radius in inches
b = coil length in inches
N = number of turns

FOR EXAMPLE:

1. Find the inductance of a 100-turn coil with a diameter of 2 in. and a winding length of 0.8 in.

Find K(diameter/length) 2/0.8 to be 2.5. Connecting 2.5 on the K scale to 100 on the N scale intersects the turning axis at 3.8. Now connect 3.8 with 2 on the D scale, and read the inductance as 600 μH.

2. Determine the number of turns required for a 290-μH coil 3 in. long with a diameter of 2.5 in. K is equal to 0.8. Connect 290 on the L scale with 2.5 on the D scale, and read 4.6 on the turning axis. Connecting 4.6 and 0.8 on the K scale gives the answer as 90 turns on the N scale.

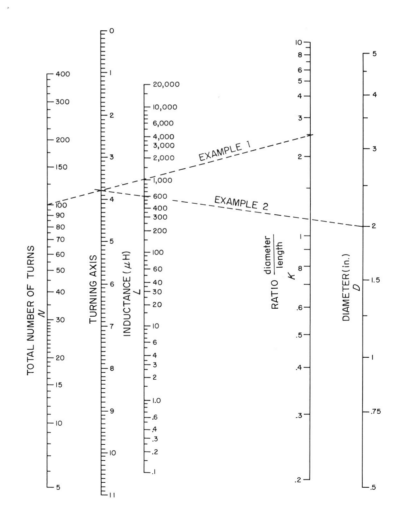

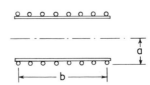

SINGLE-LAYER COIL DESIGN
NOMOGRAM (B)

This nomogram solves for the number of close-wound turns required to achieve inductances in the range of values required for television, fm, and radar if transformers. The nomogram is based on a slight modification of H.A. Wheeler's inductance formula that was used to construct nomogram A. The formula used here (with all dimensions in inches) is

$$L = \frac{a^2 N^2}{8.85a + 10b}$$

FOR EXAMPLE:
Ten turns of number 30 AWG enameled wire close-wound on a 0.25-inch diameter coil form will produce an inductance of 0.7 μH.

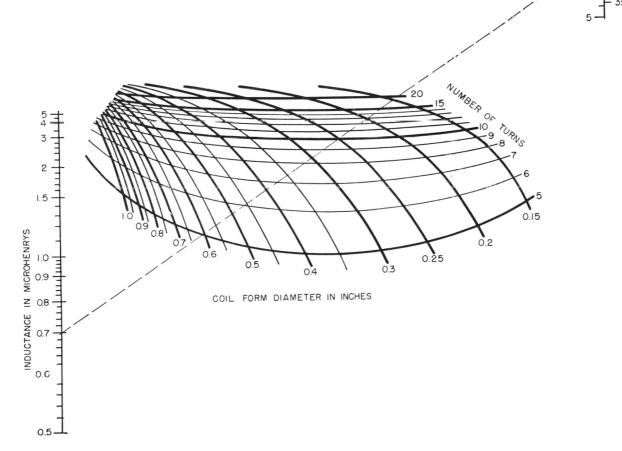

COIL FORM DIAMETER IN INCHES

INDUCTANCE OF STRAIGHT, ROUND WIRE AT HIGH FREQUENCIES

Above several megahertz the inductance of relatively short lengths of wire becomes important because of the effect on circuit performance. The chart shows the relationship between diameter, wire length, and inductance for various diameters. A more precise tabulation is also shown for short lengths of commonly used wire sizes.

FOR EXAMPLE:

A straight piece of wire 4 in. long with a diameter of 25 mil has an inductance of 0.2 μH. At a frequency of 80 MHz, this represents an inductive reactance of about 100 ohms.

Wire Size	Length (in.)	Approx. Inductance (μH)
20	1/4	0.0031
	1/2	0.0064
	3/4	0.0115
	1	0.019
	1 1/2	0.031
	2	0.04
24	1/4	0.0037
	1/2	0.0082
	3/4	0.014
	1	0.022
	1 1/2	0.036
	2	0.05

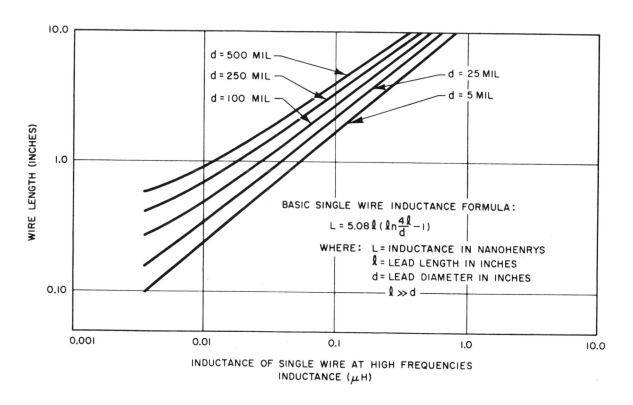

BASIC SINGLE WIRE INDUCTANCE FORMULA:

$$L = 5.08\, \ell \left(\ln \frac{4\ell}{d} - 1 \right)$$

WHERE: L = INDUCTANCE IN NANOHENRYS
ℓ = LEAD LENGTH IN INCHES
d = LEAD DIAMETER IN INCHES
$\ell \gg d$

INDUCTANCE OF SINGLE WIRE AT HIGH FREQUENCIES
INDUCTANCE (μH)

(From "Radio Engineers' Handbook" by Frederick E. Terman. Copyright 1943 by McGraw-Hill Book Company. Used with permission of McGraw-Hill Book Company.)

RF PENETRATION (SKIN RESISTANCE) OF VARIOUS MATERIALS

At very high frequencies current travels close to the outer surface of the conductor and eddy current losses increase beneath the surface. This effect is called "skin resistance" or "rf resistance." This chart shows the minimum required conductor depth related with frequency. The depth varies with the resistivity of the material and is least for silver. Therefore, a silver plating is frequently applied to conductors that are used at high frequencies so as to reduce the skin resistance.

FOR EXAMPLE:
At 200 MHz a minimum thickness of 0.81 mils of cadmium is required, whereas only 0.18 mils of silver are needed at the same frequency.

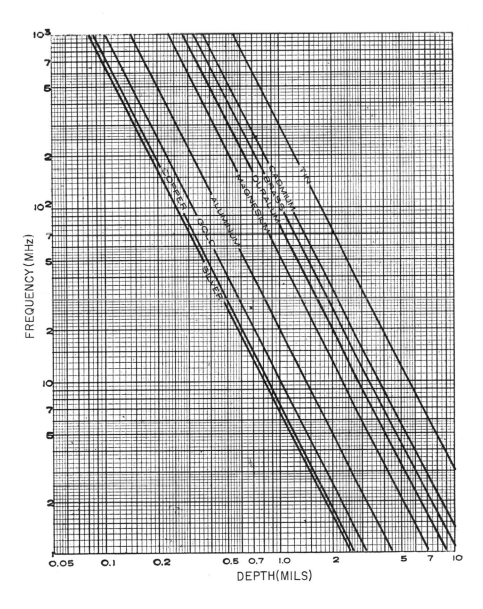

TRANSFORMER IMPEDANCE
NOMOGRAM

Tapped transformers provide standard impedances between the various taps and the common terminal. If a nonstandard impedance is required, it can often be found between the taps. This nomogram determines the impedance between terminals B and C if the impedance from A to B and A to C are known, and it is based on the following formula

$$Z_{(B-C)} = (\sqrt{Z_{(A-C)}} - \sqrt{Z_{(A-B)}})^2 \; *$$

FOR EXAMPLE:

If the impedance from A to B is 15 ohms, and the impedance from A to C is 250 ohms, then the impedance from B to C is ≈ 145 ohms.

*Derived from $Z_{(B-C)} = Z_{A-B} \left(\sqrt{\dfrac{Z_{(A-C)}}{Z_{(A-B)}}} - 1 \right)^2 .$

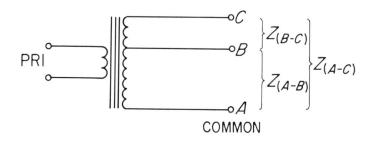

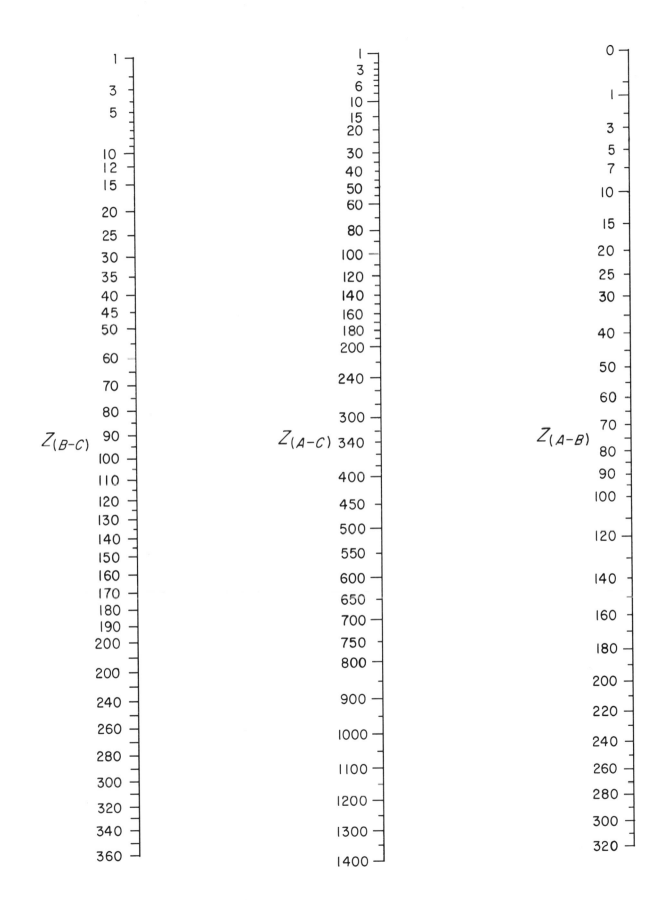

The nomogram relates capacitance, charging voltage, and stored energy in a capacitor in accordance with the formula

$$J \text{ or } W = \frac{CV^2}{2}$$

where J or W = energy in joules or watt-seconds
C = capacitance in microfarad
V = charging voltage

FOR EXAMPLE:
The energy stored in a 525-μF capacitor charged to 450 V is 53 W-sec or joules.

ENERGY STORAGE NOMOGRAM

(Continued)

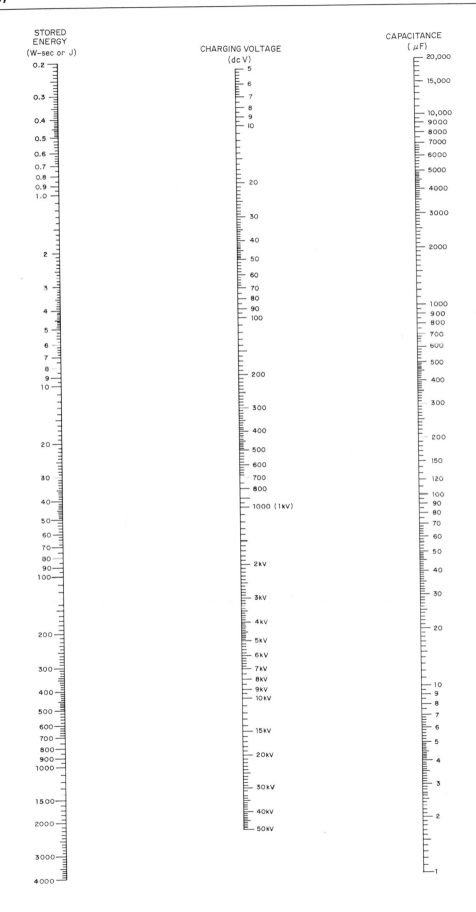

STORED
ENERGY
(W—sec or J)

CHARGING VOLTAGE
(dc V)

CAPACITANCE
(μF)

POWER-FACTOR CORRECTION

Power factor is the ratio (usually given in percent) of the actual power used in a circuit to the power apparently drawn from the line.

$$PF = \frac{actual\ power}{apparent\ power}$$

A low power factor is undesirable, and it can be raised by the addition of power-factor correction capacitors which are rated in kVAR (kilovolt-ampere reactive). To determine the kVAR of the capacitors needed to correct from an existing to a higher power factor, multiply the proper value in the table by the average power consumption, in kilowatts, of the load.

FOR EXAMPLE:
Find the kVAR of capacitors that is required to raise the power factor of a 500-kW load from 70% to 85%.

From the table select the multiplying factor 0.400 which corresponds to the existing 70% and required 85% power factor. Multiplying 0.400 by 500 shows that 200 kVAR of capacitors are required.

Existing Power Factor %	Corrected Power Factor					
	100%	95%	90%	85%	80%	75%
50	1.732	1.403	1.247	1.112	0.982	0.850
52	1.643	1.314	1.158	1.023	0.893	0.761
54	1.558	1.229	1.073	0.938	0.808	0.676
55	1.518	1.189	1.033	0.898	0.768	0.636
56	1.479	1.150	0.994	0.859	0.729	0.597
58	1.404	1.075	0.919	0.784	0.654	0.522
60	1.333	1.004	0.848	0.713	0.583	0.451
62	1.265	0.936	0.780	0.645	0.515	0.383
64	1.201	0.872	0.716	0.581	0.451	0.319
65	1.168	0.839	0.683	0.548	0.418	0.286
66	1.139	0.810	0.654	0.519	0.389	0.257
68	1.078	0.749	0.593	0.458	0.328	0.196
70	1.020	0.691	0.535	0.400	0.270	0.138
72	0.964	0.635	0.479	0.344	0.214	0.082
74	0.909	0.580	0.424	0.289	0.159	0.027
75	0.882	0.553	0.397	0.262	0.132	
76	0.855	0.526	0.370	0.235	0.105	
78	0.802	0.473	0.317	0.182	0.052	
80	0.750	0.421	0.265	0.130		
82	0.698	0.369	0.213	0.078		
84	0.646	0.317	0.161			
85	0.620	0.291	0.135			
86	0.594	0.265	0.109			
88	0.540	0.211	0.055			
90	0.485	0.156				
92	0.426	0.097				
94	0.363	0.034				
95	0.329					

kVAR-CAPACITY NOMOGRAM
FOR 60-Hz SYSTEMS

This nomogram is based on the formula

$$kVAR = \frac{2\pi f\, CE^2}{10^9}$$

where C is in microfarad E in volts, and f is 60 Hz.

FOR EXAMPLE:
To provide 5 kVAR at 460 V requires 62 μF.

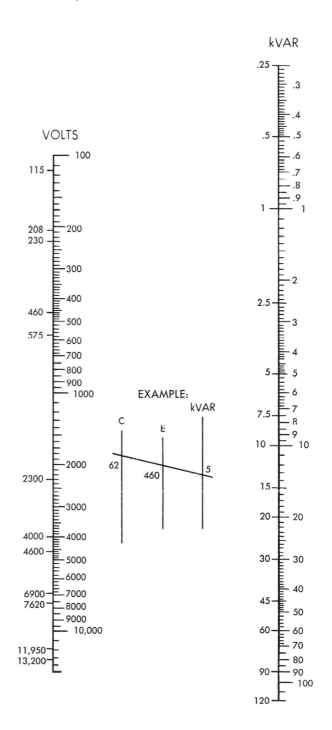

SELF-RESONANT FREQUENCY OF PARALLEL LEAD CAPACITORS

The curves show the approximate self-resonant frequency of capacitors with various lead lengths. They apply to parallel lead wires of equal length #20 to #24 AWG, spaced no further than 0.375 in. apart.

FOR EXAMPLE:
A 1000-pF capacitor with 2-in. leads resonates at about 18 MHz. The same capacitor with 0.2-in. leads will resonate at 60 MHz.

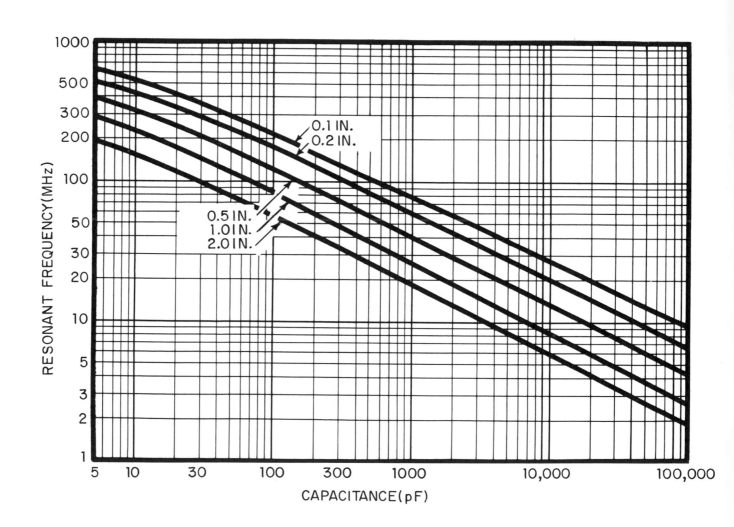

REACTANCE NOMOGRAMS

This set of three nomograms covers the frequency range of 1 Hz to 1000 MHz in three ranges which give direct answers without the need for additional calculations to locate the decimal point. These nomograms may be used to find capacitive reactance, inductive reactance, as well as resonant frequency $(X_L = X_C)$ of any combination of inductance and capacitance.

FOR EXAMPLE:
1. The reactance of a 10-mH inductor at 10-kHz is 630 ohms.
2. The reactance of a 3-pF capacitor at 5 MHz is 10,500 ohms.
3. A 5-μF capacitor and a 1.4-H inductance resonate at 60 Hz.

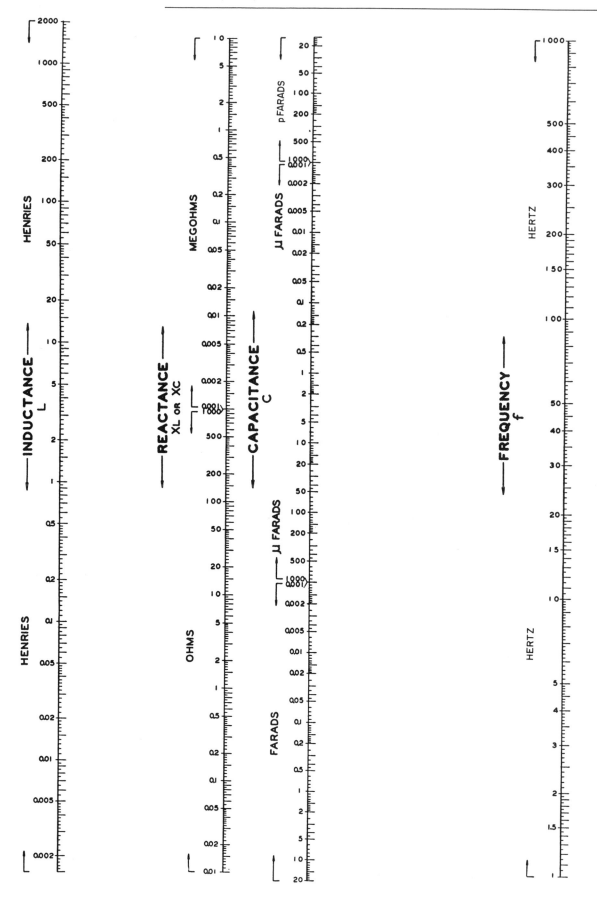

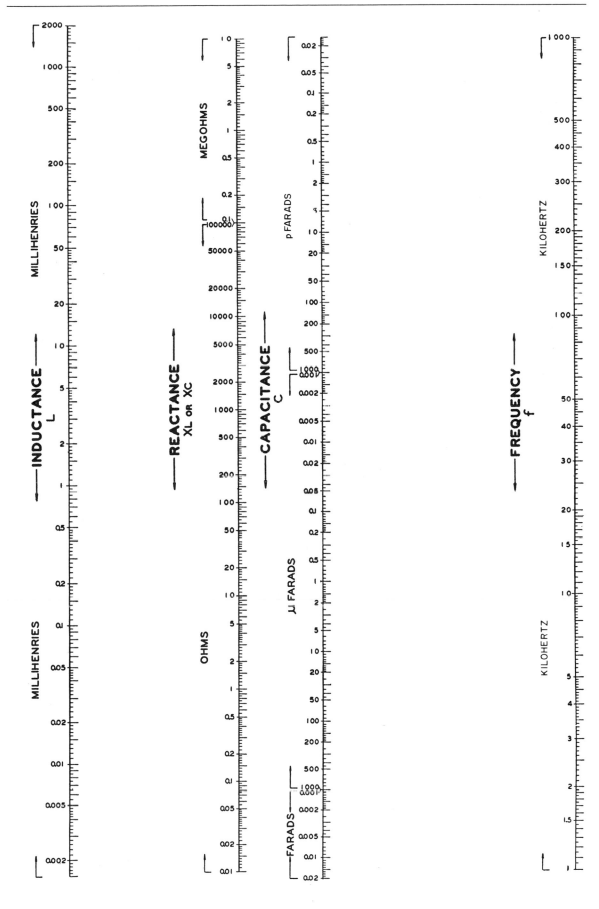

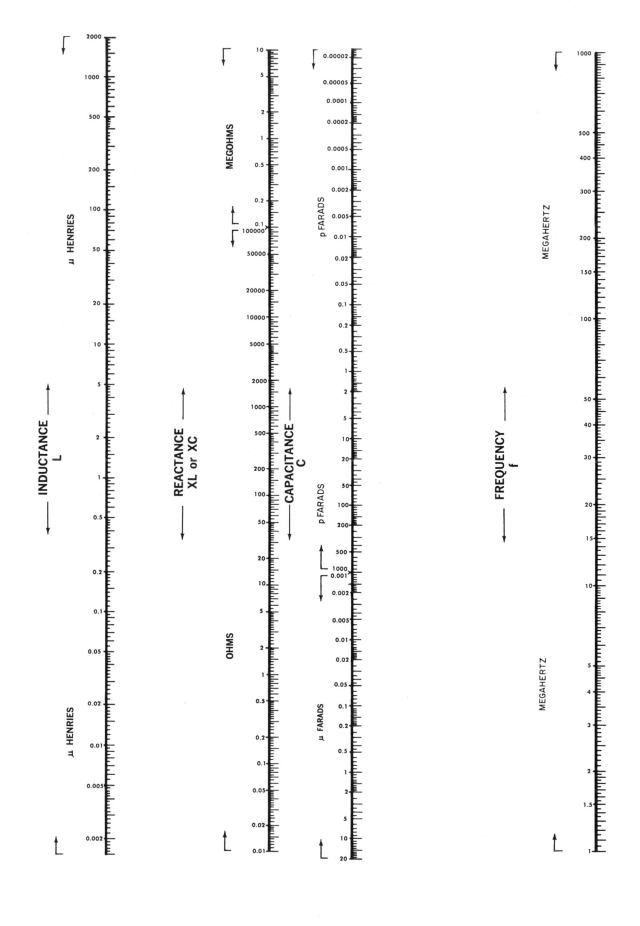

IMPEDANCE OF SERIES-CONNECTED AND PARALLEL-CONNECTED COMBINATIONS OF *L, C,* AND *R*

Circuit	Series combination	Impedance $\mathbf{Z} = R + jX$	Magnitude of impedance $|Z| = \sqrt{R^2 + X^2}$	Phase angle $\phi = \tan^{-1}(X/R)$	Admittance[a] $\mathbf{Y} = 1/\mathbf{Z}$
	R	ohms R	ohms R	radians 0	mhos $1/R$
	L	$+j\omega L$	ωL	$+\pi/2$	$-j(1/\omega L)$
	C	$-j(1/\omega C)$	$1/\omega C$	$-\pi/2$	$j\omega C$
	$R_1 + R_2$	$R_1 + R_2$	$R_1 + R_2$	0	$1/(R_1 + R_2)$
	$L_1(M)L_2$	$+j\omega(L_1 + L_2 \pm 2M)$	$\omega(L_1 + L_2 \pm 2M)$	$+\pi/2$	$-j/\omega(L_1 + L_2 \pm 2M)$
	$C_1 + C_2$	$-j\dfrac{1}{\omega}\left(\dfrac{C_1 + C_2}{C_1 C_2}\right)$	$\dfrac{1}{\omega}\left(\dfrac{C_1 + C_2}{C_1 C_2}\right)$	$-\dfrac{\pi}{2}$	$j\omega\left(\dfrac{C_1 C_2}{C_1 + C_2}\right)$
	$R + L$	$R + j\omega L$	$\sqrt{R^2 + \omega^2 L^2}$	$\tan^{-1}\dfrac{\omega L}{R}$	$\dfrac{R - j\omega L}{R^2 + \omega^2 L^2}$
	$R + C$	$R - j\dfrac{1}{\omega C}$	$\sqrt{\dfrac{\omega^2 C^2 R^2 + 1}{\omega^2 C^2}}$	$-\tan^{-1}\dfrac{1}{\omega RC}$	$\dfrac{\omega^2 C^2 R + j\omega C}{\omega^2 C^2 R^2 + 1}$
	$L + C$	$+j\left(\omega L - \dfrac{1}{\omega C}\right)$	$\left(\omega L - \dfrac{1}{\omega C}\right)$	$\pm\dfrac{\pi}{2}$	$-\dfrac{j\omega C}{\omega^2 LC - 1}$
	$R + L + C$	$R + j\left(\omega L - \dfrac{1}{\omega C}\right)$	$\sqrt{R^2 + \left(\omega L - \dfrac{1}{\omega C}\right)^2}$	$\tan^{-1}\left(\dfrac{\omega L - 1/\omega C}{R}\right)$	$\dfrac{R - j(\omega L - 1/\omega C)}{R^2 + (\omega L - 1/\omega C)^2}$

Circuit	Parallel combination	Impedance $\mathbf{Z} = R + jX$	Magnitude of impedance $|Z| = \sqrt{R^2 + X^2}$	Phase angle $\phi = \tan^{-1}(X/R)$	Admittance[a] $\mathbf{Y} = 1/\mathbf{Z}$
	R_1, R_2	ohms $\dfrac{R_1 R_2}{R_1 + R_2}$	ohms $\dfrac{R_1 R_2}{R_1 + R_2}$	radians 0	mhos $\dfrac{R_1 + R_2}{R_1 R_2}$
	C_1, C_2	$-j\dfrac{1}{\omega(C_1 + C_2)}$	$\dfrac{1}{\omega(C_1 + C_2)}$	$-\dfrac{\pi}{2}$	$+j\omega(C_1 + C_2)$
	L, R	$\dfrac{\omega^2 L^2 R + j\omega L R^2}{\omega^2 L^2 + R^2}$	$\dfrac{\omega L R}{\sqrt{\omega^2 L^2 + R^2}}$	$\tan^{-1}\dfrac{R}{\omega L}$	$\dfrac{1}{R} - \dfrac{j}{\omega L}$
	R, C	$\dfrac{R - j\omega R^2 C}{1 + \omega^2 R^2 C^2}$	$\dfrac{R}{\sqrt{1 + \omega^2 R^2 C^2}}$	$\tan^{-1}(-\omega RC)$	$\dfrac{1}{R} + j\omega C$
	L, C	$+j\dfrac{\omega L}{1 - \omega^2 LC}$	$\dfrac{\omega L}{1 - \omega^2 LC}$	$\pm\dfrac{\pi}{2}$	$j\left(\omega C - \dfrac{1}{\omega L}\right)$
	$L_1(M)L_2$	$+j\omega\dfrac{L_1 L_2 - M^2}{L_1 + L_2 \mp 2M}$	$\omega\dfrac{L_1 L_2 - M^2}{L_1 + L_2 \mp 2M}$	$\pm\dfrac{\pi}{2}$	$-j\dfrac{1}{\omega}\left(\dfrac{L_1 + L_2 \mp 2M}{L_1 L_2 - M^2}\right)$
	L, C, R	$\dfrac{\dfrac{1}{R} - j\left(\omega C - \dfrac{1}{\omega L}\right)}{\left(\dfrac{1}{R}\right)^2 + \left(\omega C - \dfrac{1}{\omega L}\right)^2}$	$\dfrac{R}{\sqrt{1 + R^2\left(\omega C - \dfrac{1}{\omega L}\right)^2}}$	$\tan^{-1} - R\left(\omega C - \dfrac{1}{\omega L}\right)$	$\dfrac{1}{R} + j\left(\omega C - \dfrac{1}{\omega L}\right)$

The power factor ($\cos\phi$) of a series RL or a parallel RC network is given by the following formulas

$$P.F.\text{ (inductive)} = \frac{R_s}{\sqrt{R_s^2 + (\omega L)^2}},$$

$$P.F.\text{ (capacitive)} = \frac{1}{\sqrt{(R_p\,\omega\,C)^2 + 1}}$$

To use the nomogram connect frequency with the desired value of L or C and note the intersect point on the turning scale. Using this intersect point, connect to the resistance, and by extending this line, read power factor and phase angle.

FOR EXAMPLE:

1. A 1-H inductance in series with 100 ohms is connected to a 60-Hz source. In this case ϕ is 75° and $\cos\phi = 0.26$.

2. An inverter operating at 2 kHz is used to supply a 100-ohm load which is in parallel with a capacitance of 0.047 μF. In this case ϕ is 3.5° and $\cos\phi = 0.998$.

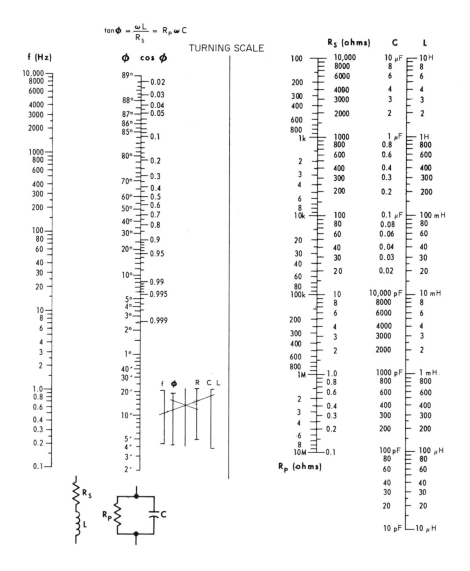

FREQUENCY CHARACTERISTICS
OF RESISTORS, CAPACITORS,
AND INDUCTORS

Tabulated here are the effects produced when potentials of increasing frequency are applied to resistors, capacitors, and inductors.

As the frequency increases from dc to above resonance, the effective "look" of the component changes as shown.

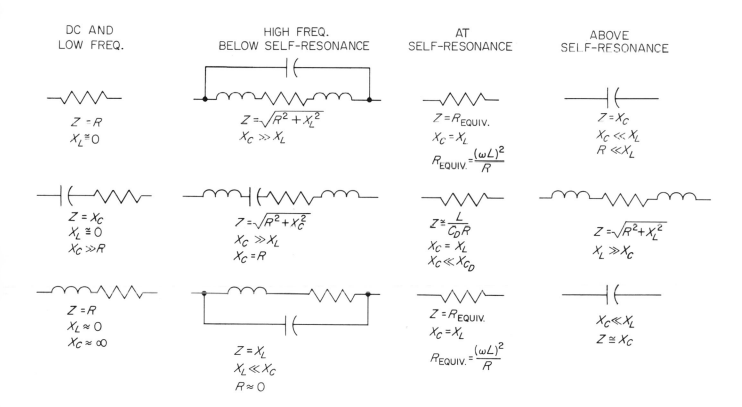

| DC AND LOW FREQ. | HIGH FREQ. BELOW SELF-RESONANCE | AT SELF-RESONANCE | ABOVE SELF-RESONANCE |

RESISTANCE–VOLTAGE–CURRENT–POWER NOMOGRAM

This nomogram is based on Ohm's law, and one straight line will determine two unknown parameters if two others are given. Preferred (±20%) resistance values are marked in addition to the ordinary resistance scale divisions. The power scale is calibrated in watts and dBm with a reference level of 0 dBm = 1 mW into 600 ohms. Thus, direct conversion between dBm and watts can be made. To cover a wide range of values and yet maintain accuracy, a dual numbering system is used. To avoid confusion, all members should be read from either the regular or the gray-barred scales.

FOR EXAMPLE:

1. The current through a 150-k resistor with a potential drop of 300 V is 2 mA, and the power dissipated is 600 mW or 0.6 W.
2. When a 12,000-ohm resistor has a current of 6 mA through it, the power dissipated is 0.43 W and the voltage across the resistor is 72 V.
3. The voltage across a 4.7 Mohm resistor with a signal level of −30 dBm is about 2.15 V rms.
4. The maximum allowable current through a 10-W 200-ohm resistor is 0.22 A. Under these operating conditions there will be 45 V across the resistor.

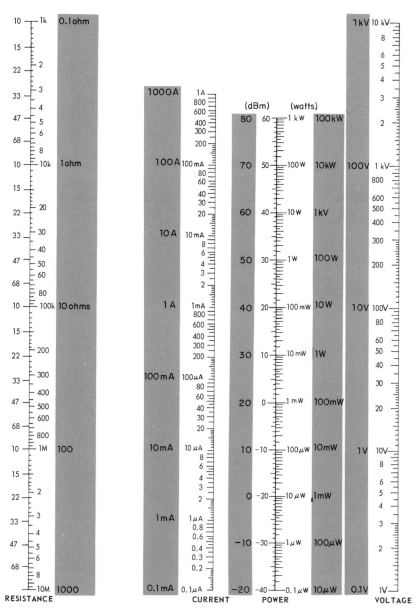

NOMOGRAM FOR CAPACITIVELY COUPLED CIRCUITS

It is often necessary to know the portion of the input voltage that will appear across the load resistor in a capacitively coupled circuit. This is a function of frequency and a factor of the ratio of R to X_c, the required ratio is shown on the center scale. It is interesting to note that any ratio of R to X_c greater than 7.4 : 1 yields over 99% output. The X_c and R scales can be multiplied by any common power of ten to extend the range of the nomogram.

FOR EXAMPLE:
For $R = 100$ k and $X_c = 10$ k, V_2 will be 99.4% of V_1.

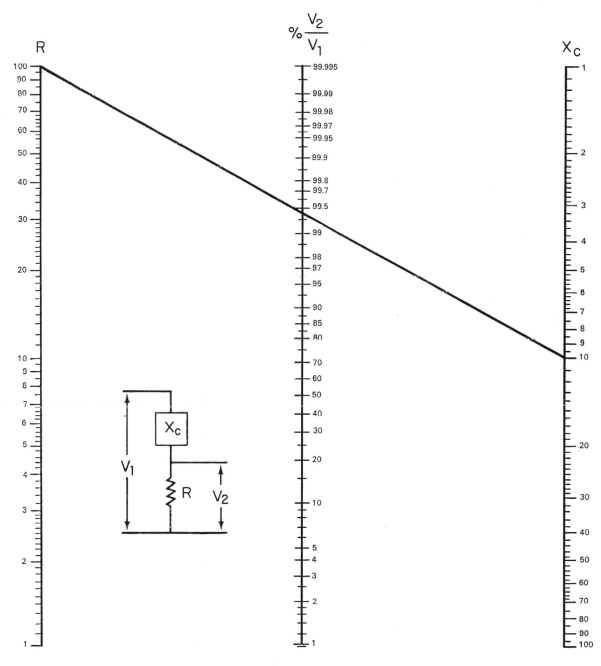

(From *Electronics and Communications*, June 1966.)

This nomogram aids in the rapid selection of component values for the simple resistive and capacitive voltage dividers illustrated, where

$$\frac{e_o}{e_i} = \frac{R_g}{R_g + R_s} \quad \text{or} \quad \frac{e_o}{e_i} = \frac{C_s + C_g}{C_s}$$

Only two decades are covered on the left and right scale to achieve maximum accuracy. The range of the nomogram can be extended by multiplying these two columns by the same power of ten without making any changes in the center column.

FOR EXAMPLE:

1. A blocking oscillator must be held at cutoff by means of a voltage divider between B- and ground. Cut-off bias is -15 V, the negative supply is 150 V, and the grid-to-ground resistor is 22,000 ohms. Thus, e_o/e_i is 0.1. Joining that value with 2.2 on the R_g scale gives 20 on the R_s scale, which makes that resistor equal to 200,000 ohms since each scale had to be multiplied by 10^4.

2. Design an rf probe with a 5:1 attenuator using standard capacitance values. Rotating about the 0.2 point on the center scale gives typical values of 30 pF for C_g and 7.5 pF for C_s.

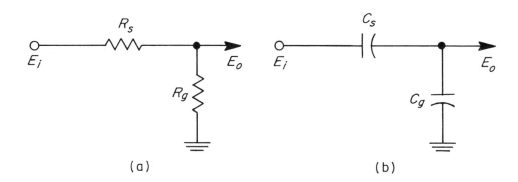

(a)   (b)

VOLTAGE DIVIDER NOMOGRAM (Continued)

Note: The longer lines outside the left and right columns locate standard ±10% values and the shorter lines locate standard ±5% values.

C_g

100
90
80
70
60
50
40
30
20
10
9
8
7
6
5
4
3
2
1

R_s

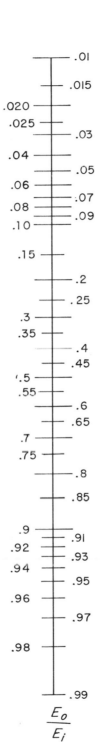

.01
.015
.020
.025
.03
.04
.05
.06
.07
.08
.09
.10
.15
.2
.25
.3
.35
.4
.45
.5
.55
.6
.65
.7
.75
.8
.85
.9
.91
.92
.93
.94
.95
.96
.97
.98
.99

$$\frac{E_o}{E_i}$$

C_s

1
2
3
4
5
6
7
8
9
10
20
30
40
50
60
70
80
90
100

R_g

95

R-C COUPLING NOMOGRAM

This nomogram is used to calculate phase shift and attenuation in R-C coupling networks. To use, connect capacitance with frequency and note the intersect point on the turning scale. Using this intersect point, connect to the resistance, and by extending this line, read attenuation and phase shift.

FOR EXAMPLE:
At 60 Hz, a 0.01-μF capacitor and 10,000-ohm resistor will exhibit a phase shift of 72° and an attenuation factor of 0.35.

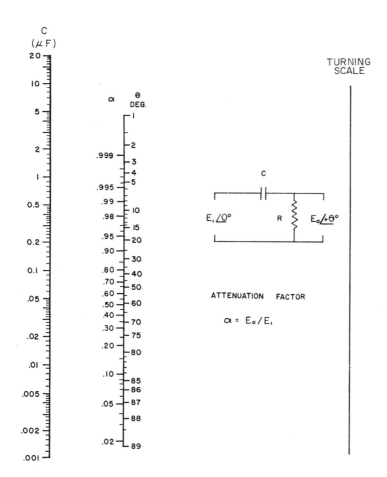

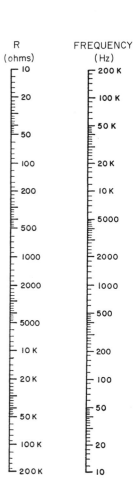

(From *Electronics and Communications*, June 1952.)

TIME-CONSTANT NOMOGRAM (A)

This nomogram is based on the formula $T = RC$ where T (the time constant) is the time required for the capacitor in an RC series circuit to reach 63.2% of the applied voltage.

FOR EXAMPLE:

The time constant of 10 msec can be achieved with a 1-Mohm resistor and a 0.01-μF capacitor.

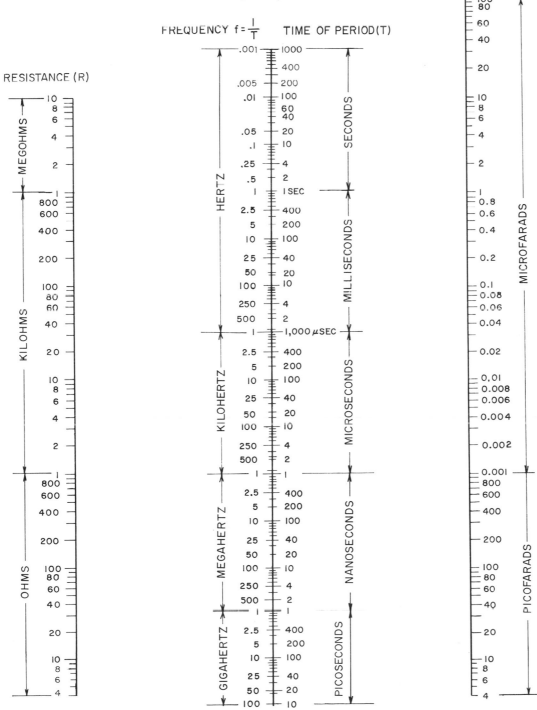

This chart is used to determine the time required in an RC series circuit to reach a given fraction of an applied step input, or to determine the percent of the applied input when the time constant is given.

The nomogram is based on the relationship

$$\frac{E_{out}}{E_{in}} = 1 - e^{-t/RC}$$

FOR EXAMPLE:

Determine the time required to charge a 5000-μF capacitor to 400 V through 1000 ohms from a 450 V supply. The percent of applied voltage is 88.5% (400/450) which requires 2.2 time constants. The time constant is 5 sec (from time-constant nomogram A), so the time required to charge to 400 V is 11 sec.

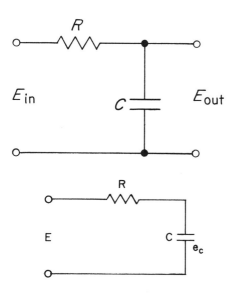

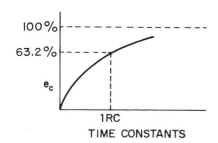

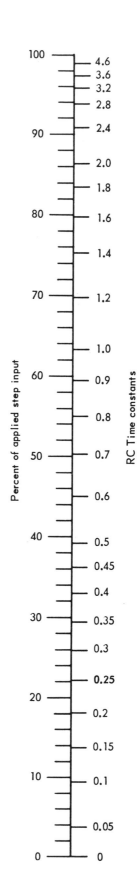

FREQUENCY SELECTIVE NETWORK NOMOGRAM

The expression $f = 1/2\pi RC$, where f is in hertz, C in farads, and R in ohms, is the expression for:

1. The 3-dB bandwidth of a single tuned circuit having parameters as shown in Figure 1.
2. The frequency at 3 dB relative attenuation of the parallel RC low-pass network shown in Figure 2.
3. The frequency at 3 dB relative transfer attenuation of the series RC high-pass network of Figure 3.
4. Wien bridge balance.

FOR EXAMPLE:

1. The circuit shown in Figure 1 is used to couple two succesive stages of an amplifier. The 3-dB bandwidth of the circuit must be 3.4 MHz and the equivalent shunt capacitance of the circuit is 25 pF. What equivalent resonant resistance will the circuit exhibit? Connect 3.4 MHz and 25 pF and find the equivalent resonant resistance as 1850 ohms.
2. The low-pass network of Figure 2 uses a 0.05-μF capacitor. What value of resistance is required for the output to drop to 0.707 of the input at 5 kHz? Connect 0.05-μF with 5 kHz and read answer as 620 ohms.

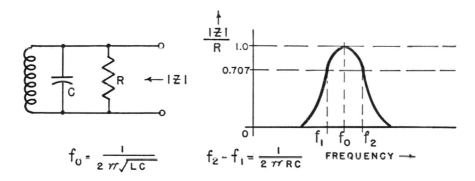

Figure 1. Characteristics of a single tuned circuit.

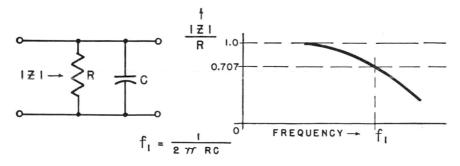

Figure 2. Characteristics of a parallel RC low-pass network.

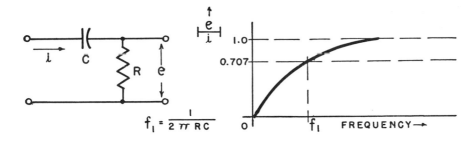

Figure 3. Transfer characteristics of an RC high-pass network.

3. It is required that the RC high-pass network in Figure 3 attenuate rapidly below 300 Hz. What value resistor must be used with a 0.1-μF capacitor? Connect 0.1-μF with 300 Hz (0.3 kHz) and read answer as 5250 ohms.

4. Figure 4 shows an RC coupled amplifier and its equivalent circuits. It is assumed that the reactance of the bypass capacitors is negligible throughout the frequency range of the amplifier. If the equivalent circuit resistance has a value of 1300 ohms and the equivalent capacitance is 25 pF, at what frequency is the amplification 0.707 of the midfrequency range of the amplifier? Connect 25 pF and 1300 ohms and read frequency of 4.75 MHz at which amplifier gain is down 3 dB.

5. The Wien bridge circuit shown in Figure 5 has R_1 and R_2 equal to 10,000 ohms and C_1 and C_2 equal to 0.1-μF. With those values the balance frequency of the circuit is 1.59 kHz.

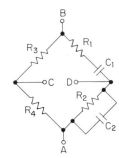

Figure 5. Conventional Wien bridge circuit.

$$R_1 = R_2 = R$$
$$C_1 = C_2 = C$$
$$\frac{R_3}{R_4} = 2$$

For the measurement of frequency, the unknown frequency is connected across A and B and a null detector, across C and D.

When used with an oscillator, the circuit is connected to a suitable amplifier with regenerative feedback.

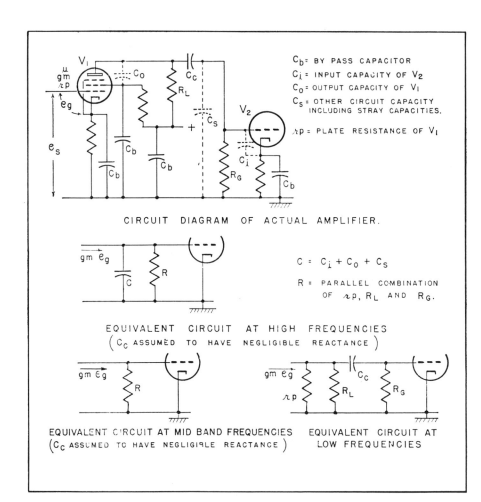

Figure 4. An RC-coupled amplifier and its equivalent circuits.

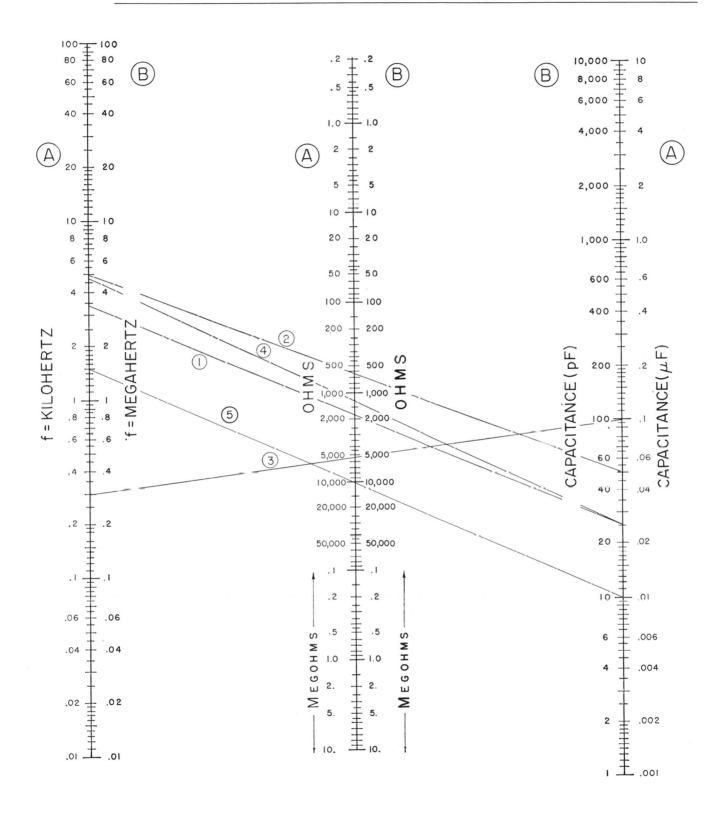

Note: Scales with corresponding letters (A or B) are used together.

BANDWIDTH NOMOGRAM

This nomogram is used to compute the bandwidth of a tuned circuit at 70.7% (-3 dB) of maximum gain. It is based on the equation

$$\Delta f = \frac{f_r}{Q}$$

where

Δf = bandwidth in kilohertz

f_r = resonant frequency in megahertz

Q = figure of merit of the inductance

FOR EXAMPLE:
1. A circuit that has a resonant frequency of 6 MHz, and uses an inductance with a Q of 140, will have a bandwidth of 43 kHz.
   Note: The range of the nomogram can be extended to cover other frequencies by multiplying or dividing both frequency scales by the same power of 10.
2. To achieve a bandwidth of 2.5 kHz at a resonant frequency of 600 kHz the inductance must have a Q of 240.

BANDWIDTH NOMOGRAM
(Continued)

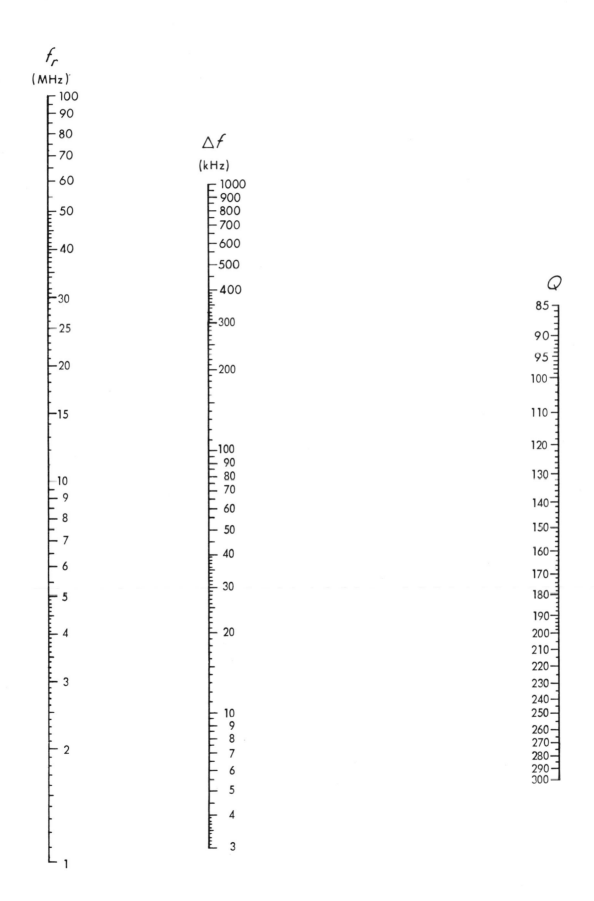

CROSSOVER NETWORKS, DESIGN EQUATIONS, AND RATE OF ATTENUATION CURVES

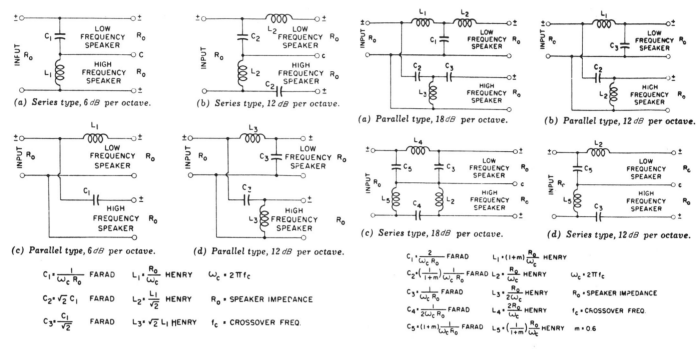

(a) Series type, 6 dB per octave.

(b) Series type, 12 dB per octave.

(a) Parallel type, 18 dB per octave.

(b) Parallel type, 12 dB per octave.

(c) Parallel type, 6 dB per octave.

(d) Parallel type, 12 dB per octave.

(c) Series type, 18 dB per octave.

(d) Series type, 12 dB per octave.

$$C_1 = \frac{1}{\omega_c R_0} \text{ FARAD} \qquad L_1 = \frac{R_0}{\omega_c} \text{ HENRY} \qquad \omega_c = 2\pi f_c$$

$$C_2 = \sqrt{2}\, C_1 \text{ FARAD} \qquad L_2 = \frac{L_1}{\sqrt{2}} \text{ HENRY} \qquad R_0 = \text{SPEAKER IMPEDANCE}$$

$$C_3 = \frac{C_1}{\sqrt{2}} \text{ FARAD} \qquad L_3 = \sqrt{2}\, L_1 \text{ HENRY} \qquad f_c = \text{CROSSOVER FREQ.}$$

m-derived crossover network.

$$C_1 = \frac{2}{\omega_c R_0} \text{ FARAD} \qquad L_1 = (1+m)\frac{R_0}{\omega_c} \text{ HENRY}$$

$$C_2 = \left(\frac{1}{1+m}\right)\frac{1}{\omega_c R_0} \text{ FARAD} \qquad L_2 = \frac{R_0}{\omega_c} \text{ HENRY} \qquad \omega_c = 2\pi f_c$$

$$C_3 = \frac{1}{\omega_c R_0} \text{ FARAD} \qquad L_3 = \frac{R_0}{2\omega_c} \text{ HENRY} \qquad R_0 = \text{SPEAKER IMPEDANCE}$$

$$C_4 = \frac{1}{2\omega_c R_0} \text{ FARAD} \qquad L_4 = \frac{2R_0}{\omega_c} \text{ HENRY} \qquad f_c = \text{CROSSOVER FREQ.}$$

$$C_5 = (1+m)\frac{1}{\omega_c R_0} \text{ FARAD} \qquad L_5 = \left(\frac{1}{1+m}\right)\frac{R_0}{\omega_c} \text{ HENRY} \qquad m = 0.6$$

Constant-resistance, constant-k network.

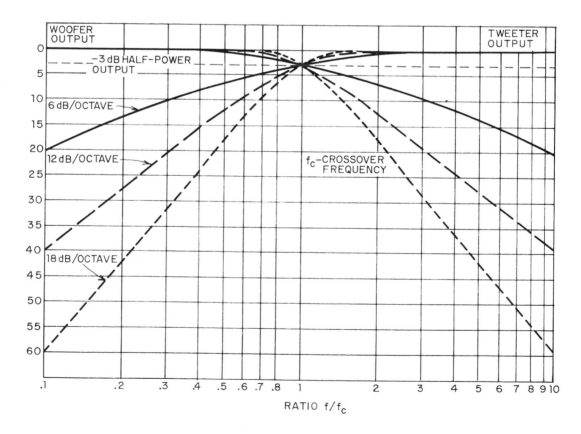

Crossover curves showing 6, 12, and 18 dB/octave crossover net-work cutoff rate.

(Reprinted from *Radio-Electronics*, copyright Gernsback Publications, Inc., March 1968.)

Calculations involved in the design of low-pass filter systems can be minimized by using the following nomograms. Practical low-pass filters can be constructed as shown in Figure 1. The intermediate portion of the filter is composed of a series of cascaded π or T sections. Each filter section works into a resistive load impedance with little or no reactive component.

Figure 1 shows unbalanced filter configurations. To construct filter sections which will work with balanced lines, divide the series reactance equally between the two sides of the line.

Either the π or T-type filter may be used in the intermediate filter section. The designer has a choice as either type will perform the same electrically.

Placement of the filter components should be considered carefully. There should be no coupling, magnetic or capacitive, between elements of the filter sections as this may deteriorate overall performance.

Figure 2 is a nomogram that solves the values of L and $L/2$ for the simple π and T-type filters. The values of L are read from the left side of the center scale for the π type while the values for $L/2$ are read from the right side of the center scale for the T type.

The nomogram of Figure 3 solves for the values of C and $C/2$ used in the simple π and T types. The values of $C/2$ are read on the right side of the center scale for the π-type filter while the values of C are read on the left side of the center scale for the T type. M-derived filter sections may be solved for after C and L have been found by using Figures 4 and 5. Figure 4 solves for values of C_2 and $\frac{1}{2}(C_2)$. as well as L_1 and $\frac{1}{2}(L_1)$. Figure 5 solves for L_2, $2L_2$, C_1 and $2C_1$.

Extension of Scales

The scales may be extended to cover other ranges of frequencies than those shown. The range of Figure 2 may be extended by shifting the decimal point of L or $L/2$ scale one place to the left for each zero added to f_c. A zero subtracted from f_c moves the decimal point one place to the right. Note that a zero added to the right or resistance scale also will move the decimal point on the center scale one place to the right.

LOW-PASS FILTER DESIGN
(Continued)

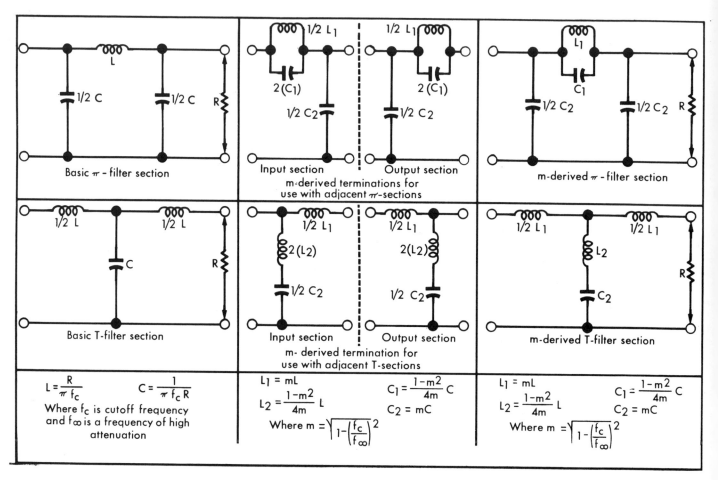

Figure 1. Basic and *m*-derived low-pass filter sections.

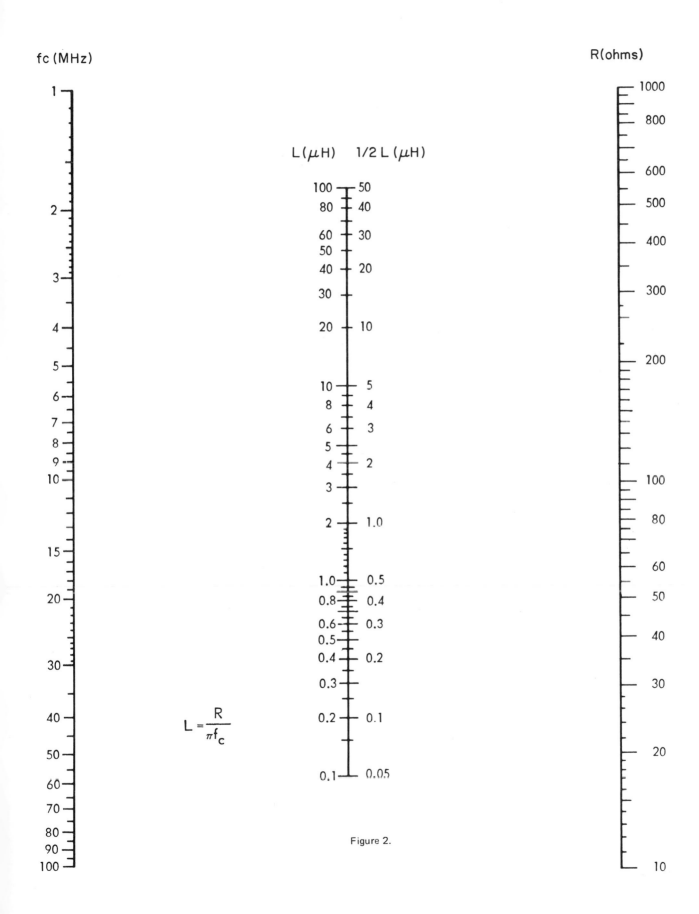

fc (MHz)

R(ohms)

L(μH)   1/2 L (μH)

$$L = \frac{R}{\pi f_c}$$

Figure 2.

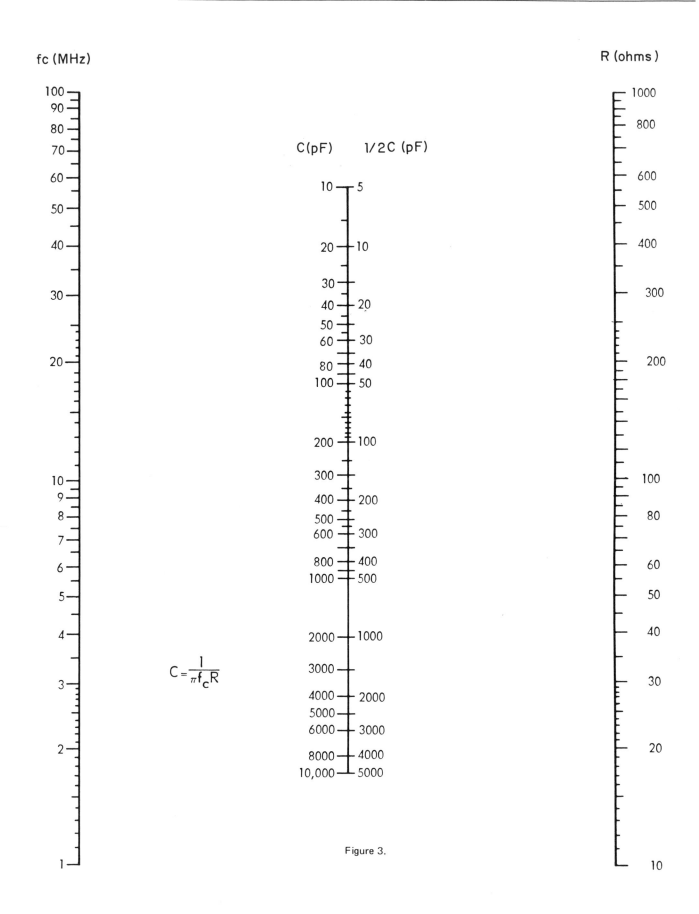

Figure 3.

$$C = \frac{1}{\pi f_c R}$$

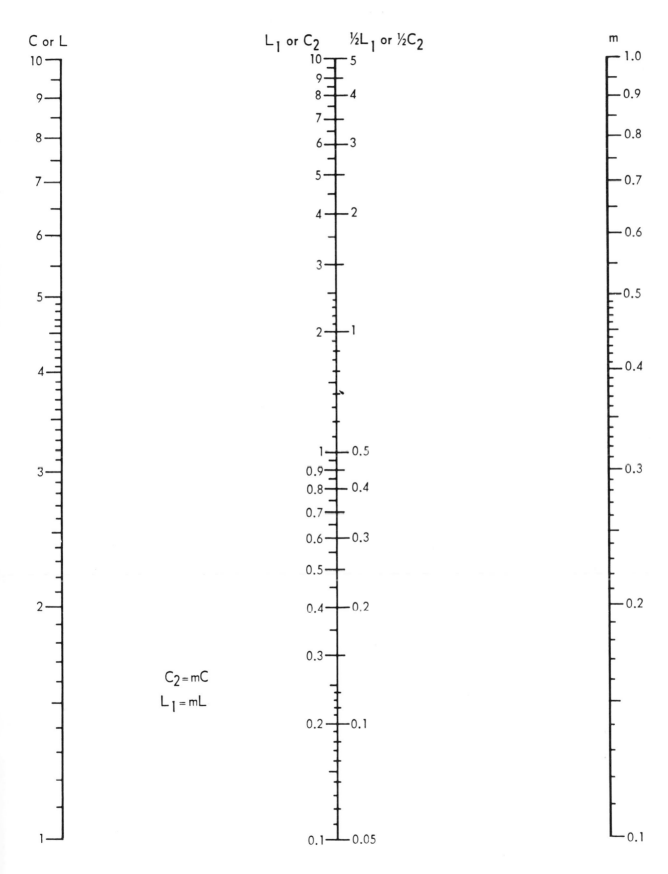

Figure 4.

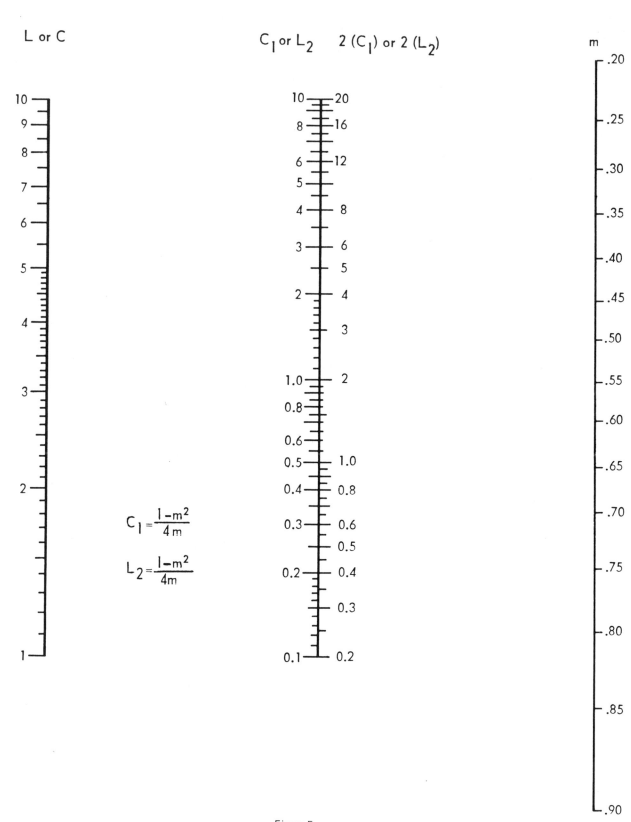

$$C_1 = \frac{1-m^2}{4m}$$

$$L_2 = \frac{1-m^2}{4m}$$

Figure 5.

LOW-PASS FILTER DESIGN
(Continued)

The range of Figure 3 may be extended by moving the decimal point on the center scale one place to the left for every zero added to either the ohms or f_c scales. Conversely, the decimal point is moved to the right for every zero subtracted from these scales.

When using the nomograms of Figures 4 and 5, it is never necessary to alter the decimal point on the m scales, and other range changes may be made by inspection.

FOR EXAMPLE:

A 52-ohm coaxial line feeds an RF signal from a transmitter to an antenna that is essentially a resistive load at the operating frequency. The transmitter frequency is 28.5 MHz. A filter is required with a cutoff frequency of 50 MHz. A π-derived

T section filter is to be used and m is chosen at 0.6. An input and output terminating section will be used with the device and a configuration as shown in Figure 6(a) is to be used. This may be modified to Figure 6(b) by lumping the capacitors marked $\frac{1}{2}C_2$ together.

L is found from Figure 2 by drawing a line from 20 on the f_c scale to 52 on the R scale. L is found to be 0.83 μH. Similarly L_1 and $\frac{1}{2}L_1$ are found from Figure 4 to be 0.5 and 0.25 μH, respectively. C in turn is found on the scales of Figure 3 while C and C_1 are found on Figures 4 and 5.

A filter using T-type sections also could be solved for using the charts in a similar manner. It is only necessary to observe the equations on the nomograms and determine if the final reading should be taken from the left or right side of the center scale.

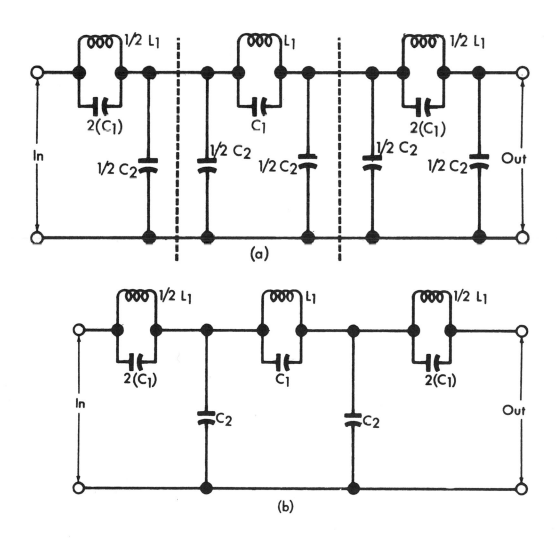

Figure 6.

LOW-PASS FILTER DESIGN
(Continued)

Formulas have been developed from which low-pass filters can be designed with sharper cutoff characteristics than those obtained from the image-parameter method shown previously.

Examples are given here for three-element, five-element, seven-element and nine-element Butterworth filters. With the values shown, all of these filters have 3-dB points at an f_c of 159 Hz, flat bandpass, and various stopband slopes as shown in the accompanying curves. The element values are based on 1000-ohm source and load impedances. Constants were selected from the table of normalized element values for Butterworth ladder low-pass filters.

The reason for the selection of 159 Hz is the fact that this frequency corresponds to 1000 rad. This in turn makes the element values correspond, without conversion, to the normalized values appearing in design tables such as the one given here.

The 3-dB point (i.e., f_c) can be moved to any other frequency by multiplying all inductance and capacitance values by the ratio

$$\frac{159}{\text{new } f_c \text{ (Hz)}}$$

For filters designed to be used at other than 1000-ohm source and load impedances, multiply inductance values and divide capacitance values by the ratio

$$\frac{R_{new}}{1000}$$

THREE-ELEMENT

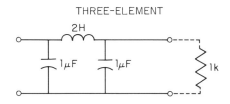

FIVE-ELEMENT

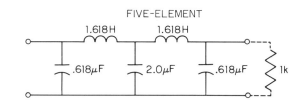

SEVEN-ELEMENT

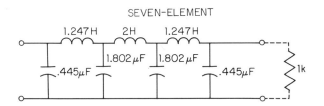

NINE-ELEMENT

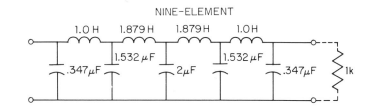

LOW-PASS FILTER DESIGN
(Continued)

(n) Number of Elements	C_1 or L'_1	L_2 or C'_2	C_3 or L'_3	L_4 or C'_4	C_5 or L'_5	L_6 or C'_6	C_7 or L'_7	L_8 or C'_8	C_9 or C'_9	L_{10} or C'_{10}
1	2.0000									
2	1.4142	1.4142								
3	1.0000	2.0000	1.0000							
4	0.7654	1.8478	1.8478	0.7654						
5	0.6180	1.6180	2.0000	1.6180	0.6180					
6	0.5176	1.4142	1.9319	1.9319	1.4142	0.5176				
7	0.4450	1.2470	1.8019	2.0000	1.8019	1.2470	0.4450			
8	0.3902	1.1111	1.6629	1.9616	1.9616	1.6629	1.1111	0.3902		
9	0.3473	1.0000	1.5321	1.8794	2.0000	1.8794	1.5321	1.0000	0.3473	
10	0.3129	0.9080	1.4142	1.7820	1.9754	1.9754	1.7820	1.4142	0.9080	0.3129

Normalized element values for Butterworth ladder low-pass filters.

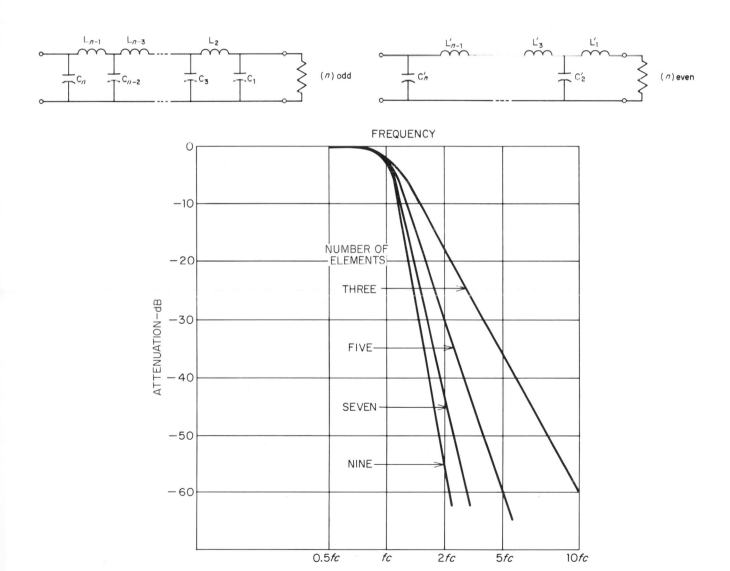

LOW-PASS FILTER DESIGN
(Continued)

Using modern computer methods of synthesis, more exotic low-pass filters with steeper rates of cutoff have been tabulated. For example, the two filters shown have five elements each—two inductors and three capacitors. The pass band is almost identical for both filters, but the stop bands are different. As before, each filter has its 3-dB point at an f_c of 159 Hz. Whole catalogs of filters of this second type have been published. They have Tchebychev-type pass bands and are called "Cauer" filters.

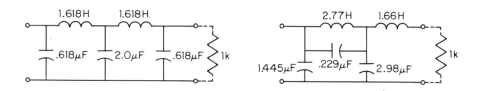

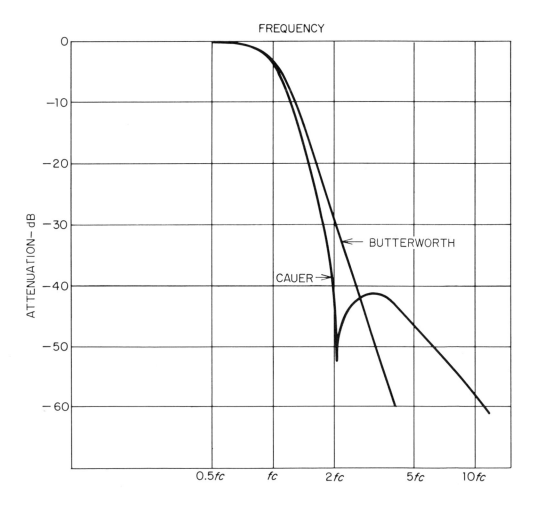

High-pass filter design can be accomplished with the following nomograms. The accuracy of the calculations will be the same as a slide rule of comparable dimensions.

Figure 1 shows schematically how the filter elements can be arranged. Precautions should be taken to prevent magnetic or capacitive coupling between the reactive components of the filter. The high-pass filter can be considered as an impedance matching device terminating in a load with no reactive component.

As shown, Figure 1 would be used with an unbalanced transmission line. The same filter sections can be modified for use with a balanced transmission line by dividing the series reactances equally between the two sides of the line. Both intermediate filter sections, π or T types, will produce the same filtering action.

Figure 2 is a nomogram that solves the values of L and $2L$ for the simple π and T-type filters. The values of L for the T-type filter are read from the left side of the middle scale while the values of $2L$ are read from the right side of this scale.

The nomogram of Figure 3 solves values of C and $2C$ used in the simple π and T-type filters. The values of C are read on the right side of the center scale for the π-type filter, while $2C$ is read on the left side of this scale for the T-type filter.

M-derived filter sections may be solved for after C and L have been found by using Figures 4 and 5. Figure 4 solves for values of C_1, $2(C_1)$, L_2, and $2(L_2)$. Figure 5 solves for L_1, $\frac{1}{2}(L_1)$, C_2, and $\frac{1}{2}(C_2)$.

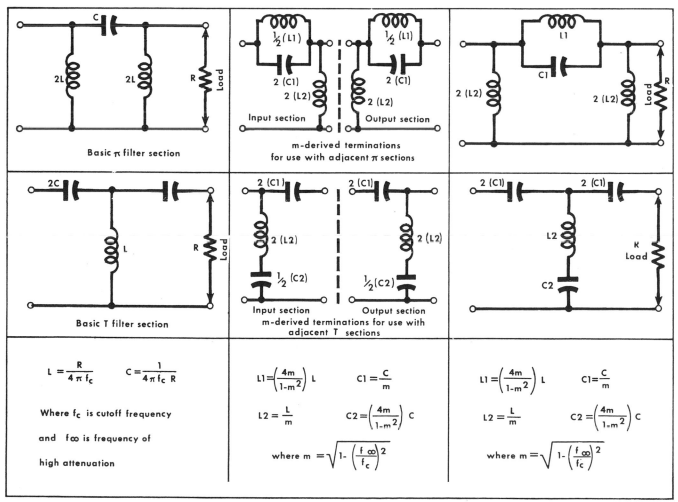

Figure 1. Basic and *m*-derived high-pass filter sections.

HIGH-PASS FILTER DESIGN
(Continued)

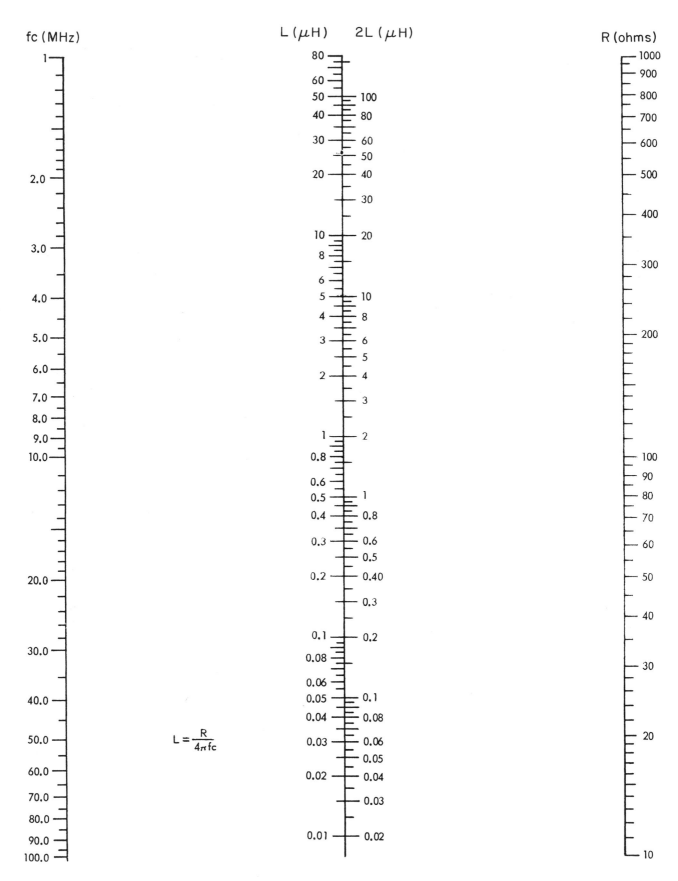

$$L = \frac{R}{4\pi f_c}$$

Figure 2

HIGH-PASS FILTER DESIGN
(Continued)

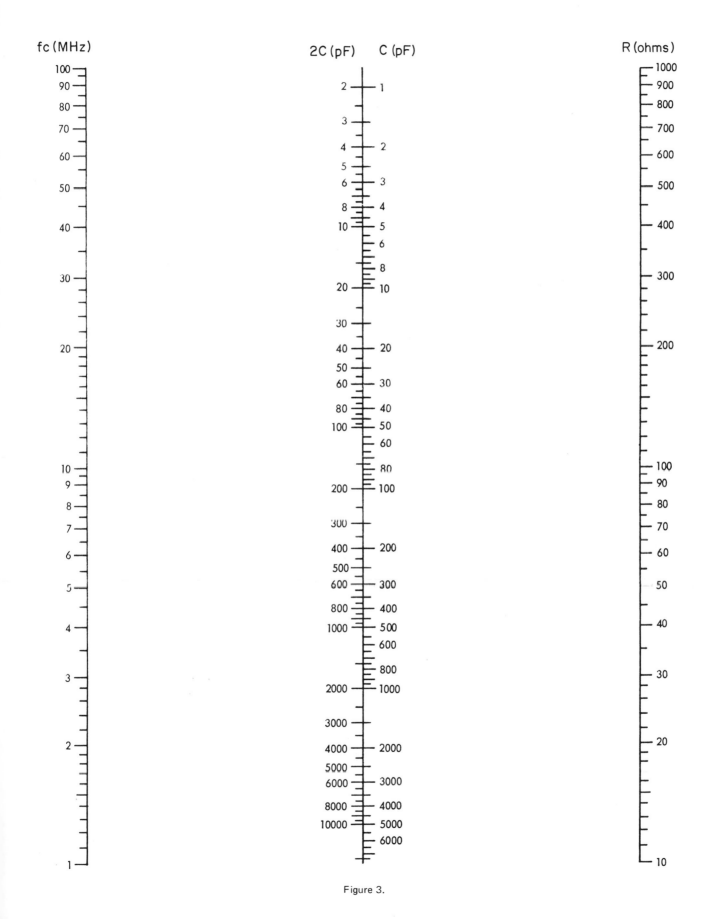

Figure 3.

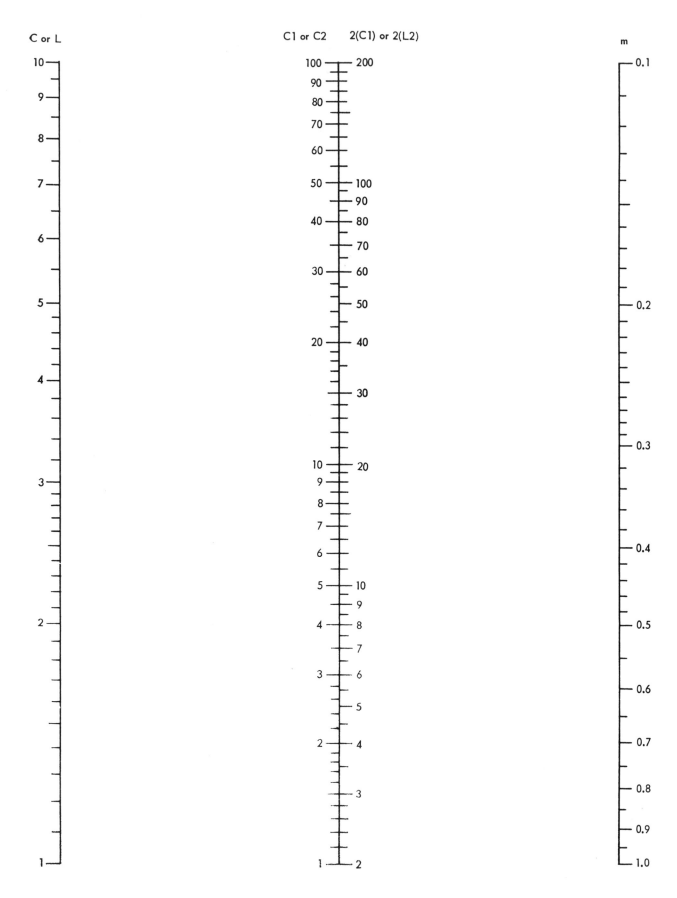

Figure 4.

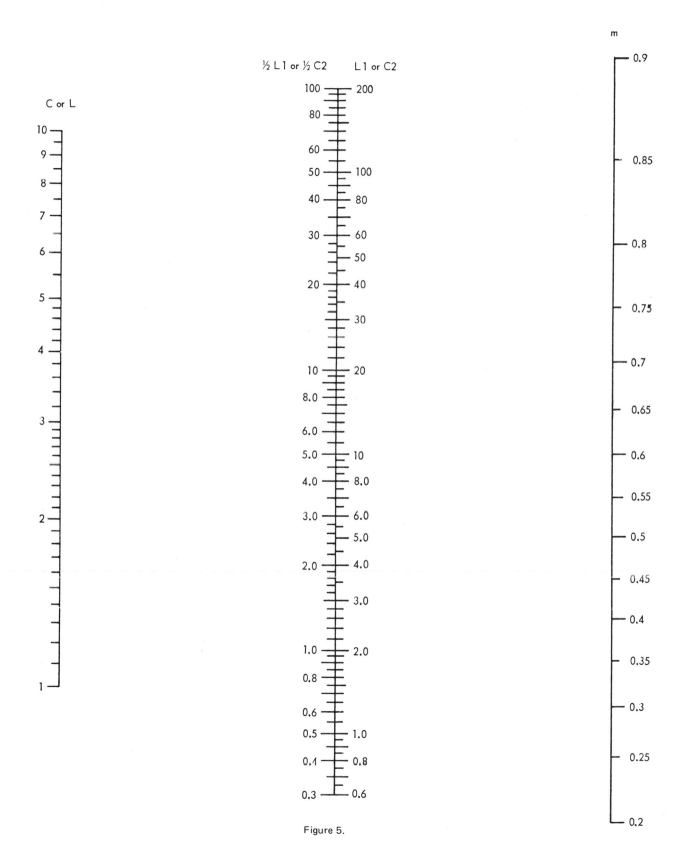

Figure 5.

HIGH-PASS FILTER DESIGN
(Continued)

Extension of Scales

The scales may be extended to cover other ranges of frequencies than those shown. The range of Figure 2 may be extended by shifting the decimal point of the L or $2L$ scale one place to the left for each zero added to f_c. A zero subtracted from f_c moves the decimal point one place to the right. Note that a zero added to the right or resistance scale will also move the decimal point on the center scale one place to the right.

The range of Figure 3 may be extended by moving the decimal point on the center scale one place to the left for every zero added to either the ohms or f_c scales. Conversely, the decimal point is moved to the right for every zero subtracted from these scales.

When the nomograms of Figures 4 and 5 are used, the decimal point position remains unaltered and extensions of the scales are unnecessary.

FOR EXAMPLE:

A 52-ohm coaxial line feeds a wide-band video signal from a camera to an amplifier. The output signal of the camera has been mixed with another frequency to produce a video signal from 20 to 26 MHz. Unwanted pulses of lower frequency information are being picked up by the cable. An M-derived filter with an f_c (cutoff frequency) of 20 MHz is needed at the input to the amplifier. M is chosen at 0.6 and a T section is chosen to isolate the input of the amplifier from dc present on the cable. The filter using two end sections and an intermediate T is shown in Figure 6a. The $2C_1$ pairs in series can be lumped to $C1$ as shown in Figure 6b.

L is shown from Figure 2 by drawing a line from $f_c = 20$ to $R = 52$. L is found on the center scale to be 0.2×10^{-6} H or $0.2~\mu$H. $2(L_2)$ and L_2 are found on Figure 4 by drawing a line from 2 on the L scale to 0.6 on the m scale. L_2 is found to be $0.335~\mu$H and $2(L_2)$ is $0.67~\mu$H.

C is found from Figure 3 to be 77×10^{-12} or 77 pF. C_1 is found from Figure 4 to be 128 pF. C_2 and $\frac{1}{2}(C_2)$ are found on Figure 5 to be 290 and 145 pF respectively. This completes the design.

High-pass filters with sharper cutoff characteristics can be designed in a manner similar to the examples previously shown for low-pass filters.

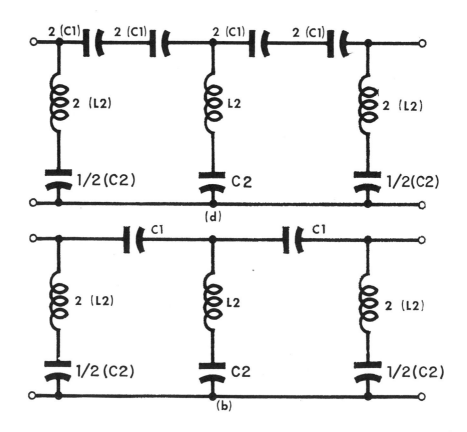

Figure 6.

FILTER CHARACTERISTICS AND DESIGN FORMULAS

Band-Pass Sections

Fundamental Relations

R = load resistance f_1 = lower frequency limit of pass band f_2 = higher frequency limit of pass band

$f_{1\infty}$ = a frequency of very high attenuation in low-frequency attenuating band $f_{2\infty}$ = a frequency of very high attenuation in high-frequency attenuating band

$$L_{1k} = \frac{R}{\pi(f_2 - f_1)} \qquad L_{2k} = \frac{(f_2 - f_1)R}{4\pi f_1 f_2} \qquad C_{1k} = \frac{f_2 - f_1}{4\pi f_1 f_2 R} \qquad C_{2k} = \frac{1}{\pi(f_2 - f_1)R}$$

Design of Sections

Type	Attenuation characteristic	A. Filters having T intermediate sections		B. Filters having π intermediate sections		Notation for both T and π sections
		Configuration	Formulas	Configuration	Formulas	
End ($m_1 = m_2$ = approximately 0.6)			$L_1 = m_1 L_{1k}$ $L_2 = a L_{1k}$ $L_2' = c L_{1k}$ $C_1 = \dfrac{C_{1k}}{m_2}$ $C_2 = \dfrac{C_{1k}}{b}$ $C_2' = \dfrac{C_{1k}}{d}$		$L_1 = \dfrac{L_{2k}}{b}$ $L_1' = \dfrac{L_{2k}}{d}$ $L_2 = \dfrac{L_{2k}}{m_2}$ $C_1 = a C_{2k}$ $C_1' = c C_{2k}$ $C_2 = m_1 C_{2k}$	$g = \sqrt{\left(1 - \dfrac{f_{1\infty}^2}{f_1^2}\right)\left(1 - \dfrac{f_{1\infty}^2}{f_2^2}\right)}$ $h = \sqrt{\left(1 - \dfrac{f_1^2}{f_{2\infty}^2}\right)\left(1 - \dfrac{f_2^2}{f_{2\infty}^2}\right)}$ $m_1 = \dfrac{\frac{f_1 f_2}{f_{2\infty}^2}g + h}{1 - \frac{f_1^2}{f_{2\infty}^2}}$ $m_2 = \dfrac{g + \frac{f_{1\infty}^2}{f_1 f_2}h}{1 - \frac{f_{1\infty}^2}{f_2^2}}$ $a = \dfrac{(1-m_1^2)f_{2\infty}^2}{4gf_1f_2}\left(1 - \dfrac{f_{1\infty}^2}{f_{2\infty}^2}\right) = \dfrac{(1-m_2^2)f_1f_2}{4gf_1f_2}\left(1 - \dfrac{f_{1\infty}^2}{f_{2\infty}^2}\right)$ $b = \dfrac{(1-m_2^2)}{4g}\left(1 - \dfrac{f_{1\infty}^2}{f_{2\infty}^2}\right)$ $c = \dfrac{(1-m_1^2)}{4h}\left(1 - \dfrac{f_{1\infty}^2}{f_{2\infty}^2}\right)$ $d = \dfrac{(1-m_1^2)f_{2\infty}^2}{4hf_1f_2}\left(1 - \dfrac{f_{1\infty}^2}{f_{2\infty}^2}\right) = \dfrac{(1-m_2^2)f_1f_2}{4hf_1f_2}\left(1 - \dfrac{f_{1\infty}^2}{f_{2\infty}^2}\right)$ when $(m_1 = m_2)$, $g = h,\ a = d,\ b = c,\ f_{1\infty} = \dfrac{f_1 f_2}{f_{2\infty}},\ m_1 = m_2 = \dfrac{h}{1 - \frac{f_1^2}{f_{2\infty}^2}}$ and $f_{2\infty}^2 = \dfrac{f_1^2 + f_2^2 - 2m^2 f_1 f_2}{2(1-m^2)} + \left[\left(\dfrac{f_1^2+f_2^2-2m^2f_1f_2}{2(1-m^2)}\right)^2 - f_1^2 f_2^2\right]^{1/2}$
I			$L_1 = m_1 L_{1k}$ $L_2 = a L_{1k}$ $L_2' = c L_{1k}$ $C_1 = \dfrac{C_{1k}}{m_2}$ $C_2 = \dfrac{C_{1k}}{b}$ $C_2' = \dfrac{C_{1k}}{d}$		$L_1 = \dfrac{L_{2k}}{b}$ $L_1' = \dfrac{L_{2k}}{d}$ $L_2 = \dfrac{L_{2k}}{m_2}$ $C_1 = a C_{2k}$ $C_1' = c C_{2k}$ $C_2 = m_1 C_{2k}$	These formulas apply for both end and Type I sections
II $f_{1\infty} = 0$ $f_{2\infty} = f_2$			$L_1 = \dfrac{f_1 R}{\pi f_2(f_2-f_1)}$ $L_2 = \dfrac{(f_1+f_2)R}{4\pi f_1 f_2}$ $C_1 = C_{1k}$		$C_1 = \dfrac{f_1+f_2}{4\pi f_1 f_2 R}$ $C_2 = \dfrac{f_1}{\pi f_2(f_2-f_1)R}$ $L_2 = L_{2k}$	
III $f_{1\infty} = f_1$ $f_{2\infty} = \infty$			$L_1 = L_{1k}$ $C_2' = \dfrac{1}{\pi(f_1+f_2)R}$ $C_1 = \dfrac{f_2-f_1}{4\pi f_1^2 R}$		$L_1' = \dfrac{R}{\pi(f_1+f_2)}$ $L_2 = \dfrac{(f_2-f_1)R}{4\pi f_1^2}$ $C_2 = C_{2k}$	
IV $f_{1\infty} = 0$ $f_{2\infty} = \infty$			$L_1 = L_{1k}$ $L_2 = L_{2k}$ $C_1 = C_{1k}$ $C_2 = C_{2k}$		$L_1 = L_{1k}$ $L_2 = L_{2k}$ $C_1 = C_{1k}$ $C_2 = C_{2k}$	
V $f_{2\infty} = f_2$			$L_1 = m_1 L_{1k}$ $L_2 = \dfrac{(1-m_1^2)}{4m_1}L_{1k}$ $C_1 = \dfrac{C_{1k}}{m_2}$ $C_2 = \dfrac{4m_2}{1-m_2^2}C_{1k}$ See notation for m_1 and m_2		$L_1 = \dfrac{4m_2}{1-m_2^2}L_{2k}$ $L_2 = \dfrac{L_{2k}}{m_2}$ $C_1 = \dfrac{(1-m_1^2)}{4m_1}C_{2k}$ $C_2 = m_1 C_{2k}$ See notation for m_1 and m_2	$m_1 = \dfrac{f_1}{f_2}m_2 \qquad m_2 = \sqrt{\dfrac{1 - \frac{f_{1\infty}^2}{f_1^2}}{1 - \frac{f_{1\infty}^2}{f_2^2}}}$
VI $f_{1\infty} = f_1$		Same circuit as above for Type V	Same formulas as above for Type V. See notation for m_1 and m_2	Same circuit as above for Type V	Same formulas as above for Type V See notation for m_1 and m_2	$m_1 = \sqrt{\dfrac{1 - \frac{f_2^2}{f_{2\infty}^2}}{1 - \frac{f_1^2}{f_{2\infty}^2}}} \qquad m_2 = \dfrac{f_1}{f_2}m_1$
VII $f_{1\infty} = 0$			$L_1 = m_1 L_{1k}$ $L_2 = a L_{1k}$ $L_2' = \dfrac{(1-m_1^2)}{4h}L_{1k}$ $C_1 = C_{1k}$ $C_2' = \dfrac{h}{a}C_{1k}$		$L_1' = \dfrac{h}{a}L_{2k}$ $L_2 = L_{2k}$ $C_1 = a C_{2k}$ $C_2 = m_1 C_{2k}$ $C_1' = \dfrac{(1-m_1^2)}{4h}C_{2k}$	$h = \sqrt{\left(1 - \dfrac{f_1^2}{f_{2\infty}^2}\right)\left(1 - \dfrac{f_2^2}{f_{2\infty}^2}\right)}$ $m_1 = \dfrac{\frac{f_1 f_2}{f_{2\infty}^2} + h}{}$ $a = \dfrac{(1-m_1^2)f_{2\infty}^2}{4f_1 f_2}$
VIII $f_{2\infty} = \infty$			$L_1 = L_{1k}$ $L_2 = \dfrac{d}{g}L_{1k}$ $C_1 = \dfrac{C_{1k}}{m_2}$ $C_2 = \dfrac{4g}{1-m_2^2}C_{1k}$ $C_2' = \dfrac{C_{1k}}{d}$		$L_1 = \dfrac{4g}{1-m_2^2}L_{2k}$ $L_2 = \dfrac{L_{2k}}{m_2}$ $L_1' = \dfrac{L_{2k}}{d}$ $C_1 = \dfrac{d}{g}C_{2k}$ $C_2 = C_{2k}$	$g = \sqrt{\left(1 - \dfrac{f_{1\infty}^2}{f_1^2}\right)\left(1 - \dfrac{f_{1\infty}^2}{f_2^2}\right)}$ $m_2 = g + \dfrac{f_{1\infty}^2}{f_1 f_2}$ $d = \dfrac{(1-m_2^2)f_1 f_2}{4f_{1\infty}^2}$

COMB-FILTER DESIGN

Comb filters consist of a chain of narrow-band filters which pass spectral lines over the frequency spectrum of the signal. They pass discrete frequency components and discriminate against noise. Such filters are used to separate a composite input signal into a number of channels before data processing in telemetry systems and radar. The spacing between channels may be expressed as a frequency ratio which depends on the number of channels needed to cover one octave, or "n." Thus $f/f_c = 2^n$, where f_c is the reference frequency, f is the unknown frequency of the adjacent channel, and n is any positive or negative real number. For $n = \pm 1$, f equals $2f_c$ and $\frac{1}{2}f_c$. These values are the center frequencies of channels, one octave away from the reference frequency.

The nomogram solves for positive or negative fractional values of n. The frequency scales, f_c and f, are normalized so that the nomogram can be used for any frequency range by shifting the decimal point. The ratio scale, n, has a decimal range as well as fractional values.

To use the nomogram, place a straight-edge from the octave fraction or decimal on the n scale to the reference frequency on the f_c scale. Read the center frequency of the next channel on the f scale. Hold the n-scale value as a pivot point and shift the straight-edge to the same frequency on the f_c scale as the first answer. Read the next bandpass center frequency on the f scale. Continue the process until all center frequencies are obtained. For negative n values, divide the reference frequency by two to obtain the lower octave. After this step, proceed as for a positive n value.

FOR EXAMPLE:
Calculate the center frequencies for 1/3 octave filters, starting at 100 Hz (see illustration).

Set the straight-edge from 1/3 or 0.33 on the n scale to the one (for 100 Hz) on the f_c scale and and read 1.26 on the f scale; the center frequency of the next channel bandpass filter is 126 Hz. Pivot at 1/3 on the n scale and shift the straight-edge to 126 on the f_c scale. Read 160 Hz on the f scale. When 1260 Hz on the f scale and 1000 Hz on f_c is reached, shift back to the lower portion of the f_c scale and continue.

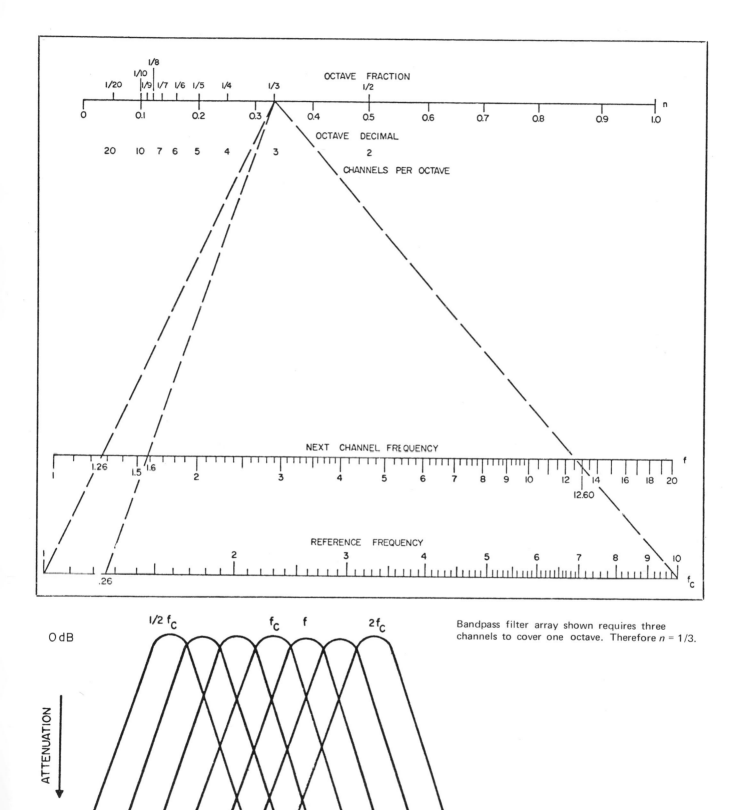

Bandpass filter array shown requires three channels to cover one octave. Therefore $n = 1/3$.

PULSE-FORMING NETWORK
NOMOGRAM

Pulse-forming networks supply high-voltage pulses to magnetrons and lasers. This nomogram relates the pulse width and characteristic impedance to the network's inductances and capacitances. It is based on the formulas:

$$Z_o = \sqrt{\frac{L}{C}}; \quad P_w = 2n\sqrt{LC};$$

and

$$n = \frac{P_w}{2r}$$

where

Z_o = characteristic impedance

L = inductance per section

C = capacitance per section

n = number of sections

P_w = pulse width

r = rise time

FOR EXAMPLE:

Design a PFN that delivers a 4-kV, 500-μsec pulse with a 25-μsec rise time into a 1-ohm load. The number of sections ($P_w/2r$) is 10. Connecting 1 ohm to 500 μsec on the left and right scales yields 250 μF and 250 μH as total capacitance C_N and total inductance L_N. Dividing by 10 gives 25 μF and 25 μH per section. The two end inductances are 1.15 the value of each section or 2.875 μH.

The nomogram solves for the total network inductance and capacitance. Each value must be divided by the number of sections to obtain the values of the individual elements.

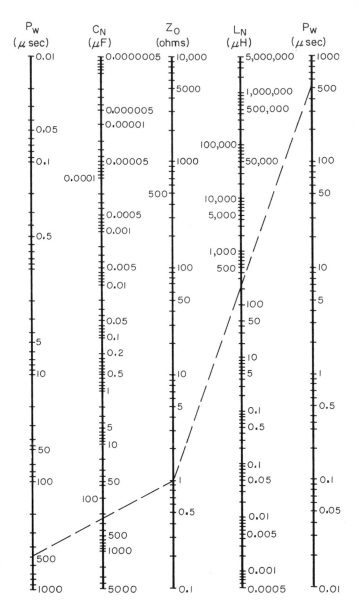

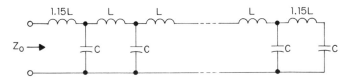

DELAY LINE DESIGN NOMOGRAM

A pulse applied to the input of a delay line is continuously delayed by a predetermined amount as it travels along the line. The artificial or lumped parameter type of delay line consists of a series of low-pass LC filters. The delay for n sections is given by the formula

$$t = n\sqrt{LC}$$

where

t = time delay in microseconds

n = number of sections

L = inductance in microhenries

C = capacitance in microfarad

The characteristic impedance Z_o must be matched to reduce reflections within the delay line and is given by the formula

$$Z_o = \sqrt{L/C} \quad \text{where } Z_o \text{ is in ohms}$$

The cutoff frequency of each section must be higher than the operating frequency

$$f_c = \frac{1}{\pi\sqrt{LC}}$$

where f_c is the cutoff frequency in megahertz

FOR EXAMPLE:

Determine the parameters for a delay line with a 1.5-μsec delay and an f_c of 5 MHz. Pivot around 5 MHz on scale 2 and select standard values of L and C on scales 1 and 4. (120 μH and 33 pF) The cutoff frequency on scale 2 corresponds to the time delay per section shown on scale 3—in this case 0.063 μsec/section. The time delay per section aligned with the required total delay (1.5 μsec) on scale 5 shows the total number of sections required as 24 on scale 1. The characteristic impedance of the line is found to be 1900 ohms as shown on scale 6 by aligning C (33 pF) on scale 5, with the previously selected value of L (120 μH) on scale 8.

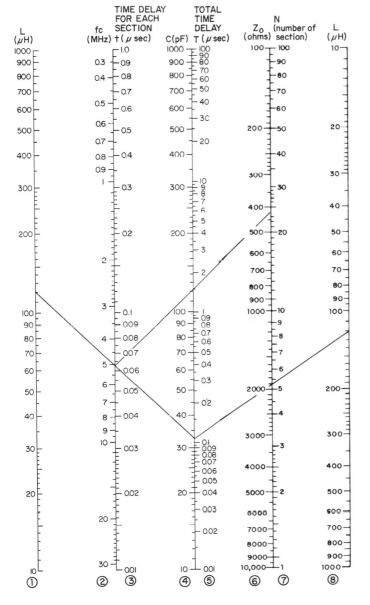

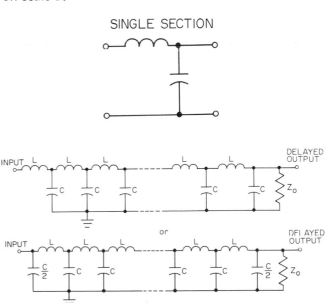

SINGLE SECTION

INPUT — DELAYED OUTPUT

or

INPUT — DELAYED OUTPUT

COAXIAL CABLE SIGNAL DELAY
NOMOGRAM

This nomogram solves for the delay per foot as well as the total delay of a coaxial cable when the relative dielectric constant of the insulation is known. The nomogram is based on the relationship

$$T = 1.108\sqrt{E} \text{ nsec/ft}$$

The relative dielectric constant and delay per foot are plotted on the left-hand index and can be related directly. The chart gives the approximate ranges of dielectric constants of commonly used insulating materials. Some dielectric properties are a function of composition, frequency, and temperature, and the values shown should be used accordingly.

FOR EXAMPLE:
A 4-ft cable with a polystyrene dielectric will produce a total delay of about 6.3 to 6.5 nsec.

DIELECTRIC CONSTANTS

Bakelite [1]	3.95
Fluorinated ethylene propylene	2.2
Irradiated polyethylene	2.3
Lucite [2]	2.7
Magnesium oxide	9.7
Nylon	3.0
Polyethylene	2.25-2.32
Polystyrene	2.4-2.6
Polytetrafluoroethylene	2.0-2.3
Polyurethane	6.4-7.6
Polyvinylchloride (nonrigid)	7.0
Rubber (natural)	2.4-4.6
Rubber (silicone)	2.9-3.7

1- TM Union Carbide Corp.

2- TM DuPont

VOLTAGE MULTIPLIER CIRCUITS

Circuit diagrams are given and the minimum voltage ratings of the capacitors are shown as related to V_m. The minimum PIV of the diodes is $2V_m$.

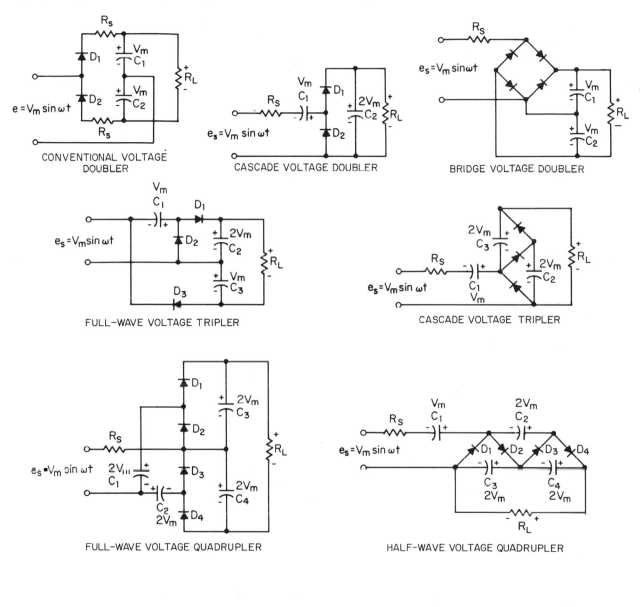

CONVENTIONAL VOLTAGE DOUBLER

CASCADE VOLTAGE DOUBLER

BRIDGE VOLTAGE DOUBLER

FULL−WAVE VOLTAGE TRIPLER

CASCADE VOLTAGE TRIPLER

FULL−WAVE VOLTAGE QUADRUPLER

HALF−WAVE VOLTAGE QUADRUPLER

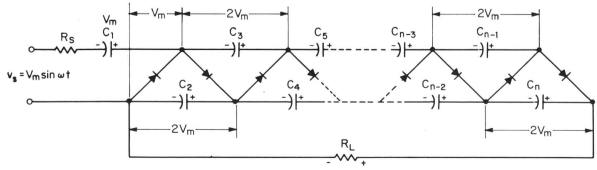

n-SECTION VOLTAGE MULTIPLIER

PERCENT REGULATION OF
POWER SUPPLIES

The percent regulation of a power supply is found by the change in output voltage between *Full Load* and *No Load* voltage as given by the formula:

% regulation =

$$\frac{No\ Load\ Voltage - Full\ Load\ Voltage}{Full\ Load\ Voltage} \times 100.$$

FOR EXAMPLE:

1. What is percent regulation if *No Load Voltage* is 500 V and *Full Load Voltage* is 492 V? The difference is 8 V. *Answer*: Connecting 492 and 8 gives a regulation of about 1.6%.
2. For 0.04% regulation what is maximum allowable change in output voltage if required *Full Load Voltage* is 15 V. *Answer*: 0.006 V.

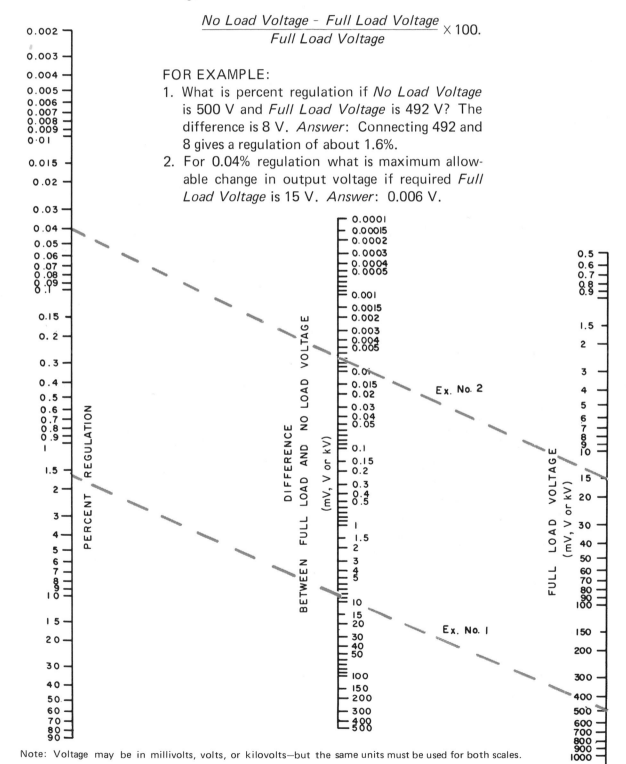

Note: Voltage may be in millivolts, volts, or kilovolts—but the same units must be used for both scales.

POWER LOSS DUE TO
IMPEDANCE MISMATCH

This chart shows the power loss resulting from in-
equality in the absolute magnitude of two impedances connected so as to transfer power from
one to the other. The figures on the curves are the
number of degrees of algebraic phase difference
between the two impedances.

FOR EXAMPLE
Find the resulting power loss when a loudspeaker
with an impedance of 10 ohms and a phase angle of
60° is fed from a generator with a 100-ohm internal
impedance. The impedance mismatch ratio is 10:1,
and at the 60° line the loss due to mismatch is
read as 5.7 dB.

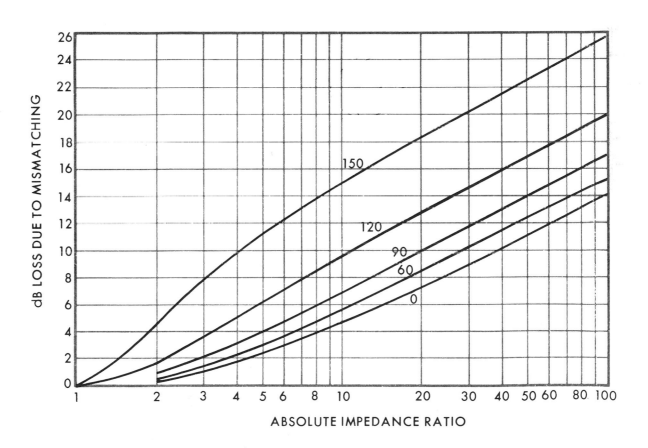

SEVEN COMMONLY USED BRIDGE CIRCUITS AND THEIR BALANCE EQUATIONS

A bridge consists essentially of four arms connected in series and so arranged, that when an electromotive force is applied across one pair of opposite junctions, the response of a detecting and/or indicating device connected between the other pair of junctions may be zeroed by adjusting one or more of the elements of the arms of the bridge. Seven commonly used bridge circuits and their balance equations are shown.

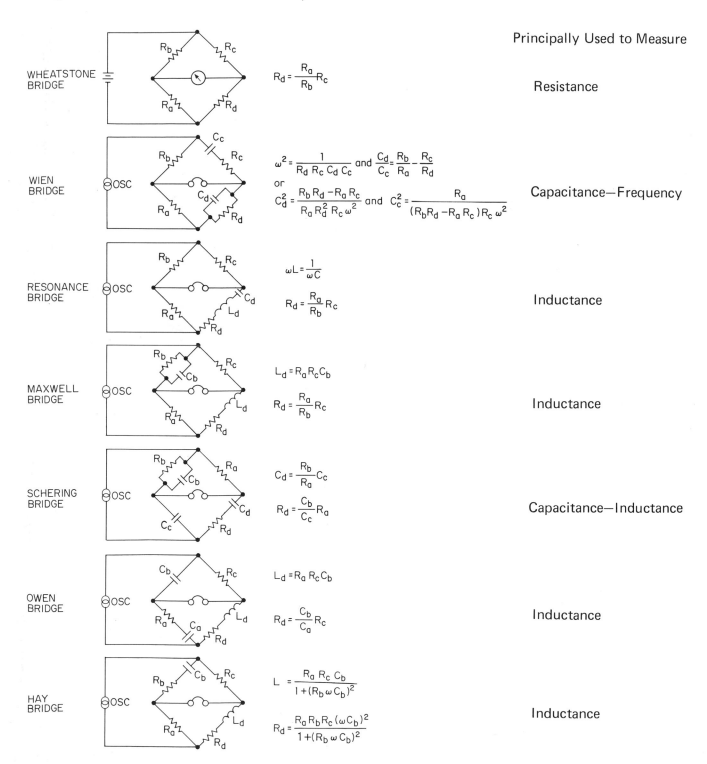

Principally Used to Measure

WHEATSTONE BRIDGE

$$R_d = \frac{R_a}{R_b} R_c$$

Resistance

WIEN BRIDGE

$$\omega^2 = \frac{1}{R_d R_c C_d C_c} \quad \text{and} \quad \frac{C_d}{C_c} = \frac{R_b}{R_a} - \frac{R_c}{R_d}$$

or

$$C_d^2 = \frac{R_b R_d - R_a R_c}{R_a R_d^2 R_c \omega^2} \quad \text{and} \quad C_c^2 = \frac{R_a}{(R_b R_d - R_a R_c) R_c \omega^2}$$

Capacitance—Frequency

RESONANCE BRIDGE

$$\omega L = \frac{1}{\omega C}$$

$$R_d = \frac{R_a}{R_b} R_c$$

Inductance

MAXWELL BRIDGE

$$L_d = R_a R_c C_b$$

$$R_d = \frac{R_a}{R_b} R_c$$

Inductance

SCHERING BRIDGE

$$C_d = \frac{R_b}{R_a} C_c$$

$$R_d = \frac{C_b}{C_c} R_a$$

Capacitance—Inductance

OWEN BRIDGE

$$L_d = R_a R_c C_b$$

$$R_d = \frac{C_b}{C_a} R_c$$

Inductance

HAY BRIDGE

$$L = \frac{R_a R_c C_b}{1 + (R_b \omega C_b)^2}$$

$$R_d = \frac{R_a R_b R_c (\omega C_b)^2}{1 + (R_b \omega C_b)^2}$$

Inductance

SECTION 4

ACTIVE COMPONENTS AND CIRCUITS

MAJOR SEMICONDUCTOR COMPONENTS

NAME OF DEVICE	CIRCUIT SYMBOL	COMMONLY USED JUNCTION SCHEMATIC	ELECTRICAL CHARACTERISTICS		MAJOR APPLICATIONS	ROUGHLY ANALOGOUS TO:
Diode or Rectifier	ANODE / CATHODE	ANODE p n CATHODE		Conducts easily in one direction, blocks in the other	Rectification, Blocking, Detecting, Steering	Check valve, Diode tube, Gas diode
Avalanche (Zener) Diode	ANODE / CATHODE	ANODE p n CATHODE		Constant voltage characteristic in negative quadrant	Regulation, Reference, Clipping	V-R tube
Integrated Voltage Regulator (IVR)	IVR		$\frac{R_{31}}{R_{31}+R_{21}}$	Programmed to desired V_{21} by two resistors	Shunt voltage regulator, Reference element, Error modifier, Level sensing, Level shifting	Avalanche Diode
Tunnel Diode	POSITIVE ELECTRODE / NEGATIVE ELECTRODE	POSITIVE ELECTRODE p n NEGATIVE ELECTRODE		Displays negative resistance when current exceeds peak point current I_p	UHF converter, Logic circuits, Microwave circuits, Level sensing	None
Back Diode	ANODE / CATHODE	ANODE n p CATHODE		Similar characteristics to conventional diode except very low forward voltage drop	Microwave mixers and low power oscillators	None
Thyrector		p n p n		Rapidly increasing current above rated voltage in either direction	Transient voltage suppression and arc suppression	Thyrite, Two avalanche diodes in inverse-series connection
n-p-n Transistor	COLLECTOR, BASE, EMITTER	COLLECTOR n p n EMITTER		Constant collector current for given base drive	Amplification, Switching, Oscillation	Pentode Tube
p-n-p Transistor	COLLECTOR, BASE, EMITTER	COLLECTOR p n p EMITTER		Complement to n-p-n transistor	Amplification, Switching, Oscillation	None
Photo Transistor	COLLECTOR, BASE, EMITTER	COLLECTOR n p n EMITTER		Incident light acts as base current of the photo transistor	Tape readers, Card readers, Position sensor, Tachometers	None
Unijunction Transistor (UJT)	EMITTER, BASE 2, BASE 1	BASE 2 p n BASE 1		Unijunction emitter blocks until its voltage reaches V_p; then conducts	Interval timing, Oscillation, Level Detector, SCR Trigger	None

NAME OF DEVICE	CIRCUIT SYMBOL	COMMONLY USED JUNCTION SCHEMATIC	ELECTRICAL CHARACTERISTICS		MAJOR APPLICATIONS	ROUGHLY ANALOGOUS TO:
Complementary Unijunction Transistor (CUJT)				Functional complement to UJT	High stability timers Oscillators and level detectors	None
Programmable Unijunction Transistor (PUT)				Programmed by two resistors for V_p, I_p, I_v. Function equivalent to normal UJT.	Low cost timers and oscillators Long period timers SCR trigger Level detector	UJT
Silicon Controlled Rectifier (SCR)				With anode voltage (+), SCR can be triggered by I_g, remaining in conduction until anode I is reduced to zero	Power switching Phase control Inverters Choppers	Gas thyratron or ignitron
Complementary Silicon Controlled Rectifier (CSCR)				Polarity complement to SCR	Ring counters Low speed logic Lamp driver	None
Light Activated SCR* (LASCR)				Operates similar to SCR, except can also be triggered into conduction by light falling on junctions	Relay Replacement Position controls Photoelectric applications Slave flashes	None
Silicon Controlled Switch* (SCS)				Operates similar to SCR except can also be triggered on by a negative signal on anode-gate. Also several other specialized modes of operation	Logic applications Counters Nixie drivers Lamp drivers	Complementary transistor pair
Silicon Unilateral Switch (SUS)				Similar to SCS but zener added to anode gate to trigger device into conduction at ~ 8 volts. Can also be triggered by negative pulse at gate lead.	Switching Circuits Counters SCR Trigger Oscillator	Shockley or 4-layer diode
Silicon Bilateral Switch (SBS)				Symmetrical bilateral version of the SUS. Breaks down in both directions as SUS does in forward.	Switching Circuits Counters TRIAC Phase Control	Two inverse Shockley diodes
Triac				Operates similar to SCR except can be triggered into conduction in either direction by (+) or (-) gate signal	AC switching Phase control Relay replacement	Two SCR's in inverse parallel
Diac Trigger				When voltage reaches trigger level (about 35 volts), abruptly switches down about 10 volts.	Triac and SCR trigger Oscillator	Neon lamp

LETTER SYMBOLS AND ABBREVIATIONS FOR SEMICONDUCTOR DEVICES

TABLE I

General Semiconductor Symbols

I, i	region of a device which is intrinsic and in which neither holes nor electrons predominate
N, n	region of a device where electrons are the majority carriers
NF	noise figure
P, p	region of a device where holes are the majority carriers
K_θ	thermal derating factor
T	temperature
T_A	ambient temperature
T_C	case temperature
T_J	junction temperature
T_{STG}	storage temperature
θ, or R_θ	thermal resistance
θ_{J-A}	thermal resistance, junction to ambient
θ_{J-C}	thermal resistance, junction to case
$\theta_{(t)}$	transient thermal impedance
$\theta_{J-A(t)}$	transient thermal impedance, junction to ambient
$\theta_{J-C(t)}$	transient thermal impedance, junction to case
t_d	delay time
t_f	fall time
t_{fr}	forward recovery time (diodes)
t_p	pulse time
t_r	rise time
t_{rr}	reverse recovery time (diodes)
t_s	storage time

TABLE 2

Signal Diode and Rectifier Diode Symbols

$V_{(BR)}$ or $V_{(BR)R}$	reverse breakdown voltage, dc
$v_{(BR)}$ or $v_{(BR)R}$	reverse breakdown voltage, instantaneous total value
I_F	forward current, dc
$I_{F(AV)}$	forward current, average value
i_F	forward current, instantaneous total value
I_f	forward current, rms value of alternating component
$I_{F(RMS)}$	forward current, rms total value
I_{FM}	forward current, maximum (peak) total value
$I_{FM(rep)}$	forward current, repetitive, maximum (peak), total value
$I_{FM(surge)}$	forward current, maximum (peak), total value of surge
I_o	output current, average rectified
I_R	reverse current, dc
i_R	reverse current, instantaneous total value
$I_{R(AV)}$	reverse current, average value
I_{RM}	reverse current, maximum (peak) total value
I_r	reverse current, rms value of alternating component
$I_{R(RMS)}$	reverse current, rms total value
L_c	conversion loss (microwave diodes)
P_F	forward power dissipation, dc
$P_{F(AV)}$	forward power dissipation, average value
P_{FM}	forward power dissipation, maximum (peak) total value
p_F	forward power dissipation, instantaneous total value
P_R	reverse power dissipation, dc
$P_{R(AV)}$	reverse power dissipation, average value
P_{RM}	reverse power dissipation, maximum (peak) total value
p_R	reverse power dissipation, instantaneous total value
V_F	forward voltage drop, dc
v_F	forward voltage drop, instantaneous total value
$V_{F(AV)}$	forward voltage drop, average value
V_{FM}	forward voltage drop, maximum (peak) total value
$V_{F(RMS)}$	forward voltage drop, total rms value
V_f	forward voltage drop, rms value of alternating component
V_R	reverse voltage, dc
v_R	reverse voltage, instantaneous total value
$V_{R(AV)}$	reverse voltage, average value
V_{RM}	reverse voltage, maximum (peak) total value
$V_{RM(wkg)}$	working peak reverse voltage, maximum (peak) total value
$V_{RM(rep)}$	repetitive peak reverse voltage, maximum (peak) total value
$V_{RM(nonrep)}$	nonrepetitive peak reverse voltage, maximum (peak) total value
$V_{R(RMS)}$	reverse voltage, total rms value
V_r	reverse voltage, rms value of alternating component

LETTER SYMBOLS AND ABBREVIATIONS FOR SEMICONDUCTOR DEVICES (Continued)

TABLE 3
Transistor Symbols

BV_{CBO}	obsolete—see $V_{(BR)CBO}$
BV_{CEO}	obsolete—see $V_{(BR)CEO}$
BV_{CER}	obsolete—see $V_{(BR)CER}$
BV_{CES}	obsolete—see $V_{(BR)CES}$
BV_{CEX}	obsolete—see $V_{(BR)CEX}$
BV_{EBO}	obsolete—see $V_{(BR)EBO}$
BV_R	obsolete—see $V_{(BR)R}$
C_{ibo}	open-circuit input capacitance, common base
C_{ibs}	short-circuit input capacitance, common base
C_{ieo}	open-circuit input capacitance, common emitter
C_{ies}	short-circuit input capacitance, common emitter
C_{obo}	open-circuit output capacitance, common base
C_{obs}	short-circuit output capacitance, common base
C_{oeo}	open-circuit output capacitance, common emitter
C_{oes}	short-circuit output capacitance, common emitter
f_{hfb}	small-signal short-circuit forward current transfer ratio cutoff frequency (common base)
f_{hfc}	small-signal short-circuit forward current transfer ratio cutoff frequency (common collector)
f_{hfe}	small-signal short-circuit forward current transfer ratio cutoff frequency (common emitter)
f_{max}	maximum frequency of oscillation
f_T	frequency at which small-signal forward current transfer ratio (common emitter) extrapolates to unity
g_{ME}	static transconductance (common emitter)
g_{me}	small-signal transconductance (common emitter)
G_{PB}	large-signal average power gain (common base)
G_{pb}	small-signal average power gain (common base)
G_{PC}	large-signal average power gain (common collector)
G_{pc}	small-signal average power gain (common collector)
G_{PE}	large-signal average power gain (common emitter)
G_{pe}	small-signal average power gain (common emitter)

h_{FB}	static forward current transfer ratio (common base)
h_{fb}	small-signal short-circuit forward current transfer ratio (common base)
h_{FC}	static forward current transfer ratio (common collector)
h_{fc}	small-signal short-circuit forward current transfer ratio (common collector)
h_{FE}	static forward current transfer ratio (common emitter)
h_{fe}	small-signal short-circuit forward current transfer ratio (common emitter)
h_{IB}	static input resistance (common base)
h_{ib}	small-signal short-circuit input impedance (common base)
h_{IC}	static input resistance (common collector)
h_{ic}	small-signal short-circuit input impedance (common collector)
h_{IE}	static input resistance (common emitter)
h_{ie}	small-signal short-circuit input impedance (common emitter)
h_{ob}	small-signal open-circuit output admittance (common base)
h_{oc}	small-signal open-circuit output admittance (common collector)
h_{oe}	small-signal open-circuit output admittance (common emitter)
h_{rb}	small-signal open-circuit reverse voltage transfer ratio (common base)
h_{rc}	small-signal open-circuit reverse voltage transfer ratio (common collector)
h_{re}	small-signal open-circuit reverse voltage transfer ratio (common emitter)
I_B	base current, dc
I_b	base current, rms value of alternating component
i_B	base current, instantaneous total value
I_C	collector current, dc
I_c	collector current, rms value of alternating component
i_C	collector current, instantaneous total value
I_{CBO}	collector cutoff current, dc, emitter open
I_{CEO}	collector cutoff current, dc, base open
I_{CER}	collector cutoff current, dc, with specified resistance between base and emitter
I_{CEV}	collector cutoff current, dc, with specified voltage between base and emitter
I_{CEX}	collector current, dc, with specified circuit between base and emitter
I_{CES}	collector cutoff current, dc, with base short circuited to emitter
I_{DSS}	drain current, dc, with gate shorted to emitter
I_E	emitter current, dc
I_e	emitter current, rms value of alternating component
I_{EBO}	emitter cutoff current (dc), collector open

P_{BE}	power input (dc) to the base (common emitter)
p_{BE}	power input (instantaneous total) to the base (common emitter)
P_{CB}	power input (dc) to the collector (common base)
p_{CB}	power input (instantaneous total) to the collector (common base)
P_{CE}	power input (dc) to the collector (common emitter)
p_{CE}	power input (instantaneous total) to the collector (common emitter)
P_{EB}	power input (dc) to the emitter (common base)
p_{EB}	power input (instantaneous total) to the emitter (common base)
P_{IB}	large-signal input power (common base)
p_{ib}	small-signal input power (common base)
P_{IC}	large-signal input power (common collector)
p_{ic}	small-signal input power (common collector)
P_{IE}	large-signal input power (common emitter)
p_{ie}	small-signal input power (common emitter)
P_{OB}	large-signal output power (common base)
p_{ob}	small-signal output power (common base)
P_{OC}	large-signal output power (common collector)
p_{oc}	small-signal output power (common collector)
P_{OE}	large-signal output power (common emitter)
p_{oe}	small-signal output power (common emitter)
P_T	total nonreactive power input (dc) to all terminals
p_T	nonreactive power input (instantaneous total) to all terminals
R_B	external base resistance
R_C	external collector resistance
$r_{CE(sat)}$	collector-to-emitter saturation resistance
R_E	external emitter resistance
$Re(h_{ie})$	real part of the small-signal short-circuit input impedance (common emitter)
$V_{(BR)CBO}$	breakdown voltage, collector-to-base, emitter open
$V_{(BR)CEO}$	breakdown voltage, collector-to-emitter, base open
$V_{(BR)CER}$	breakdown voltage, collector-to-emitter, with specified resistance between base and emitter
$V_{(BR)CES}$	breakdown voltage, collector-to-emitter, with base short-circuited to emitter
$V_{(BR)CEX}$	breakdown voltage, collector-to-emitter, with specified circuit between base and emitter
$V_{(BR)DGO}$	breakdown voltage, drain-to-gate, source open

$V_{(BR)EBO}$	breakdown voltage, emitter-to-base, collector open
$V_{(BR)R}$	breakdown voltage, reverse
V_{BB}	base supply voltage
V_{BC}	base-to-collector voltage, dc
V_{bc}	base-to-collector voltage, rms value of alternating component
v_{bc}	base-to-collector voltage, instantaneous value of ac component
V_{BE}	base-to-emitter voltage, dc
V_{be}	base-to-emitter voltage, rms value of alternating component
v_{be}	base-to-emitter voltage, instantaneous value of ac component
V_{CB}	collector-to-base voltage, dc
V_{cb}	collector-to-base voltage, rms value of alternating component
v_{cb}	collector-to-base voltage, instantaneous value of ac component
$V_{CB(fl)}$	dc open-circuit voltage (floating potential) between the collector and base, with the emitter biased with respect to the base
V_{CC}	collector supply voltage, dc
V_{CE}	collector-to-emitter voltage, dc
V_{ce}	collector-to-emitter voltage, rms value of alternating component
v_{ce}	collector-to-emitter voltage, instantaneous value of ac component
$V_{CE(fl)}$	dc open-circuit voltage (floating potential) between the collector and emitter, with the base biased with respect to the emitter
V_{CEO}	collector-to-emitter voltage, dc, with base open
$V_{CEO(sus)}$	collector-to-emitter (breakdown) sustaining voltage with base open
V_{CER}	collector-to-emitter voltage, dc with specified resistor between base emitter
$V_{CER(sus)}$	collector-to-emitter (breakdown) sustaining voltage with specified resistor between base and emitter
V_{CES}	collector-to-emitter voltage, dc with base short circuited to emitter
$V_{CES(sus)}$	collector-to-emitter (breakdown) sustaining voltage with base short-circuited to emitter
V_{CEX}	collector-to-emitter voltage, dc with specified circuit between base and emitter
$V_{CEX(sus)}$	collector-to-emitter (breakdown) sustaining voltage with specified circuit between base and emitter
$V_{CE(sat)}$	collector-to-emitter saturation voltage, dc
V_{EB}	emitter-to-base voltage, dc
$V_{EB(fl)}$	dc open-circuit voltage (floating potential) between the emitter and base, with the collector biased with respect to the base
V_{eb}	emitter-to-base voltage, rms value of alternating component
v_{eb}	emitter-to-base voltage, instantaneous value of ac component
V_{EC}	emitter-to-collector voltage, dc
$V_{EC(fl)}$	dc open-circuit voltage (floating potential) between the emitter and collector, with the base biased with respect to the collector
V_{ec}	emitter-to-collector voltage, rms value of alternating component
V_{ec}	emitter-to-collector voltage, instantaneous value of ac component
V_{EE}	emitter supply voltage
V_{RT}	reach-through voltage

LETTER SYMBOLS AND ABBREVIA-TIONS FOR SEMICONDUCTOR DEVICES (Continued)

TABLE 4
Tunnel Diode Symbols

I_I inflection point current
I_P peak point current
I_V valley point current
r_i dynamic resistance at inflection point
V_{PP} projected peak point voltage
        [forward voltage point (greater than the peak voltage), at which the current
        is equal to the peak current]
V_I inflection point voltage
V_P peak point voltage
V_V valley point voltage

Typical Characteristics

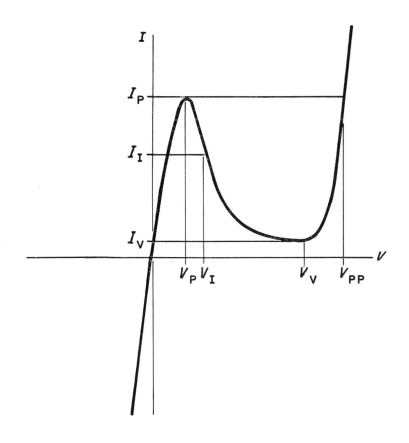

COMPARATIVE CHARACTERISTICS
OF ACTIVE DEVICES

Characteristic	Vacuum Tube	Small-Signal Transistor	High-Power Transistor	Junction Fet	Mosfet
Input impedance	High	a	Very low	High	Very high
Output impedance	High	a	Low/moderate	High	High
Noise	Low	Low	Moderate	Low	Unpredictable
Warm-up time	Long	Short	Short	Short	Short
Power consumption	Large	Small	Moderate	Very Small	Very small
Aging	Appreciable	Low	Low	Low	Moderate
Reliability	Poor	Excellent	Very good	Excellent	Very good
Overload sensitivity	Excellent	Good	Fair	Good	Poor
Size	Large	Small	Moderate	Small	Small

[a]Impedances depend on circuit arrangement:

	Input Impedance	Output impedance
For common base	Low (10's of ohms)	High (megohms)
For common emitter	Medium (kilohms)	Medium (10's of kilohms)
For common collector	High (100's of kilohms)	Low (100's of ohms)

ANALOGY BETWEEN THE THREE BASIC JUNCTION TRANSISTOR CIRCUITS AND THEIR EQUIVALENT ELECTRON TUBE CIRCUITS

A transistor can be operated with the input signal applied to the base and the output taken from the collector (common emitter), with the input signal applied to the emitter and the output taken from the collector (common base), or with the input signal applied to the base and the output taken from the emitter (common collector or emitter follower). The performance characteristics of these three connections correspond roughly to the three tube connections shown below, with the exception that the input impedance is generally lower in the transistor circuit. General characteristics of these three connections are given in the table.

Common Emitter	Common Base	Common Collector
Large current gain	Approximate unity current gain	Large current gain
Large voltage gain	Large voltage gain	Approximate unity voltage gain
Highest power gain	Intermediate power gain	Lowest power gain
Low input resistance	Very low input resistance	High input resistance
High output resistance	Very high output resistance	Low output resistance
Analogous to grounded cathode	Analogous to grounded grid	Analogous to cathode follower generally

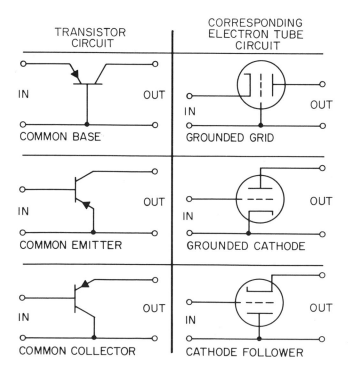

TRANSISTOR CIRCUIT

CORRESPONDING ELECTRON TUBE CIRCUIT

IN   OUT   IN   OUT

COMMON BASE   GROUNDED GRID

IN   OUT   IN   OUT

COMMON EMITTER   GROUNDED CATHODE

IN   OUT   IN   OUT

COMMON COLLECTOR   CATHODE FOLLOWER

DEFINITIONS OF EQUIVALENT CIRCUIT PARAMETERS

Parameter	Common Base	Common Emitter	Common Collector	Definition
z	$z_{11}, z_{11b},$ or z_{ib}	z_{11e} or z_{ie}	z_{11c} or z_{ic}	Input impedance with open-circuit output
	$z_{12}, z_{12b},$ or z_{rb}	z_{12e} or z_{re}	z_{12c} or z_{rc}	Reverse transfer impedance with open-circuit input
	$z_{21}, z_{21b},$ or z_{fb}	z_{21e} or z_{fe}	z_{21c} or z_{fc}	Forward transfer impedance with open-circuit output
	z_{22}, z_{22b} or z_{ob}	z_{22e} or z_{oe}	z_{22c} or z_{oc}	Output impedance with open-circuit input
y	$y_{11}, y_{11b},$ or y_{ib}	y_{11e} or y_{ie}	y_{11c} or y_{ic}	Input admittance with short-circuit output
	$y_{12}, y_{12b},$ or y_{rb}	y_{12e} or y_{re}	y_{12c} or y_{rc}	Reverse transfer admittance with short-circuit input
	$y_{21}, y_{21b},$ or y_{fb}	y_{21e} or y_{fe}	y_{21c} or y_{fc}	Forward transfer admittance with short-circuit output
	$y_{22}, y_{22b},$ or y_{ob}	y_{22e} or y_{oe}	y_{22c} or y_{oc}	Output admittance with short-circuit input
h	$h_{11}, h_{11b},$ or h_{ib}	h_{11e} or h_{ie}	h_{11c} or h_{ic}	Input impedance with short-circuit output
	$h_{12}, h_{12b},$ or h_{rb}	h_{12e} or h_{re}	h_{12c} or h_{rc}	Reverse open-circuit voltage amplification factor
	$h_{21}, h_{21b},$ or h_{fb}	h_{21e} or h_{fe}	h_{21c} or h_{fc}	Forward short-circuit current amplification factor
	$h_{22}, h_{22b},$ or h_{ob}	h_{22e} or h_{oe}	h_{22c} or h_{oc}	Output admittance with open-circuit input

Note: $h_{11} = 1/y_{11}$ and $h_{22} = 1/z_{22}$.

Typical Transistor Parameters

Common Base

$h_{11} = 39$ ohms
$h_{12} = 380 \times 10^{-6}$
$h_{21} = -0.98$
$h_{22} = 0.49 \ \mu\text{mho}$

Common Emitter

$h_{11} = 2,000$ ohms
$h_{12} = 600 \times 10^{-6}$
$h_{21} = 50$
$h_{22} = 25 \ \mu\text{mhos}$

Common Collector

$h_{11} = 2,000$ ohms
$h_{12} - 1$
$h_{21} = -51$
$h_{22} = 25 \ \mu\text{mhos}$

EQUIVALENT CIRCUITS FOR SMALL-SIGNAL LOW-FREQUENCY TRANSISTOR STAGES

Small-signal, low-frequency, T-equivalent circuits for transistor stages

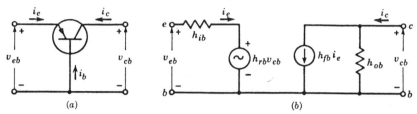

Common-base configuration (a) and hybrid equivalent circuit (b).

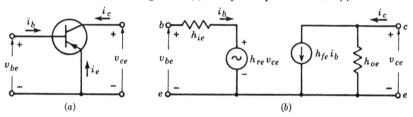

Common-emitter configuration (a) and hybrid equivalent circuit (b).

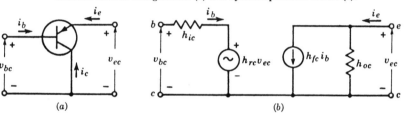

Common-collector configuration (a) and hybrid equivalent circuit (b).

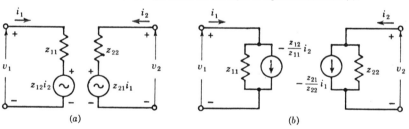

z-Parameter equivalent circuit.

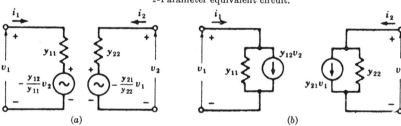

y-Parameter equivalent circuit.

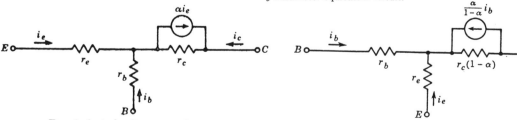

T-equivalent circuit, common base.   T-equivalent circuit, common emitter.

TRANSISTOR PARAMETER CONVERSION TABLES

(A) Common-base h parameters in terms of common-emitter, common-collector, and T parameters.
(B) Common-collector h parameters in terms of common-emitter, common-base, and T parameters.
(C) Common-emitter h parameters in terms of common-base, common-collector, and T parameters.
(D) T parameters in terms of common-emitter, common-base, and common-collector parameters.

(A)

h parameter	Common emitter	Common collector	T-equivalent circuit
h_{ib}	$\dfrac{h_{ie}}{(1+h_{fe})(1-h_{re})+h_{ie}h_{oe}} \cong \dfrac{h_{ie}}{1+h_{fe}}$	$\dfrac{h_{ic}}{h_{ic}h_{oc}-h_{fc}h_{rc}} \cong -\dfrac{h_{ic}}{h_{fc}}$	$r_e+(1-\alpha)r_b$
h_{rb}	$\dfrac{h_{ie}h_{oe}-h_{re}(1+h_{fe})}{(1+h_{fe})(1-h_{re})+h_{ie}h_{oe}} \cong \dfrac{h_{ie}h_{oe}}{1+h_{fe}}-h_{re}$	$\dfrac{h_{fc}(1-h_{rc})+h_{ic}h_{oc}}{h_{ic}h_{oc}-h_{fc}h_{rc}} \cong h_{re}-1-\dfrac{h_{ic}h_{oc}}{h_{fc}}$	$\dfrac{r_b}{r_c+r_b} \cong \dfrac{r_b}{r_c}$
h_{fb}	$\dfrac{-h_{fe}(1-h_{re})-h_{ie}h_{oe}}{(1+h_{fe})(1-h_{re})+h_{ie}h_{oe}} \cong \dfrac{h_{fe}}{1+h_{fe}}$	$\dfrac{h_{rc}(1+h_{fc})-h_{ic}h_{oc}}{h_{ic}h_{oc}-h_{fc}h_{rc}} \cong \dfrac{1+h_{fc}}{h_{fc}}$	α
h_{ob}	$\dfrac{h_{oe}}{(1+h_{fe})(1-h_{re})+h_{ie}h_{oe}} \cong \dfrac{h_{oe}}{1+h_{fe}}$	$\dfrac{h_{oc}}{h_{ic}h_{oc}-h_{fc}h_{rc}} \cong \dfrac{h_{oc}}{h_{fc}}$	$\dfrac{1}{r_c+r_b} \cong \dfrac{1}{r_c}$

(B)

h parameter	Common emitter	Common base	T-equivalent circuit
h_{ic}	h_{ie}	$\dfrac{h_{ib}}{(1+h_{fb})(1-h_{rb})+h_{ob}h_{ib}} \cong \dfrac{h_{ib}}{1+h_{fb}}$	$r_b+\dfrac{r_e r_c}{r_e+r_c-ar_c} \cong r_b+\dfrac{r_e}{1-\alpha}$
h_{rc}	$1-h_{re}$	$\dfrac{1+h_{fb}}{(1+h_{fb})(1-h_{rb})+h_{ob}h_{ib}} \cong 1$	$\dfrac{r_c-ar_c}{r_e+r_c-ar_c} \cong 1-\dfrac{r_e}{(1-\alpha)r_c}$
h_{fc}	$-(1+h_{fe})$	$\dfrac{h_{rb}-1}{(1+h_{fb})(1-h_{rb})+h_{ob}h_{ib}} \cong -\dfrac{1}{1+h_{fb}}$	$-\dfrac{r_c}{r_e+r_c-ar_c} \cong \dfrac{-1}{1-\alpha}$
h_{oc}	h_{oe}	$\dfrac{h_{ob}}{(1+h_{fb})(1-h_{rb})+h_{ob}h_{ib}} \cong \dfrac{h_{ob}}{1+h_{fb}}$	$\dfrac{1}{r_e+r_c-ar_c} \cong \dfrac{1}{(1-\alpha)r_c}$

(C)

h parameter	Common base	Common collector	T-equivalent circuit
h_{ie}	$\dfrac{h_{ib}}{(1+h_{fb})(1-h_{rb})+h_{ob}h_{ib}} \cong \dfrac{h_{ib}}{1+h_{fb}}$	h_{ic}	$r_b+\dfrac{r_e r_c}{r_e+r_c-ar_c} \cong r_b+\dfrac{r_e}{1-\alpha}$
h_{re}	$\dfrac{h_{ib}h_{ob}-h_{rb}(1+h_{fb})}{(1+h_{fb})(1-h_{rb})+h_{ob}h_{ib}} \cong \dfrac{h_{ib}h_{ob}}{1+h_{fb}}-h_{rb}$	$1-h_{rc}$	$\dfrac{r_e}{r_e+r_c-ar_c} \cong \dfrac{r_e}{(1-\alpha)r_c}$
h_{fe}	$\dfrac{-h_{fb}(1-h_{rb})-h_{ob}h_{ib}}{(1+h_{fb})(1-h_{rb})+h_{ob}h_{ib}} \cong \dfrac{-h_{fb}}{1+h_{fb}}$	$-(1+h_{fc})$	$\dfrac{ar_c-r_e}{r_e+r_c-ar_c} \cong \dfrac{\alpha}{1-\alpha}$
h_{oe}	$\dfrac{h_{ob}}{(1+h_{fb})(1-h_{rb})+h_{ob}h_{ib}} \cong \dfrac{h_{ob}}{1+h_{fb}}$	h_{oc}	$\dfrac{1}{r_e+r_c-ar_c} \cong \dfrac{1}{(1-\alpha)r_c}$

(D)

T parameter	Common emitter	Common base	Common collector
α	$\dfrac{h_{fe}(1-h_{re})+h_{ie}h_{oe}}{(1+h_{fe})(1-h_{re})+h_{ie}h_{oe}} \cong \dfrac{h_{fe}}{1+h_{fe}}$	$-h_{fb}$	$\dfrac{h_{ic}h_{oc}-h_{rc}(1+h_{fc})}{h_{ic}h_{oc}-h_{fc}h_{rc}} \cong \dfrac{1+h_{fc}}{h_{fc}}$
r_c	$\dfrac{h_{fe}+1}{h_{oe}}$	$\dfrac{1-h_{rb}}{h_{ob}}$	$-\dfrac{h_{fc}}{h_{oc}}$
r_e	$\dfrac{h_{re}}{h_{oe}}$	$h_{ib}-(1+h_{fb})\dfrac{h_{rb}}{h_{ob}}$	$\dfrac{1-h_{rc}}{h_{oe}}$
r_b	$h_{ie}-\dfrac{h_{re}(1+h_{fe})}{h_{oe}}$	$\dfrac{h_{rb}}{h_{ob}}$	$h_{ic}+\dfrac{h_{fc}(1-h_{rc})}{h_{oc}}$
a	$\dfrac{h_{fe}+h_{re}}{1+h_{fe}}$	$-\dfrac{h_{fb}+h_{rb}}{1-h_{rb}}$	$\dfrac{h_{fc}+h_{rc}}{h_{fc}}$

TRANSISTOR PARAMETER
CONVERSION TABLES (Continued)

(E) Input impedance and output impedance in terms of h and T parameters.
(F) Insertion power gain and transducer power gain in terms of h parameters.
(G) Current gain and voltage gain in terms of h and T parameters.
(H) Available power gain and operating power gain in terms of h parameters.

	Input impedance	Output impedance
h parameter	$Z_i = \dfrac{v_i}{i_i} = h_i - \dfrac{h_f h_r Z_L}{1 + h_o Z_L}$	$Z_o = \dfrac{v_o}{i_o} = \dfrac{1}{h_o - \dfrac{h_f h_r}{h_i + Z_g}}$
(E) Common base T-equivalent circuit	$r_e + r_b\left(\dfrac{r_c - ar_c + R_L}{r_c + r_b + R_L}\right) \cong r_e + r_b(1-\alpha)$	$r_c + r_b\left(1 - \dfrac{ar_c + r_b}{r_e + r_b + R_g}\right) \cong r_c$
Common emitter T-equivalent circuit	$r_b + \dfrac{r_e(r_c + R_L)}{r_c - ar_c + r_e + R_L} \cong r_b + \dfrac{r_e}{1-\alpha}$	$r_c - ar_c + r_e\left(1 + \dfrac{ar_c - r_e}{r_e + r_b + R_g}\right) \cong \dfrac{r_c}{1-\alpha}$
Common collector T-equivalent circuit	$r_b + \dfrac{r_c(r_e + R_L)}{r_c - ar_c + r_e + R_L} \cong r_b + \dfrac{r_e + R_L}{1-\alpha}$	$r_e + (r_b + R_g)\dfrac{r_c - ar_c}{r_c + r_b + R_g}$

	Insertion power gain $\left(\dfrac{\text{power into load}}{\text{power generator would deliver directly}}\right)$	Transducer power gain $\left(\dfrac{\text{power into load}}{\text{maximum available generator power}}\right)$
(F) h parameter where Z_g and Z_L are pure resistance	$G_i = \dfrac{h_f^2(R_g + R_L)^2}{[(h_i + R_g)(1 + h_o R_L) - h_f h_r R_L]^2}$	$G_t = \dfrac{4h_f^2 R_g R_L}{[(h_i + R_g)(1 + h_o R_L) - h_f h_r R_L]^2}$

	Current gain	Voltage gain
h parameter	$A_i = \dfrac{i_o}{i_i} = \dfrac{h_f}{1 + h_o Z_L}$	$A_v = \dfrac{v_o}{v_i} = \dfrac{1}{h_r - \dfrac{h_i}{Z_L}\left(\dfrac{1 + h_o Z_L}{h_f}\right)}$
(G) Common base T-equivalent circuit	$\dfrac{ar_c + r_b}{r_c + r_b + R_L} \cong \alpha$	$\dfrac{(ar_c + r_b)R_L}{r_e(r_c + r_b + R_L) + r_b(r_c - ar_c + R_L)} \cong \dfrac{\alpha R_L}{r_e + r_b(1-\alpha)}$
Common emitter T-equivalent circuit	$\dfrac{-(ar_c - r_e)}{r_c - ar_c + r_e + R_L} \cong \dfrac{\alpha}{1-\alpha}$	$\dfrac{-(ar_c - r_e)R_L}{r_e(r_e + R_L) + r_b(r_c - ar_c + r_e + R_L)} \cong -\dfrac{\alpha R_L}{r_e + r_b(1-\alpha)}$
Common collector T-equivalent circuit	$\dfrac{r_c}{r_c - ar_c + r_e + R_L} \cong \dfrac{1}{1-\alpha}$	$\dfrac{r_c R_L}{r_c(r_e + R_L) + r_b(r_c - ar_c + r_e + R_L)} \cong \dfrac{1}{1 + r_e + r_b\dfrac{1-\alpha}{R_L}}$

	Available power gain $\left(\dfrac{\text{maximum available output power}}{\text{maximum available generator power}}\right)$	Operating power gain $\left(\dfrac{\text{power into load}}{\text{power into transistor}}\right)$
(H) h parameter where Z_g and Z_L are pure resistance	$G_a = \dfrac{h_f^2 R_g}{(h_i + R_g)[h_o(h_i + R_g) - h_f h_r]}$	$G_1 = A_v A_i = \dfrac{v_o i_o}{v_i i_i} = \dfrac{\left(\dfrac{h_f}{1 + h_o R_L}\right)}{h_r - \dfrac{h_i}{R_L}\left(\dfrac{1 + h_o R_L}{h_f}\right)}$

(I) z parameters in terms of h parameters.

(J) y parameters in terms of h parameters.

(K) Common emitter z parameters in terms of common collector and common base z parameters and T parameters.

(L) Common emitter y parameters in terms of common collector and common base y parameters and T parameters.

		Common emitter	Common base	Common collector
(I)	z_{11b}	$\dfrac{\Delta h}{h_{oe}}$	$\dfrac{\Delta h}{h_{ob}}$	$\dfrac{1}{h_{oc}}$
	z_{12b}	$\dfrac{\Delta h - h_{re}}{h_{oe}}$	$\dfrac{h_{rb}}{h_{ob}}$	$\dfrac{1 + h_{fc}}{h_{oc}}$
	z_{21b}	$\dfrac{\Delta h + h_{fe}}{h_{oe}}$	$\dfrac{-h_{fb}}{h_{ob}}$	$\dfrac{1 - h_{rc}}{h_{oc}}$
	z_{22b}	$\dfrac{d}{h_{oe}}$	$\dfrac{1}{h_{ob}}$	$\dfrac{d}{h_{oc}}$
(J)	y_{11b}	$\dfrac{d}{h_{ie}}$	$\dfrac{1}{h_{ib}}$	$\dfrac{d}{h_{ic}}$
	y_{12b}	$\dfrac{h_{re} + h_{fe}}{h_{ie}}$	$-\dfrac{h_{rb}}{h_{ib}}$	$-\dfrac{1 + h_{fe}}{h_{ic}}$
	y_{21b}	$-\dfrac{\Delta h + h_{fe}}{h_{ie}}$	$\dfrac{h_{fb}}{h_{ib}}$	$\dfrac{h_{rc} - 1}{h_{ic}}$
	y_{22b}	$\dfrac{\Delta h}{h_{ie}}$	$\dfrac{\Delta h}{h_{ib}}$	$\dfrac{1}{h_{ic}}$

$\Delta h = h_i h_o - h_r h_f$

$d = (1 + h_f)(1 - h_r) + h_i h_o \cong 1 + h_f$

z parameter		Common collector	Common base	T equivalent-circuit
(K)	z_{11e}	$z_{11} - z_{12} - z_{21} + z_{22}$	z_{11}	$r_e + r_b$
	z_{12e}	$z_{22} - z_{12}$	$z_{11} - z_{12}$	r_e
	z_{21e}	$z_{22} - z_{21}$	$z_{11} - z_{21}$	$r_e - ar_c$
	z_{22e}	z_{22}	$z_{11} - z_{12} - z_{21} + z_{22}$	$r_e + r_c(1 - a)$

y parameter		Common collector	Common base	T equivalent-circuit
(L)	y_{11e}	y_{11}	$y_{11} + y_{12} + y_{21} y_{22}$	$\dfrac{r_e + r_c(1 - a)}{\Delta}$
	y_{12e}	$-(y_{11} + y_{12})$	$-(y_{12} + y_{22})$	$-\dfrac{r_e}{\Delta}$
	y_{21e}	$-(y_{11} + y_{21})$	$-(y_{21} + y_{22})$	$-\dfrac{r_e - ar_c}{\Delta}$
	y_{22e}	$y_{11} + y_{12} + y_{21} + y_{22}$	y_{22}	$\dfrac{r_e + r_b}{\Delta}$

$\Delta = r_e r_b + r_c [r_e + r_b(1 - a)]$

TRANSISTOR PARAMETER
CONVERSION TABLES (Continued)

(M) Common base z parameters in terms of common emitter and common collector z parameters and T parameters.

(N) Common base y parameters in terms of common emitter and common collector y parameters and T parameters.

(O) Common collector z parameters in terms of common emitter and common base z parameters and T parameters.

(P) Common collector y parameters in terms of common emitter and common base y parameters and T parameters.

(Q) Input impedance, output impedance, voltage gain, and current gain in terms of z and y parameters.

	z parameter	Common emitter	Common collector	T-equivalent circuit
(M)	z_{11b}	z_{11}	$z_{11} - z_{12} - z_{21} + z_{22}$	$r_e + r_b$
	z_{12b}	$z_{11} - z_{12}$	$z_{11} - z_{21}$	r_b
	z_{21b}	$z_{11} - z_{21}$	$z_{11} - z_{12}$	$r_b + ar_c$
	z_{22b}	$z_{11} - z_{12} - z_{21} + z_{22}$	z_{11}	$r_b + r_c$

	y parameter	Common emitter	Common collector	T-equivalent circuit
(N)	y_{11b}	$y_{11} + y_{12} + y_{21} + y_{22}$	y_{22}	$\dfrac{r_b + r_c}{\Delta}$
	y_{12b}	$-(y_{12} + y_{22})$	$-(y_{21} + y_{22})$	$-\dfrac{r_b}{\Delta}$
	y_{21b}	$-(y_{21} + y_{22})$	$-(y_{12} + y_{22})$	$-\dfrac{r_b + ar_c}{\Delta}$
	y_{22b}	y_{22}	$y_{11} + y_{12} + y_{21} + y_{22}$	$\dfrac{r_e + r_b}{\Delta}$

$\Delta = r_e r_b + r_c [r_e + r_b (1 - a)]$

	z parameter	Common emitter	Common base	T-equivalent circuit
(O)	z_{11c}	$z_{11} - z_{12} - z_{21} + z_{22}$	z_{22}	$r_b + r_c$
	z_{12c}	$z_{22} - z_{12}$	$z_{22} - z_{21}$	$r_c (1 - a)$
	z_{21c}	$z_{22} - z_{21}$	$z_{22} - z_{12}$	r_c
	z_{22c}	z_{22}	$z_{11} - z_{12} - z_{21} + z_{22}$	$r_e + r_c (1 - a)$

	y parameter	Common emitter	Common base	T-equivalent circuit
(P)	y_{11c}	y_{11}	$y_{11} + y_{12} + y_{21} + y_{22}$	$\dfrac{r_e + r_c (1 - a)}{\Delta}$
	y_{12c}	$-(y_{11} + y_{12})$	$-(y_{11} + y_{21})$	$\dfrac{-r_c (1 - a)}{\Delta}$
	y_{21c}	$-(y_{11} + y_{21})$	$-(y_{11} + y_{12})$	$-\dfrac{r_c}{\Delta}$
	y_{22c}	$y_{11} + y_{12} + y_{21} + y_{22}$	y_{11}	$\dfrac{r_b + r_c}{\Delta}$

$\Delta = r_e r_b + r_c [r_e + r_b (1 - a)]$

	Parameter	Input impedance	Output impedance	Voltage gain	Current gain
(Q)	z	$\dfrac{\Delta z + z_{11} Z_L}{z_{22} + Z_L}$	$\dfrac{\Delta z + z_{22} Z_g}{z_{11} + Z_g}$	$\dfrac{z_{21} Z_L}{\Delta z + z_{11} Z_L}$	$\dfrac{-z_{21}}{z_{22} + Z_L}$
	y	$\dfrac{y_{22} + Y_L}{\Delta y + y_{11} Y_L}$	$\dfrac{y_{11} + Y_g}{\Delta y + y_{22} Y_g}$	$\dfrac{y_{21}}{y_{22} + Y_L}$	$\dfrac{y_{21} Y_L}{\Delta y + y_{11} Y_L}$

$\Delta z = z_{11} z_{22} - z_{12} z_{21}$

$\Delta y = y_{11} y_{22} - y_{12} y_{21}$

(From "Transistor Circuit Design," Texas Instruments, Inc. Copyright 1963 by Texas Instruments Incorporated. Used with permission of McGraw-Hill Book Company.)

EUROPEAN SEMICONDUCTOR
NUMBERING SYSTEM
(Pro Electron code)

First Letter	Second Letter	Third, Fourth, and Fifth Character
Material	Type	Serial Code
A Germanium B Silicon C Compound materials, such as cadmium sulfide or gallium arsenide used in semiconductor devices. (Energy gap band of 1.3 or more electron-volts) D Materials with an energy gap band of less than 0.6 electron-volts such as indium antimonide R Radiation detectors, photoconductive cells. Hall effect generators, etc.	A Low-power diode, voltage-variable capacitor B Varicap C Small-signal audio transistor D Audio power transistor E Tunnel diode F Small-signal rf transistor G Miscellaneous H Field probe K Hall generator L Rf-power transistor M Hall modulators and multipliers P Photodiode, phototransistor, photoconductive cell (LDR), radiation device R Low-power controlled rectifier S Low-power switching transistor T Breakdown devices, high-power controlled rectifier, Shockley diode, Thyristor, pnpn diodes U High-power switching transistor X Multiplier diode Y High-power rectifier (diode) Z Zener diode	Three figures—serial codes used on devices for domestic and commercial applications One letter and two figures—serial codes used on devices for use in military, industrial, scientific, and pulse, equipment

The third letter—if there is one—indicates industrial device and is a Y. If there is no third letter, the device is for consumer or entertainment use. The digits that follow the letters for industrial units indicate how many devices of that particular type have been registered. The digits start at 10 and go up to 99. When 99 is reached—i.e. after 89 devices—the last letter changes from a Y to an X and the numbering begins anew, working back towards A. There is no Z. For consumer devices, the numbers that follow the two letters start with 100, allowing registration of 899 similar devices.

FOR EXAMPLE: The designation BLY 80 means the device uses silicon (B) is for high r-f power use (L), and is used in industrial applications, (Y); the 80 means that it is the 71st device of its type to be registered with Pro Electron.

The accompanying curves permit an easy and rapid determination of the frequency of oscillation of a symmetrical-astable (free-running) multivibrator, and the pulse duration (t_p) of a monostable (one-shot) multivibrator. The pulse duration of the astable multivibrator output also can be read from the curve.

The expressions on which the curves are based are derived readily. The expression for the voltage at the base of the "off" transistor is

$$e_b = E_{cc} (1 - 2\epsilon^{-t/RC}) + V_{be}$$

where V_{be} is the base-to-emitter voltage of an "on" transistor. The above equation assumes that base-to-emitter breakdown is prevented by using transistors whose base-to-emitter breakdown voltage is greater than E_{cc} volts, or by connecting a diode in either the base or emitter lead.

The "off" transistor turns on when $e_b = V_{be}$, or $\epsilon^{-t/RC} = 1/2$ where t is the "off" time (t_p) at the end of which time $e_b = V_{be}$. Solving the equation yields $t_p = 0.69$ RC. The curves in graph (A) are plots of this equation. For the monostable multivibrator, t_p is the pulse duration. The period of the symmetrical-astable multivibrator is equal to $2t_p$.

Graph (B) is a family of curves of frequency of the symmetrical-astable multivibrator versus capacitance C for various values of resistance R. Since the period of the output wave is $2t_p$, the equation for frequency is given as $f = 1/1.38RC$, from which the curves were plotted.

FOR EXAMPLE:
Find the value of C required to generate a frequency of 500 Hz from a free-running multivibrator, or a 1 msec pulse from a monostable. In both cases the value of R is limited to 100,000 ohms by the beta of the transistor selected. The curves indicate a value of 0.0145 μF for the capacitor.

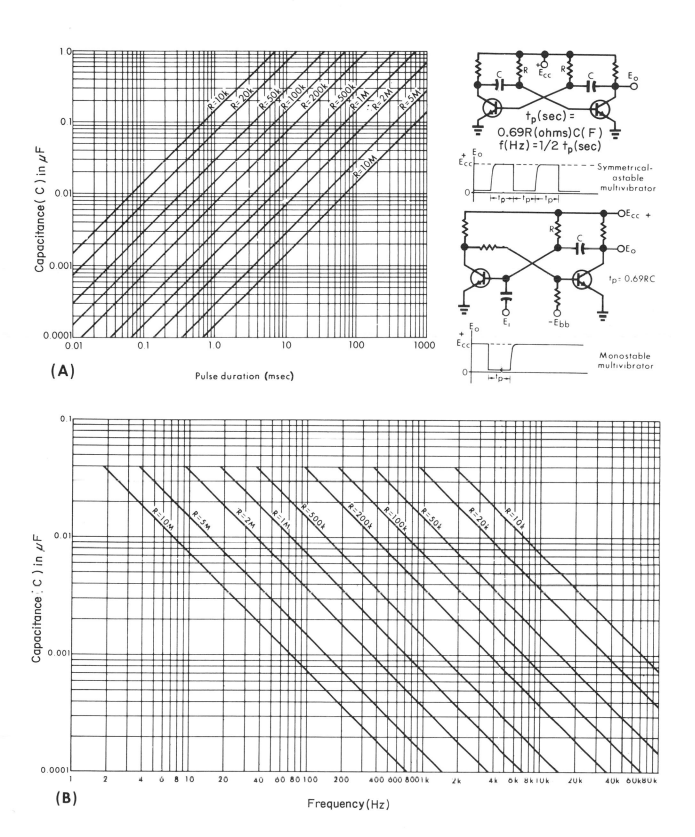

(A)

Pulse duration (msec)

$$t_p(sec) = 0.69R(ohms)C(F)$$
$$f(Hz) = 1/2 \, t_p(sec)$$

Symmetrical-astable multivibrator

$$t_p = 0.69RC$$

Monostable multivibrator

(B)

Frequency (Hz)

OPERATIONAL AMPLIFIERS

An operational amplifier is essentially a very high gain dc amplifier whose open-loop gain is generally high enough when compared with the closed-loop gain so that the closed-loop characteristics depend solely on the feedback element. Circuit applications for which operational amplifiers can be used are illustrated below.

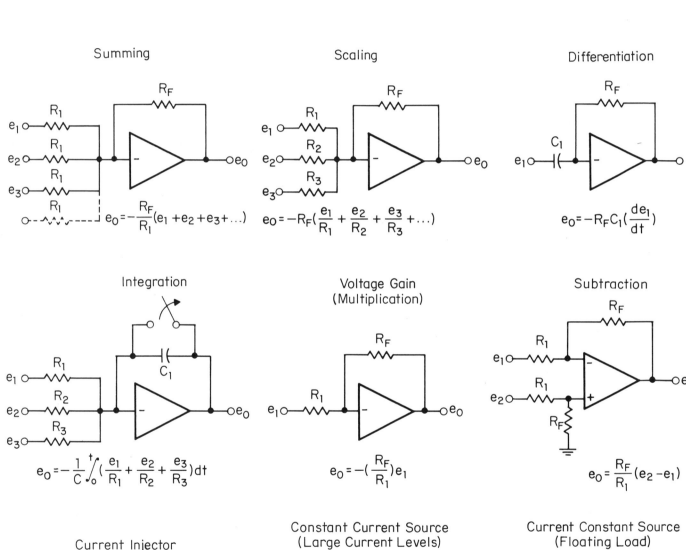

Summing

$$e_0 = -\frac{R_F}{R_1}(e_1 + e_2 + e_3 + \ldots)$$

Scaling

$$e_0 = -R_F\left(\frac{e_1}{R_1} + \frac{e_2}{R_2} + \frac{e_3}{R_3} + \ldots\right)$$

Differentiation

$$e_0 = -R_F C_1\left(\frac{de_1}{dt}\right)$$

Integration

$$e_0 = -\frac{1}{C}\int_0^t \left(\frac{e_1}{R_1} + \frac{e_2}{R_2} + \frac{e_3}{R_3}\right)dt$$

Voltage Gain (Multiplication)

$$e_0 = -\left(\frac{R_F}{R_1}\right)e_1$$

Subtraction

$$e_0 = \frac{R_F}{R_1}(e_2 - e_1)$$

Current Injector

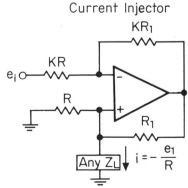

$$i = -\frac{e_1}{R}$$

Constant Current Source (Large Current Levels)

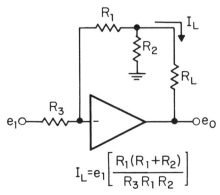

$$I_L = e_1\left[\frac{R_1(R_1 + R_2)}{R_3 R_1 R_2}\right]$$

Current Constant Source (Floating Load)

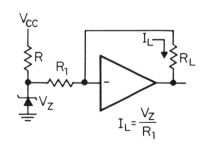

$$I_L = \frac{V_Z}{R_1}$$

Current-to-Voltage Converter

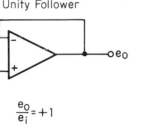

$$e_O = iR$$

Voltage Source

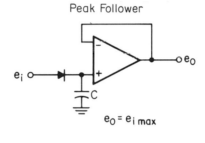

$$e_O = -V_Z\left(\frac{R_F}{R_1}\right)$$

Voltage Follower with Gain

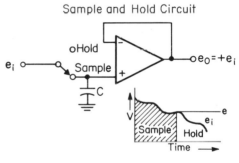

$$e_O = \left(\frac{R_F + R_1}{R_1}\right)e_i$$

Unity Follower

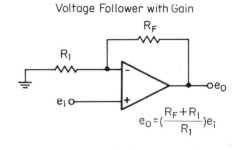

$$\frac{e_O}{e_i} = +1$$

Peak Follower

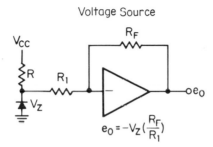

$$e_O = e_{i\,max}$$

Sample and Hold Circuit

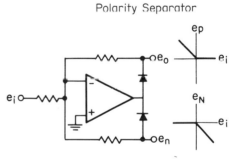

$$e_O = +e_i$$

Voltage Comparator Circuit

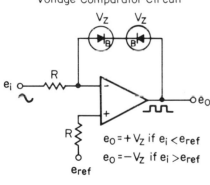

$$e_O = +V_Z \text{ if } e_i < e_{ref}$$
$$e_O = -V_Z \text{ if } e_i > e_{ref}$$

High-impedance Low-voltage Voltmeter

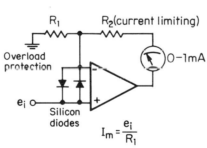

$$I_m = \frac{e_i}{R_1}$$

Polarity Separator

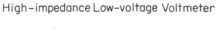

Logarithmic Transconductor

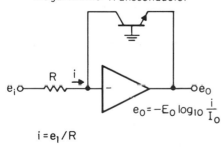

$$e_O = -E_O \log_{10} \frac{i}{I_O}$$

$$i = e_1/R$$

Simple Overvoltage Clamp

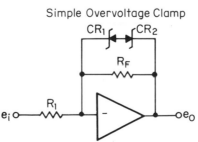

Absolute Value Amplifier

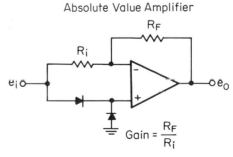

$$\text{Gain} = \frac{R_F}{R_i}$$

Automatic Gain Control Amplifier

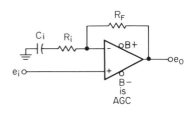

Adjustable Lag (0 to −180°) Amplifier
Adjustable Lead (0 to +180°) Amplifier

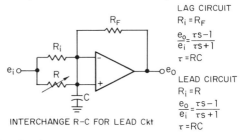

LAG CIRCUIT
$R_i = R_F$

$\dfrac{e_0}{e_i} = \dfrac{\tau s - 1}{\tau s + 1}$

$\tau = RC$

LEAD CIRCUIT
$R_i = R$

$\dfrac{e_0}{e_i} = \dfrac{\tau s - 1}{\tau s + 1}$

$\tau = RC$

INTERCHANGE R–C FOR LEAD Ckt

D/A Converter

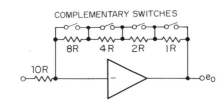

COMPLEMENTARY SWITCHES

Active Filter (Two-Pole)

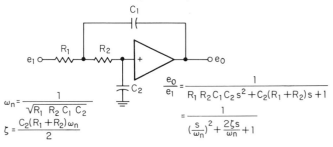

$\omega_n = \dfrac{1}{\sqrt{R_1 R_2 C_1 C_2}}$

$\zeta = \dfrac{C_2(R_1 + R_2)\omega_n}{2}$

$\dfrac{e_0}{e_1} = \dfrac{1}{R_1 R_2 C_1 C_2 s^2 + C_2(R_1 + R_2)s + 1}$

$= \dfrac{1}{\left(\dfrac{s}{\omega_n}\right)^2 + \dfrac{2\zeta s}{\omega_n} + 1}$

Second-Order Transfer Function Amplifier

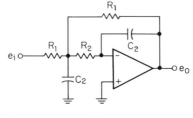

$\dfrac{e_0}{e_i} = \dfrac{1}{(R_1 R_2 C_1 C_2)s^2 + C_1(2R_2 + R_1)s + 1}$

Second-Order High-Pass Active Filter

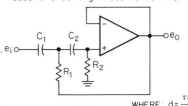

$\dfrac{e_0}{e_i} = \dfrac{\left(\dfrac{S}{\omega_n}\right)^2}{\left(\dfrac{S}{\omega_n}\right)^2 + d\left(\dfrac{S}{\omega_n}\right) + 1}$

WHERE: $d = \dfrac{\tau_1 + R_1 C_2}{\sqrt{\tau_1 \tau_2}}$

AND: $\omega_n^2 = \dfrac{1}{\tau_1 \tau_2}$

$\tau_1 = R_1 C_1$
$\tau_2 = R_2 C_2$

Second-Order Low-Pass Active Filter

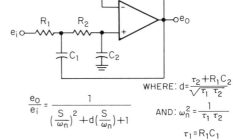

$\dfrac{e_0}{e_i} = \dfrac{1}{\left(\dfrac{S}{\omega_n}\right)^2 + d\left(\dfrac{S}{\omega_n}\right) + 1}$

WHERE: $d = \dfrac{\tau_2 + R_1 C_2}{\sqrt{\tau_1 \tau_2}}$

AND: $\omega_n^2 = \dfrac{1}{\tau_1 \tau_2}$

$\tau_1 = R_1 C_1$
$\tau_2 = R_2 C_2$

Notch Filter (Active)

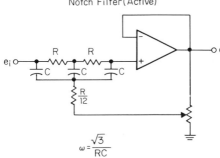

$\omega = \dfrac{\sqrt{3}}{RC}$

Bandpass Amplifier

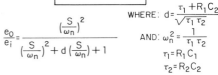

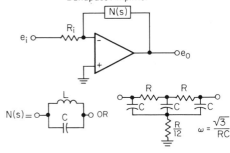

$N(s) =$

OR

$\omega = \dfrac{\sqrt{3}}{RC}$

Frequency Divider

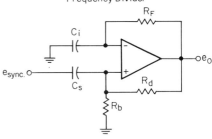

Schmitt Trigger

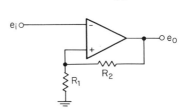

Square-wave Multivibrator

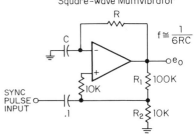

$f \cong \dfrac{1}{6RC}$

SYNC PULSE INPUT

Wien Bridge Oscillator

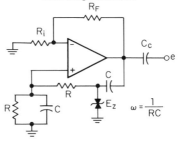

$\omega = \dfrac{1}{RC}$

Crystal Oscillator (Square Wave)

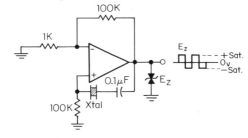

E_z --- +Sat.
0v
--- −Sat.

OPERATIONAL AMPLIFIERS
(Continued)

Glossary of Operational Amplifier Terms

Common-mode gain—Ratio of output voltage over input voltage applied to (+) and (−) terminal in parallel.

Common-mode rejection ratio (CMRR)—Ratio of an op amp's open-loop gain to its common-mode gain.

Differential-input voltage range—Range of voltages that may be applied between input terminals without forcing the op amp to operate outside its specifications.

Differential Input Impedance (Z_{in}diff)—Impedance measured between (+) and (−) input terminals.

Drift, input voltage—Change in output voltage divided by open-loop gain, as a function of temperature or time.

Input voltage offset—Dc potential required at the differential input to produce an output voltage of zero.

Input bias current—Input current required by (+) and (−) inputs for normal operation.

Input offset current—Difference between (+) and (−) input bias currents.

Offset—Measure of unbalance between halves of a symmetrical circuit.

Open-loop bandwidth—Without feedback, frequency at which amplifier gain falls 3 dB below its low-frequency value.

Open-loop voltage gain (A_{vol})—Differential gain of an op amp with no external feedback.

Slew rate—Maximum rate at which output voltage can change with time; usually given in volts per microsecond.

Modulator–Demodulator (Half-Wave)

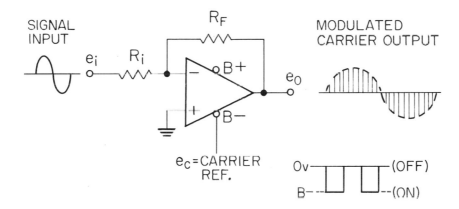

Floating Load

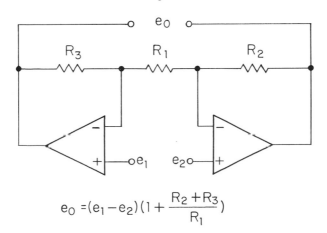

$$e_O = (e_1 - e_2)\left(1 + \frac{R_2 + R_3}{R_1}\right)$$

CHARACTERISTICS OF
INTEGRATED CIRCUIT LOGIC
FAMILIES

	Typical Circuit Diagram	Logic Type	Relative Cost Per Gate	Propagation Time Per Gate (nsec)	Power Dissipation Per Gate (mW)	Typical Noise Margin (V)	Typical Fanin	Typical Fanout	Remarks
RTL Resistor-Coupled Transistor Logic		NOR	Low	15	10	0.2	3	3	Variations in input characteristics result in base-current "hogging" problem. Proper operation not always guaranteed. More susceptible to noise because of low operating and signal voltages
RCTL Resistor-Capacitor Transistor Logic		NOR	Medium	50	10	0.2	3	4	Very similar to DCTL. Resistors resolve current "hogging" problem and reduce power dissipation. However, operating speed is reduced.
DCTL Direct-Coupled Transistor Logic		NOR	Med-high	30	10	0.2	3	4	Though capacitors can increase speed capability, noise immunity is affected by capacitive coupling of noise signals.
DTL Diode-Transistor Logic		NAND	Medium	25	15	0.7	8	8	Use of pull-up resistor and charge-control technique improves speed capabilities. Many variations of this circuit exist, each having specific advantages.

CHARACTERISTICS OF
INTEGRATED CIRCUIT LOGIC
FAMILIES (Continued)

	Typical. Circuit Diagram	Logic Type	Relative Cost Per Gate	Propagation Time Per Gate (nsec)	Power Dissipation Per Gate (mW)	Typical Noise Margin (V)	Typical Fanin	Typical Fanout	Remarks
TTL Transistor-Transistor Logic		NAND	Medium	10	20	1	8	12	Very similar to DTL. Has lower parasitic capacity at inputs. With the many existing variations, this logic family is very popular.
CTL Complimentary Transistor Logic		OR/NOR	High	5	50	0.4	5	25	Similar to a differential amplifier, the reference voltage sets the threshold voltage. High-speed, high-fanout operation is possible with associated high power dissipation. Also known as emitter-coupled logic (ECL).
CML Current-Mode Logic (ECL Emitter-Coupled Logic)		AND/OR	High	5	50	0.4	5	25	More difficult manufacturing process results in compromises of active device characteristics and higher cost.
MOSL Metal-Oxide Semiconductor Logic		NOR	Very low	250	<1	2.5	10	5	Limited in switching speed compared to bipolar transistor circuits because the MOS transistor is a high-impedance device and cannot charge the stray circuit capacitance quickly.

159

DEFINITIONS OF INTEGRATED CIRCUIT TERMS

Access time Time required in a computer to move information from memory to the computing mechanism.

Active elements Those components in a circuit which have gain or which direct current flow: diodes, transistors, SCR's, etc.

Adder Switching circuits which combine binary bits to generate the SUM and CARRY of these bits. Takes the bits from the two binary numbers to be added (ADDEND and AUGEND) plus the CARRY from the preceding less significant bit and generates the SUM and the CARRY.

Address Noun: a location, either name or number, where information is stored in a computer. Verb: to select or pick out the location of a stored information set for access.

"AND" A boolean logic expression used to identify the logic operation wherein given two or more variables, all must be logical "1" for the result to be logical "1." The AND function is graphically represented by the dot ($\cdot$) symbol.

Anticipated carry adder A parallel ADDER in which each stage is capable of looking back at all ADDEND and AUGEND bits of less significant stages and deciding whether the less significant bits provide a "0" or a "1" CARRY IN. Having determined the CARRY IN it combines it with its own ADDEND and AUGEND to give the SUM for that bit or stage. Also called FAST ADDER or look ahead CARRY ADDER.

Arrays Integrated circuits designed to perform near or actual subsystem operations. They are characterized by high complexity and component density. Each array package replaces a number of conventional I/Cs. Arrays are classified as medium-scale or larger-scale according to function performed. They can be monolithic or fabricated on a silicon wafer with interconnections between circuits.

Asynchronous inputs Those terminals in a flip-flop which can affect the output state of the flip-flop independent of the clock. Called Set, Preset, Reset or DC Set and Reset, or clear.

Basic logic diagram A logic diagram that depicts logic functions with no reference to physical implementations. It consists primarily of logic symbols and is used to depict all logic relationships as simply and understandably as possible. Nonlogic functions are not normally shown.

Binary coded decimal (BCD) A binary numbering system for coding decimal numbers in groups of 4 bits. The binary value of these 4-bit groups ranges from 0000 to 1001 and codes the decimal digits "0" through 9. To count to 9 takes 4 bits; to count to 99 takes two groups of 4 bits; to count to 999 takes three groups of 4 bits.

Binary logic Digital logic elements which operate with two distinct states. The two states are variously called true and false, high and low, on and off, or "1" and "0." In computers they are represented by two different voltage levels. The level which is more positive (or less negative) than the other is called the high level, the other the low level. If the true ("1") level is the most positive voltage, such logic is referred to as positive true or positive logic.

Bistable element Another name for flip-flop. A circuit in which the output has two stable states (output levels "0" or "1") and can be caused to go to either of these states by input signals, but remains in that state permanently after the input signals are removed. This differentiates the bistable element from a gate also having two output states but which requires the retention of the input signals to stay in a given state. The characteristic of two stable states also differentiates it from a monostable element which keeps returning to a specific state, and an astable element which keeps changing from one state to the other.

Bit A synonym for binary numeral. Also refers to a single binary numeral in a binary word.

Boolean algebra The mathematics of logic which uses alphabetic symbols to represent logical variables and "1" and "0" to represent states. There are three *basic* logic operations in this algebra: AND, OR, and NOT. (Also see NAND, NOR, Invert which are combinations of the three *basic* operations.)

Buffer A circuit element, which is used to isolate between stages to handle a large fanout or to convert input and output circuits for signal level compatibility.

Buried layer A heavily doped (N+) region directly under the N doped epitaxial collector region of transistors in a monolithic integrated circuit used to lower the series collector resistance.

Chip (die) A single piece of silicon which has been cut from a slice by scribing and breaking. It can contain one or more circuits but is packaged as a unit.

Clear An asynchronous input. Also called Reset. To restore a memory element or flip-flop to a "standard" state, forcing the Q terminal to logic "0."

Clock A pulse generator which controls the timing of computer switching circuits and memory states and regulates the speed at which the computer central processor operates. It serves to synchronize all operations in a digital system.

Clock input That terminal on a flip-flop whose condition or change of condition controls the admission of data into a flip-flop through the synchronous inputs and thereby controls the output state of the flip-flop. The clock signal performs two functions: (1) It permits data signals to enter the flip-flop; (2) after entry, it directs the flip-flop to change state accordingly.

CML (Current Mode Logic) Logic in which transistors operate in the unsaturated mode as distinguished from most other logic types which operate in the saturation region. This logic has very fast switching speeds and low logic swings. Also called ECL or MECL.

Counter A device capable of changing states in a specified sequence upon receiving appropriate input signals. The output of the counter indicates the number of pulses which have been applied. (See also Divider.) A counter is made from flip-flops and some gates. The output of all flip-flops are accessible to indicate the exact count at all times.

Counter, binary An interconnection of flip-flops having a single input so arranged to enable binary counting. Each time a pulse appears at the input, the counter changes state and tabulates the number of input pulses for readout in binary form. It has a 2^n possible counts where n is the number of flip-flops.

Counter, ring A special form of counter sometimes called a Johnson or shift counter which has very simple wiring and is fast. It forms a loop or circuits of interconnected flip-flops so arranged that only one is "0" and that as input signals are received, the position of the "0" state moves in sequence from one flip-flop to another around the loop until they are all "0," then the first one goes to "1" and this moves in sequence from one flip-flop to another until all are "1." It has $2 \times n$ possible counts where n is the number of flip-flops.

Data Term used to denote facts, numbers, letters, symbols, binary bits presented as voltage levels in a computer. In a binary system data can only be "0" or "1."

DCTL (Direct-Coupled Transistor Logic) Logic employing only transistors as active circuit elements.

Debug To remove malfunctions from a system or device.

Decimal A system of numerical representation which uses ten numerals 0, 1, 2, 3, . . . , 9. Each numeral is called a digit. A numbering system to the radix 10.

Delay The slowing up of the propagation of a pulse either

intentionally, such as to prevent inputs from changing while clock pulses are present, or unintentionally as caused by transistor rise and fall time pulse response effects.

Detailed logic diagram A diagram that depicts all logic functions and also shows nonlogic functions, socket locations, pin numbers, test points, and other physical elements necessary to describe the physical and electrical aspects of the logic. The detailed logic diagram is used primarily to facilitate the rapid diagnosis and localization of equipment malfunctions. It also is used to verify the physical consistency of the logic and to prepare fabrication instructions. The symbols are connected by lines that represent signal paths.

Die See Chip.

Diffusion A process, used in the production of semiconductors, which introduces minute amounts of impurities into a substrate material such as silicon or germanium and permits the impurity to spread into the substrate. The process is very dependent on temperature and time.

Digital circuit A circuit which operates in the manner of a switch, that is, it is either "on" or "off." More correctly should be called a binary circuit.

Diode A device permitting current to flow in one direction only. Diodes are used in logic circuits to control the passage or nonpassage of a signal from one element to another.

Discrete circuits Electronic circuits built of separate, individually manufactured, tested, and assembled diodes, resistors, transistors, capacitors, and other specific electronic components.

Divider (Frequency) A counter which has a gating structure added which provides an output pulse after receiving a specified number of input pulses. The outputs of all flip-flops are not accessible.

Dot "AND" Externally connecting separate circuits or functions so that the combination of their outputs results in an "AND" function. The point at which the separate circuits are wired together will be a "1" if all circuits feeding into this point are "1" (also called WIRED "OR").

Dot "OR" Externally connecting separate circuits or functions so that the combination of their outputs results in an "OR" function. The point at which the separate circuits are wired together will be a "1" if any of the circuits feeding into this point are "1."

Driver An element which is coupled to the output stage of a circuit in order to increase its power or current handling capability or fanout; for example, a clock driver is used to supply the current necessary for a clock line.

DTL (Diode-Transistor Logic) Logic employing diodes with transistors used only as inverting amplifiers.

Enable To permit an action or the acceptance or recognition of data by applying appropriate signals (generally a logic "1" in a positive logic) to the appropriate input. (See Inhibit.)

Epitaxial growth A chemical reaction in which silicon is precipitated from a gaseous solution and grows in a very precise manner, that is, monocrystalline, upon the surface of a silicon wafer placed in the solution.

Exclusive "OR" A logical function whose output is "1" if either of the two variables is "1" but whose output is "0" if both inputs are "1" or both are "0."

Fall time A measure of the time required for the output voltage of a circuit to change from a high voltage level to a low voltage level once a level change has started. Current could also be used as the reference, that is, from a high current to a low current level.

Fanin The number of inputs available to a specific logic stage or function.

Fanout The number of input stages that can be driven by a circuit output.

Fast ADDER (See Anticipated CARRY ADDER.)

FEB (Functional Electronic Block) Another name for a monolithic integrated circuit of thick-film circuit.

Feedback When part of the output of a circuit is channeled back to an input, it is said to have feedback. When part of the output of an amplifier is routed back to augment the input signal, the amplifier has positive feedback or if this rechanneling is employed to diminish the input it is called negative feedback.

Flip-flop (storage element) A circuit having two stable states and the capability of changing from one state to another with the application of a control signal and remaining in that state after removal of signals. (See Bistable element.)

Flip-flop, "D" D stands for delay. A flip-flop whose output is a function of the input which appeared one pulse earlier; for example, if a "1" appeared at the input, the output after the next clock pulse will be a "1."

Flip-flop, "J-K" A flip-flop having two inputs designated J and K. At the application of a clock pulse, a "1" on the "J" input and a "0" on the "K" input will set the flip-flop to the "1" state; a "1" on the "K" input and a "0" on the "J" input will reset it to the "0" state; and "1's" simultaneously on both inputs will cause it to change state regardless of the previous state. J = 0 and K = 0 will prevent change.

Flip-flop, "R-S" A flip-flop consisting of two cross-coupled NAND gates having two inputs designated "R" and "S." A "1" on the "S" input and "0" on the "R" input will reset (clear) the flip-flop to the "0" state, and "1" on the "R" input and "0" on the "S" input will set it to the "1." It is assumed that "0's" will never appear simultaneously at both inputs. If both inputs have "1's" it will stay as it was. "1" is considered nonactivating. A similar circuit can be formed with NOR gates.

Flip-flop, "R-S-T" A flip-flop having three inputs, "R," "S," and "T." This unit works as the "R-S" flip-flop except that the "T" input is used to cause the flip-flop to change states.

Flip-flop, "T" A flip-flop having only one input. A pulse appearing on the input will cause the flip-flop to change states. Used in ripple counters.

FULL ADDER See Adder

Gate definitions
below assume positive logic.

Gate, AND All inputs must have "1" level signals at the input to produce a "1" level output.

Gate, NAND All inputs must have "1" level signals at the input to produce a "0" level output.

Gate, NOR Any one input or more than one input having a "1" level signal will produce a "0" level output.

Gate, OR Any one input or more than one input having a "1" level signal will produce a "1" level output.

Gates (decision elements) A circuit having two or more inputs and one output. The output depends upon the combination of logic signals at the input.

Half ADDER A switching circuit which combines binary bits to generate the SUM and the CARRY. It can only take in the two binary bits to be added and generate the SUM and CARRY (see also ADDER).

Half shift register Another name for certain types of flip-flops when used in a shift register. It takes two of these to make one stage in a shift register.

High See Binary logic.

Hybrid A method of manufacturing integrated circuits by using a combination of monolithic, thin-film and thick-film techniques.

Inhibit To prevent an action, or acceptance of data, by applying an appropriate signal to the appropriate input (generally a logic "0" in positive logic). (See Enable.)

Integrated circuit (EIA definition) "The physical realization of a number of electrical elements inseparably associated on or within a continuous body of semiconductor material to perform the functions of a circuit." (See Slice and Chip.)

Inverter A circuit whose output is always in the opposite state from the input. This is also called a NOT circuit. (A teeter-totter is a mechanical inverter.)

Linear circuit A circuit whose output is an amplified version of its input, or whose output is a predetermined variation of its input.

Logic diagram A picture representation for the logical functions of AND, OR, NAND, NOR, NOT.

Logic function A combinational, storage, delay, or sequential function expressing a relationship between variable signal input(s) to a system or device and the resultant output(s).

Logic swing The voltage difference between the two logic levels "1" and "0."

Logic symbol The graphic representation of the aggregate of all the parts implementing a logic function.

Low See Binary logic.

Memory The semi-permanent storage of numbers, in digital form, in a circuit or system. With reference to computers, the term also describes peripheral storage such as tape or disc memory.

Microelectronics The entire spectrum of electronic art dealing with the fabrication of sophisticated, practical systems using miniaturized electronic components.

Monolithic Refers to the single silicon substrate in which an integrated circuit is constructed. (See Integrated circuit.)

"NAND" A Boolean logic operation which yields a logic "0" output when all logic input signals are logic "1."

Negative logic Logic in which the more negative voltage represents the "1" state; the less negative voltage represents the "0" state. (See Binary logic.)

Noise immunity A measure of the insensitivity of a logic circuit to triggering or reaction to spurious or undesirable electrical signals or noise, largely determined by the signal swing of the logic. Noise can be either of two directions, positive or negative.

"NOR" A Boolean logic operation which yields a logic "0" output with one or more true "1" input signals.

"NOT" A Boolean logic operation indicating negation, not "1." Actually an inverter. If input is "1" output is NOT "1" but "0." If the input is "0" output is NOT "0" but "1." Graphically represented by a bar over a Boolean symbol such as A. A means "when A is not 1."

Offset The change in input voltage required to produce a zero output voltage in a linear amplifier circuit. In digital circuits it is the dc voltage on which a signal is impressed.

One ("1") See Binary logic.

"OR" A Boolean logic operation used to identify the logic operation wherein two or more true "1" inputs only add to one true "1" output. Only one input needs to be "true" to produce a "true" output. The graphical symbol for "OR" is a plus sign (+).

Parameter Any specific characteristic of a device. When considered together, all the parameters of a device describe its operational and physical characteristics.

Parallel This refers to the technique for handling a binary data word which has more than one bit. All bits are acted upon simultaneously. It is like the line of a football team. Upon a signal all line men act. (See also Serial.)

Parallel ADDER A conventional technique for adding where the two multibit numbers are presented and added simultaneously (parallel). A ripple adder is still a parallel adder; the carry is rippled from the least significant to the most significant bit. Another type of parallel adder is the "Look Ahead," or "Anticipated Carry" adder. (See Ripple ADDER and Fast ADDER.)

Parallel operation The organization of data manipulation within computer circuitry where all the digits of a word are transmitted simultaneously on separate lines in order to speed up operation, as opposed to serial operation.

Passive elements Resistors, inductors, or capacitors, elements without gain.

Positive logic Logic in which the more positive voltage represents the "1" stage. (See Binary logic.)

Preset An input like the Set input and which works in parallel with the Set.

Propagation delay A measure of the time required for a change in logic level to be transmitted through an element or a chain of elements.

Propagation time The time necessary for a unit of binary information (high voltage or low) to be transmitted or passed from one physical point in a system or subsystem to another. For example, from input of a device to output.

Pulse A signal of very short duration.

Q output The reference output of a flip-flop. When this output is "1" the flip-flop is said to be in the "1" state; when it is "0" the output is said to be in the "0" state. (See also State and Set.)

Q̄ output The second output of a flip-flop. It is always opposite in logic level to the Q output.

RCTL (Resistor-Capacitor-Transistor-Logic) Same as RTL except that capacitors are used to enhance switching speed.

Register An interconnection of computer circuitry, made up of a number of storage devices (usually flip-flops) to store a certain number of digits, usually one computer word. For example, a 4-bit register requires 4 flip-flops.

Reset Also called clear. Similar to Set except it is the input through which the Q output can be made to go to "0."

Ripple The transmission of data serially. It is a serial reaction analogous to a bucket brigade or a row of falling dominoes.

Ripple ADDER A binary adding system similar to the system most people use to add decimal numbers—that is, add the "units" column, get the carry, add it to the "10's" column, get the carry, add it to the "100's" column, and so on. Again it is necessary to wait for the signal to propagate through all columns even though all columns are present at once (parallel). Note that the carry is rippled.

Ripple counter A binary counting system in which flip-flops are connected in series. When the first flip-flop changes it effects the second which effects the third and so on. If there are ten in a row, the signal must go sequentially from the first flip-flop to the tenth.

Rise time A measure of the time required for the output voltage of a state to go from a low voltage level ("O") to a high voltage level ("1") once a level change has been started.

RTL (Resistor-Transistor-Logic) Logic is performed by resistors. Transistors are used to produce an inverted output.

Semiconductor The name applied to materials which exhibit relatively high resistance in a pure state but much lower resistance when minute amounts of impurities are added. The word is commonly used to describe electronic devices made from semiconductor materials.

Serial The technique for handling a binary data word which has more than one bit. The bits are acted upon one at a time. It is like a parade going by a review point.

Serial operation The organization of data manipulation within computer circuitry where the digits of a word are transmitted one at a time along a single line. The serial mode of operation is slower than parallel operation, but utilizes less complex circuitry.

Set An input on a flip-flop not controlled by the clock (see Asynchronous inputs), and used to effect the Q output. It is this input through which signals can be entered to get the Q output to go to "1." Note it cannot get Q to go to "0."

Shift The process of moving data from one place to another. Generally many bits are moved at once. Shifting is done synchronously and by command of the clock. An 8-bit word can be shifted sequentially (serially)—that is, the 1st bit goes out, 2nd bit takes 1st bit's place, 3rd bit takes 2nd bit's place, and so on, in the manner of a bucket brigade. Generally referred to as shifting left or right. It takes 8 clock pulses to shift an 8-bit word or all bits of a word can be shifted simultaneously. This is called parallel load or parallel shift.

Shift register An arrangement of circuits, specifically flip-flops, which is used to shift serially or in parallel. Binary words are generally parallel loaded and then held temporarily or serially shifted out.

Skewing Refers to time delay or offset between any two signals in relation to each other.

Slewing rate Rate at which the output can be driven from limit to limit over the dynamic range.

Slice A single wafer cut from a silicon ingot forming a thin substrate on which all active and passive elements for multiple integrated circuits have been fabricated utilizing semiconductor epitaxial growth, diffusion, passivation, masking, photo resist, and metallization technologies. A completed slice generally contains hundreds of individual circuits. (See Chip.)

Solid state The electronic properties of crystalline materials (usually semiconductor in type). The interaction of light, heat, magnetic fields, and electric currents in these crystalline materials are involved in solid state devices. Less power is required to operate solid state devices and a greater variety of effects can be obtained.

Stability The specific ability of electronic circuits or other devices to withstand use and environmental stresses without changing. Also continued operation according to specifications despite adverse conditions.

State This refers to the condition of an input or output of a circuit as to whether it is a logic "1" or a logic "0." The state of a circuit (gate or flip-flop) refers to its output. The flip-flop is said to be in the "1" state when its Q output is "1." A gate is in the "1" state when its output is "1."

Substrate The physical material upon which a circuit is fabricated. It serves as a support while performing useful thermal and/or electrical functions.

Synchronous Operation of a switching network by a clock pulse generator. All circuits in the network switch simultaneously. All actions take place synchronously with the clock.

Synchronous inputs Those terminals on a flip-flop through which data can be entered but only upon command of the clock. These inputs do not have direct control of the output such as those of a gate but only when the clock permits and commands. Called JK inputs or AC set and reset inputs.

System A group of integrated circuits or other components interconnected to perform a single function or number of related functions. If further interconnected into a larger system, the individual elements are referred to as subsystems.

Thick film A method of manufacturing integrated circuits by depositing thin layers of materials on an insulated substate (often ceramic) to perform electrical functions.

Thin films Films with a thickness on the order of one-millionth of an inch are classified as thin films. They are produced through evaporation and sputtering processes. Thin films are used in fabricating microminiature components for hybrid and monolithic integrated circuits as well as some discrete devices.

Toggle To switch between two states as in a flip-flop.

Trigger A timing pulse used to initiate the transmission of logic signals through the appropriate circuit signal paths.

Truth table A chart which tabulates and summarizes all the combinations of possible states of the inputs and outputs of a circuit. It tabulates what will happen at the output for a given input combination.

TTL, T^2L (Transistor-Transistor-Logic) A logic system which evolved from DTL wherein the multiple diode cluster is replaced by a multiple-emitter transistor. A circuit which has a multiple emitter input and an active pullup network.

Turn-on time The time required for an output to turn on (sink current, to ground output, to go to 0-V). It is the propagation time of an appropriate input signal to cause the output to go to 0 V.

Turn-off time Same as Turn-on time except the output stops sinking current, goes off and/or goes to a high voltage level (logic "1").

Wired "OR" Externally connected separate circuits or functions arranged so that the combination of their outputs results in an "AND" function. The point at which separate circuits are wired together will be an "0" if any one of the separate outputs is an "0." The same as a dot "AND."

Word A group of bits treated as an entity in a computer.

Zero ("0") See Binary logic.

CLASSIFICATION OF AMPLIFIERS

The definitions of class A, B, or C operation apply to vacuum tube as well as to transistor circuits. Bias voltage on the emitter junction of a transistor determines collector current just as grid voltage determines plate current in a vacuum tube.

Class A allows for 360° operation of a sine wave.

Class B operation is with zero bias (cutoff) and allows 180° conduction.

Class C operation is with bias beyond cutoff which allows less than 180° conduction.

Class AB operation allows small-signal class A operation, and large-signal class B operation.

The above classes of operation are defined and illustrated for transistors and vacuum tubes.

Class	Bias Setting	Input-signal Voltage Swing	Plate or Collector Current Flow	Performance Characteristic
A_1	Center point of characteristic curve	Confined to linear portion of characteristic curve.	Complete cycle	Undistorted output. High gain. Low power conversion efficiency. (25% maximum)
A_2	Above center point of characteristic curve	Extends into upper (saturation) bend of characteristic curve	Complete cycle	Almost undistorted output. Lower gain but higher efficiency than class A_1.
AB_1	Below center point of characteristic curve	Extends into lower (cutoff) bend of characteristic curve	Cuts off for a small portion of negative half-cycle	In push-pull operation output is practically undistorted. Lower gain but higher efficiency than class A_2.
AB_2	Center point of characteristic curve	Extends into lower (cutoff) and upper (saturation) bends of characteristic curve	Cuts off for small portion of negative half-cycle	Slight harmonic distortion in push-pull operation. Lower gain but higher efficiency than class AB_1.
B_1	Near lower bend of characteristic curve	Extends beyond lower (cutoff) bend of characteristic curve	Cuts off for greater part of negative half-cycle	Little harmonic distortion in push-pull operation. Gain less than class AB_2. Maximum efficiency 78.5%.
B_2	Near lower bend of characteristic curve	Extends into lower (cutoff) and upper (saturation) bend of characteristic curve	Cuts off for greater part of negative half-cycle and small portion of positive half-cycle	Some harmonic distortion in push-pull operation. Lower gain but higher efficiency than class B_1.
C	Beyond lower bend of characteristic curve	Extends well beyond lower (cutoff) and upper (saturation) bends of characteristic curve	Cuts off all of negative and part of positive half-cycles	Considerable harmonic distortion. Low gain. High power conversion efficiency (80% maximum).

Subscript 1 denotes that no grid current flows during any part of the cycle.
Subscript 2 denotes that grid current flows at least for a portion of the cycle.
In class C amplifiers, grid current always flows, and a subscript is therefore unnecessary.

CLASSIFICATION OF AMPLIFIERS
(Continued)

TRANSISTORS

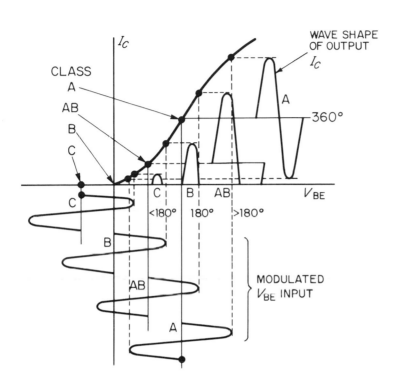

VACUUM TUBES

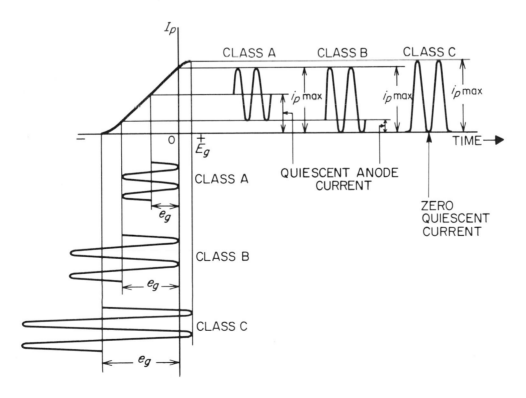

NEGATIVE FEEDBACK NOMOGRAM

In negative-feedback amplifier considerations, β (expressed as a percentage) has a negative value. A line across the β and μ scales will intersect the center scale to indicate resulting change in gain. It also indicates amount (in decibels) by which input must be increased to maintain original output. Original amplification may be expressed as voltage ratio or in decibels by using appropriate scale at right.

FOR EXAMPLE:

For a β of 10% and an amplifier μ of 30, the nomogram yields a change in μ of 0.25.

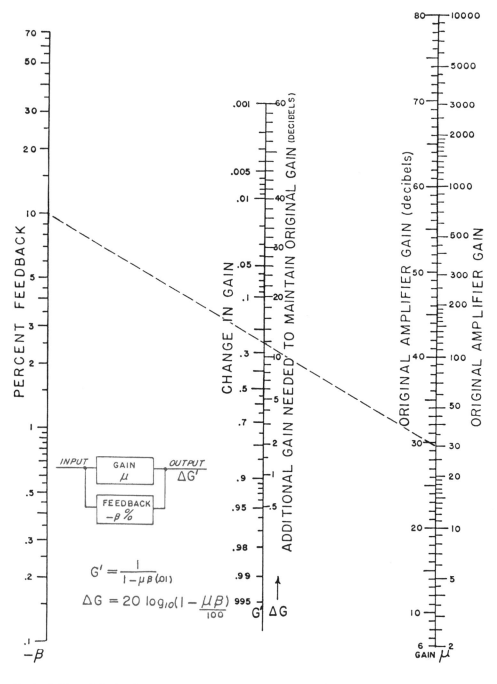

$$G' = \frac{1}{1 - \mu\beta\,(.01)}$$

$$\Delta G = 20\,\log_{10}\left(1 - \frac{\mu\beta}{100}\right)$$

RISETIME OF CASCADED AMPLIFIERS

Two cascaded amplifying devices will have an overall risetime given by:

$$T_{r_t} = \sqrt{T_{r_1}^2 + T_{r_2}^2}$$

where T_{r_1}, T_{r_2}, and T_{r_t} are the first stage, second stage, and total risetimes respectively.

The above relation is presented in the accompanying graph.

FOR EXAMPLE:

A system incorporating two cascaded amplifiers having risetimes of 100 μsec and 25 μsec (a ratio of 4:1), would have an overall risetime of 103 μsec.

Note: The Y-axis is the percentage increase in the risetime above the risetime of the slower of two cascaded devices.

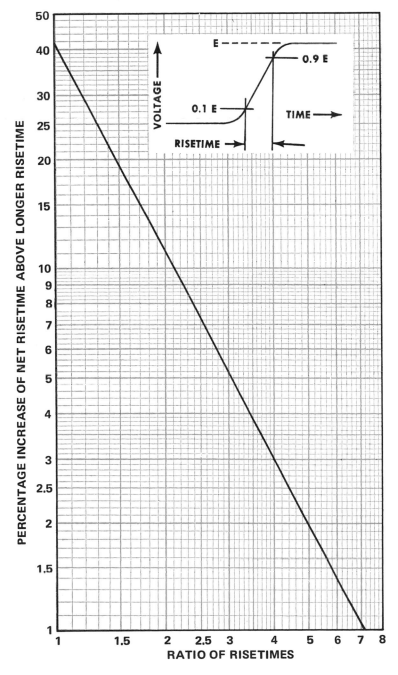

(From *Electronics and Communications*, December 1968.)

BANDWIDTH—RISETIME
NOMOGRAM

This nomogram facilitates calculations for tuned amplifiers having several stages where each stage has the same resonant frequency and bandwidth. For the case of double-tuned stages, equal primary Q's with critical coupling are assumed.

The nomogram can also be used for R-C coupled amplifiers based on the formula.

$$(\text{risetime}) \times (\text{bandwidth}) = 0.35$$

The overall risetime of several stages is

overall risetime =

$$\frac{}{(\text{risetime per stage}) \sqrt{\text{number of stages}}}$$

FOR EXAMPLE:

1. The overall bandwidth of four single-tuned circuits, each with a bandwidth of 1 MHz, is 0.44 MHz.
2. If five double-tuned transformers are to be used to obtain an overall bandwidth of 0.3 MHz, each must have a bandwidth of 0.34 MHz.
3. If the bandwidth of a single stage is 4 MHz, then the risetime of five cascaded stages is 235 nsec.
4. The risetime of ten stages, each with a rise time of 10 μsec, is 27 μsec.

Note: For R-C coupled circuits, the "single-tuned circuit" scale is used.

RC CIRCUITS

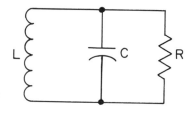

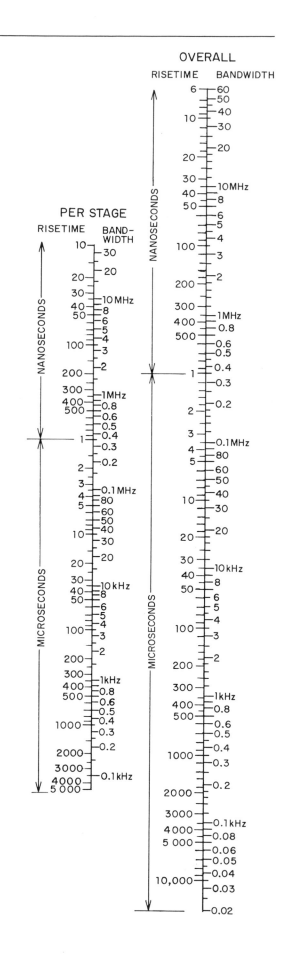

CLASS B PUSH-PULL AMPLIFIER
NOMOGRAM

This nomogram determines the available power from the output of class B vacuum tube or transistor push-pull stage operating under the following conditions: The output is a sine wave, the collector or plate swing is twice the supply voltage, and the available output power is determined by the formula

$$P = \frac{(\sqrt{2}V)^2}{Z}$$

FOR EXAMPLE:

A transistor amplifier with a 12-V supply and a collector-to-collector impedance of 400 ohms could produce 720 mW of undistorted output power.

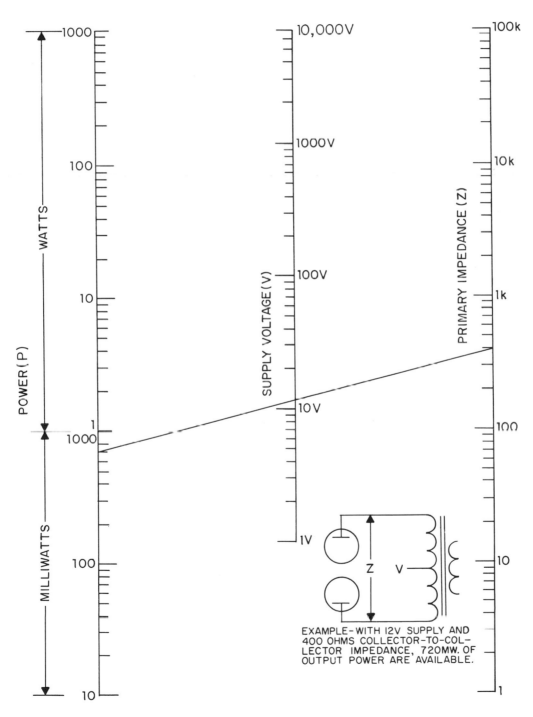

EXAMPLE-WITH 12V SUPPLY AND 400 OHMS COLLECTOR-TO-COLLECTOR IMPEDANCE, 720MW. OF OUTPUT POWER ARE AVAILABLE.

CATHODE FEEDBACK NOMOGRAM

This nomogram shows the reduction in the gain of an amplifier as a result of negative feedback that is introduced if the cathode resistor is not bypassed.

FOR EXAMPLE:
What will be the gain of an amplifier that has an initial stage gain of 20, a cathode resistor of 22 K, and a dynamic plate load resistor of 220 K if the cathode bypass capacitor is removed. The ratio of R_L to R_K is 10, thus the resultant "actual" stage gain is 7.

The range of the nomogram can be extended by multiplying all three scales by the same power of 10.

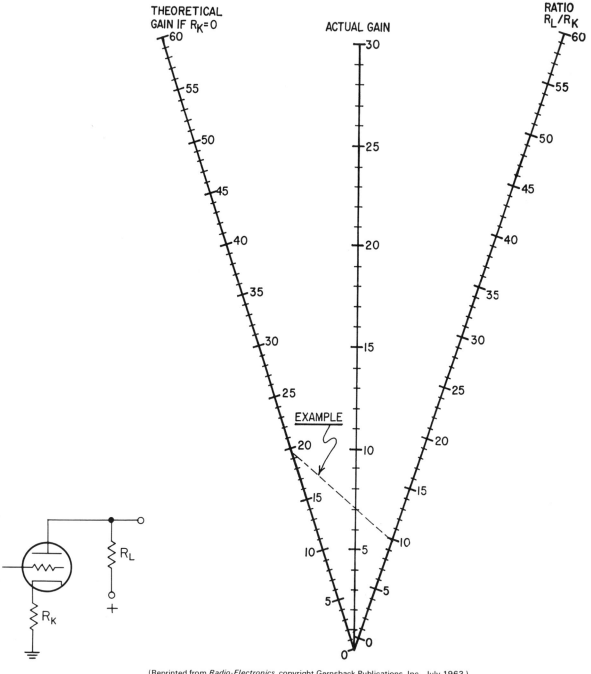

(Reprinted from *Radio-Electronics*, copyright Gernsback Publications, Inc., July 1963.)

CATHODE FOLLOWER NOMOGRAM

A cathode follower is useful for properly terminating transmission lines and coaxial cables. It provides high Z_{in} and low Z_{out}, good frequency and phase response, ground common to the input and output, reduced input capacitance, power gain and in-phase input and output. To match a transmission line, R_o should equal the impedance of the line (A). If R_o is less, add a series resistor (B), if R_o is greater use a resistor (C) so that $R = R_o Z_o / (R_o - Z_o)$.

FOR EXAMPLE:

To drive a 52-ohm line using a tube with a g_m of 5000 requires an R_o of 70 ohms. To provide proper cathode bias, determine the required cathode resistance from the tube manual or by calculation, and subtract R_o to determine R_K. Assuming that 220 ohms is required for proper bias, the R_K is 150 ohms and R_o is 70 ohms. If fixed bias is used, R_K is not needed.

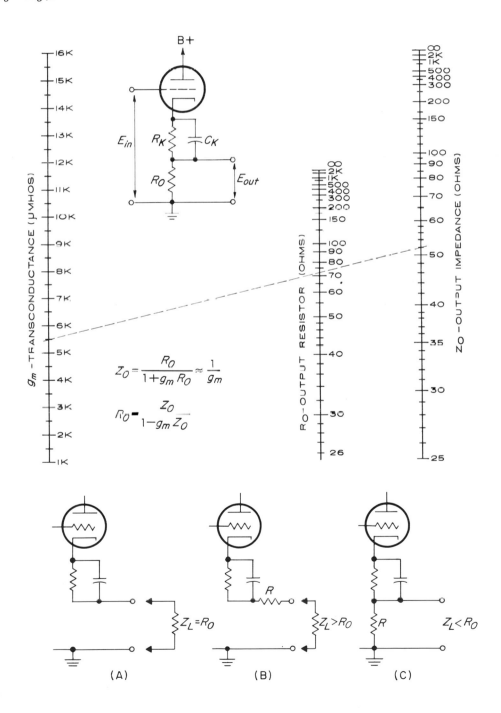

$$Z_O = \frac{R_O}{1 + g_m R_O} \approx \frac{1}{g_m}$$

$$R_O = \frac{Z_O}{1 - g_m Z_O}$$

173

EUROPEAN TUBE NUMBERING SYSTEM

Receiving and Amplifying Tubes

First Letter	Second and Subsequent Letter	Numbers
Type of Filament or Heater	Electrode Structure— Class of Tube	Type of Base
A 4 V ac (parallel)	A Single diode	1 Base indicated by second number
C 200 mA heater	B Dual diode	
D 0.5–1.5 V dc	C Triode, small-signal	2 Loctal
E 6.3 V ac (parallel)	D Triode, large-signal	3 Octal
G 5 V heater	E Tetrode, small-signal	4 European rim-lock
H 12.6 V 150 mA heater (parallel)	F Pentode, small-signal	5 Miscellaneous special bases
K 2V dc (parallel)	H Hexode or heptode	6, 7 Subminiature tube
M 2.5 V	K Octode, pentagrid converter	8 Nine-pin miniature (noval)
O no filament	L Pentode or tetrode, large-signal	9 Seven-pin minature
P 300 mA heater (series)	M Electron-beam indicator	Second and third digits differentiate between tubes that have the same general description but different characteristics. If the first number is a 1, then the second number indicates the type of base.
U 100 mA heater (series)	N Thyratron	
Z cold cathode	P Secondary emission tube	
	Q Nonode (9 electrodes)	
	T Miscellaneous	
	X Gas-filled full-wave rectifier	
	Y Vacuum half-wave rectifier	
	Z Vacuum full-wave rectifier	
	Two or more of these letters may be combined. Thus AC indicates a diode and a triode in one envelope.	

FOR EXAMPLE:

Type ECH81 Triode-heptode oscillator converter, with noval socket and 6.3 V heater

Type EL34 Power pentode with octal base and 6.3-V heater

Type GZ34 Full-wave rectifier with octal base and 5-V heater

Note: For special tubes (ruggedized, long-life, etc.) the numbers are placed between the letters. For example: E80F, E90CC, E80CF.

Transmitting Tubes

First Letter	Second Letter	Third Letter	Numbers
Tube Type	Filament	Cooling Type	Characteristic
D Rectifier	A Tungsten, directly heated	G Mercury filled	No uniform notation used
M Triode	B Thoriated tungsten, directly heated	L Forced air	
P Pentode	C Oxide coated, directly heated	W Water cooled	
Q Tetrode	E Heater/cathode	X Xenon filled	
T Triode			

FOR EXAMPLE:
Type QQE-04-20 Dual tetrode with indirectly heated cathode

NOISE FIGURE NOMOGRAM FOR TWO CASCADED STAGES

The cascade noise figure of two noise sources is given by the equation

$$F_T = F_1 + \frac{(F_2 - 1)}{G_1}$$

where F_1, F_2, and F_T are the first-stage, second-stage, and overall noise figures respectively, and G is the gain of the first stage—all expressed as power ratios. The nomogram has all scales calibrated in decibels. To use the nomogram connect F_2 and G and note the intersect point on the turning scale. That point is then connected to F_T or F_1 depending on which of these figures is given. Two ranges (high and low) are given for all three "F" scales and they must be used together. Only one "G" scale is necessary.

FOR EXAMPLE:

A first-stage noise figure of 3 dB, a second-stage noise figure of 7 dB, and a first-stage gain of 8 dB, results in an overall noise figure of 4.2 dB.

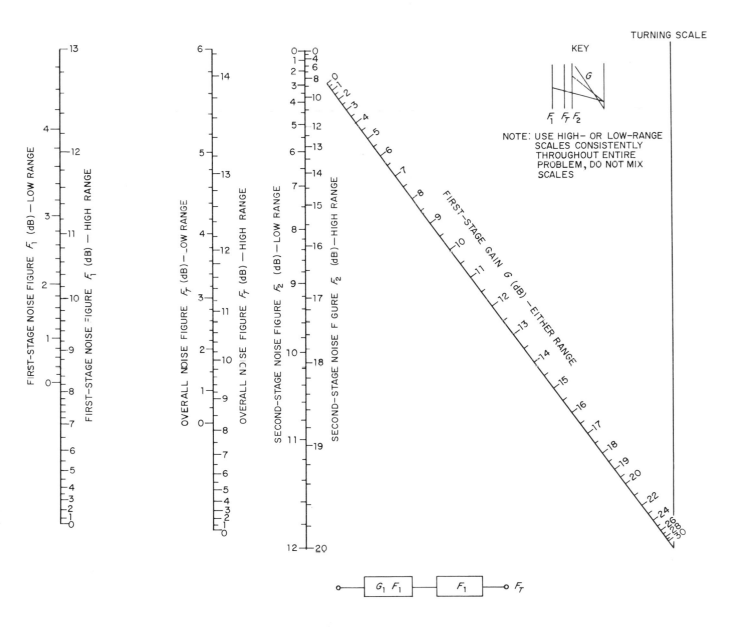

SECTION 5

MATHEMATICAL DATA, FORMULAS, SYMBOLS

RELIABILITY CHARTS

This chart relates system MTBF (Mean-Time-Between-Failures) with the number of components per system and the component MTBF.

FOR EXAMPLE:
A system using 10,000 components with a component MTBF of 30 years will have a system MTBF of 1 day.

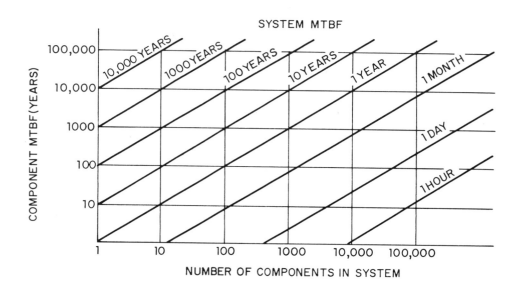

This chart relates system reliability in percent with the number of serial parts, that is, the critical parts that must function in order for the system to perform its function.

FOR EXAMPLE:
10,000 critical parts with a 99.99% parts reliability provide a system reliability of only 37%.

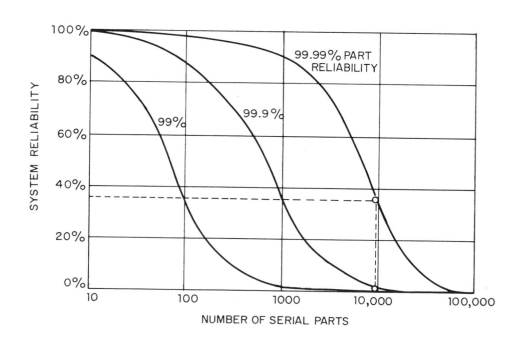

RELIABILITY NOMOGRAM

Reliability is a dependent function of operating time and failure rate. It is generally given as a percentage or a decimal that states the probability that an equipment will perform its function satisfactorily during a mission. Reliability is based on the formula

$$P_o = e^{-t/T} = e^{-\lambda t}$$

where

$T = 1/\lambda$
P_o = probability of success, i.e. reliability
e = base of natural logarithm
t = operating time in hours
T = mean time between failures
λ = failure rate (% per 1000 hr)

FOR EXAMPLE:
A circuit that has a failure rate of 100%/1000 hr (an hourly failure rate of 0.001 or an MTBF of 1000) has a reliability of 99.8% when operated for 2 hr. That means that the circuit will not operate properly an average of 2 times out of 1000 operations, or out of 1000 circuits an average of 2 will fail in 2 hr.

Note: An equipment or circuit with an MTBF of one hour will have a reliability of only 33.788% (100/e) when operated for one hour.

Note: For more detailed treatment of MTBF see *MIL-Handbook-217A*.

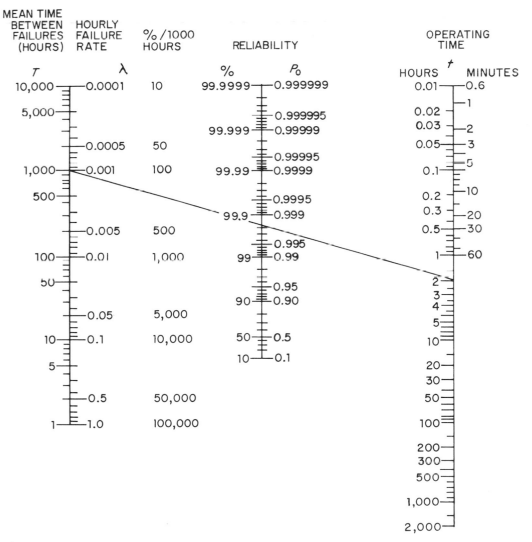

(From *Electronics and Communications*, March 1965.)

For certain critical applications, such as manned space flights, the required reliability is often greater than what can be achieved with a single system. Under these conditions it is necessary to resort to redundancy where two or more identical systems are paralleled. The required redundancy is based on the following equation:

$$P_N = 1 - (1 - P_o)^N$$

where

P_N = probability of success of N paralleled systems

P_o = probability of success of one system

N = number of paralleled systems

FOR EXAMPLE:

A subsystem for a two-week moon exploration flight has a specified reliability of 99.99% and an MTBF of 2000 hr. What is the required redundancy? On reliability nomogram (A) connect 2000 on the T scale with 336 (2 weeks) on the t scale to determine subsystem reliability to be 0.845. On redundancy nomogram connect 0.845 on the P_o scale with 0.9999 on the P_N scale to determine that a redundancy of five is required.

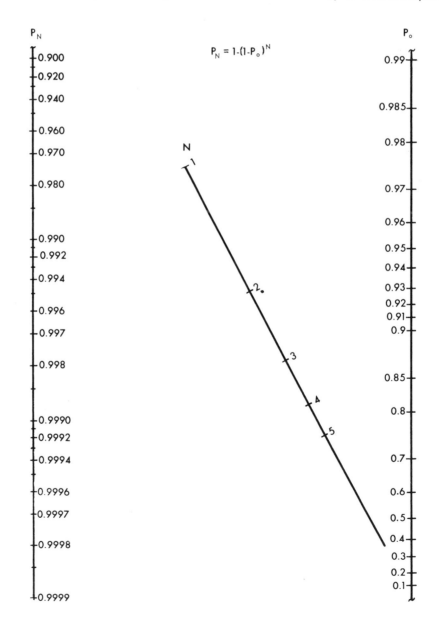

ANGULAR RESOLUTION TABLE

The shaft angle corresponding to an integral binary fraction is required wherever shaft angle encoders are used. This resolution table aids in determining accurately the angle represented by a specific number of counts or conversely, the precise number of counts which equals a given angle.

n	2^n	2^{-n}	Angular Resolution Corresponding to Integral-Exponent Binary Fraction				n
			$1\,296\,000/2^n$ (seconds)	$21\,600/2^n$ (minutes)	$360/2^n$ (degrees)	$2\pi/2^n$ (radians)	
0	1	1	1 296 000	21 600	360.0	6.283 185 307 179 586 476 92	0
1	2	.5	648 000	10 800	180.0	3.141 592 653 589 793 238 46	1
2	4	.25	324 000	5 400	90.0	1.570 796 326 794 896 619 23	2
3	8	.125	162 000	2 700	45.0	.785 398 163 397 448 309 615	3
4	16	.062 5	81 000	1 350	22.5	.392 699 081 698 724 154 808	4
5	32	.031 25	40 500	675	11.25	.196 349 540 849 362 077 404	5
6	64	.015 625	20 250	337.5	5.625	.098 174 770 424 681 038 701 9	6
7	128	.007 812 5	10 125	168.75	2.812 5	.049 087 385 212 340 519 350 9	7
8	256	.003 906 25	5 062.5	84.375	.1.406 25	.024 543 692 606 170 259 675 5	8
9	512	.001 953 125	2 531.25	42.187 5	.703 125	.012 271 846 303 085 129 837 7	9
10	1 024	.000 976 562 5	1 265.625	21.093 75	.351 562 5	.006 135 923 151 542 564 918 87	10
11	2 048	.000 488 281 25	632.812 5	10.546 875	.175 781 25	.003 067 961 575 771 282 459 43	11
12	4 096	.000 244 140 625	316.406 25	5.273 437 5	.087 890 625	.001 533 980 787 885 641 229 72	12
13	8 192	.000 122 070 312 5	158.203 125	2.636 718 75	.043 945 312 5	.000 766 990 393 942 820 614 863	13
14	16 384	.000 061 035 156 25	79.101 562 5	1.318 359 375	.021 972 656 25	.000 383 495 196 971 410 307 431	14
15	32 768	.000 030 517 578 125	39.550 781 25	.659 179 687 5	.010 986 328 125	.000 191 747 598 485 705 153 715	15
16	65 536	.000 015 258 789 062 5	19.775 390 625	.329 589 843 75	.005 493 164 062 5	.000 095 873 799 242 852 576 857 9	16
17	131 072	.000 007 629 394 531 25	9.887 695 312 5	.164 794 921 875	.002 746 582 031 25	.000 047 936 899 621 426 288 428 9	17
18	262 144	.000 003 814 697 265 625	4.943 847 656 25	.082 397 460 937 5	.001 373 291 015 625	.000 023 968 449 810 713 144 214 4	18
19	524 288	.000 001 907 348 632 812 5	2.471 923 828 125	.041 198 730 468 75	.000 686 645 507 812 5	.000 011 984 224 905 356 572 107 2	19
20	1 048 576	.000 000 953 674 316 406 25	1.235 961 914 062 5	.020 599 365 234 375	.000 343 322 753 906 25	.000 005 992 112 452 678 286 053 62	20
21	2 097 152	.000 000 476 837 158 203 125	.617 980 957 031 25	.010 299 682 617 187 5	.000 171 661 376 953 125	.000 002 996 056 226 339 143 026 81	21
22	4 194 304	.000 000 238 418 579 101 562 5	.308 990 478 515 625	.005 149 841 308 593 75	.000 085 830 688 476 562 5	.000 001 498 028 113 169 571 513 40	22
23	8 388 608	.000 000 119 209 289 550 781 25	.154 495 239 257 812 5	.002 574 920 654 296 875	.000 042 915 344 238 281 25	.000 000 749 014 056 584 785 756 702	23
24	16 777 216	.000 000 059 694 644 775 390 625	.077 247 619 628 906 25	.001 287 460 327 148 437 5	.000 021 457 672 119 140 625	.000 000 374 507 028 292 392 878 351	24
25	33 554 432	.000 000 029 802 322 387 695 312 5	.038 623 809 814 453 125	.000 643 730 163 574 218 75	.000 010 728 836 059 570 312 5	.000 000 187 253 514 146 196 439 176	25

TRUTH TABLES (1 = truth, 0 — falsity)

WORDS (English)	MATHEMATICS (Set Theory)	LOGIC	ENGINEERING
1. THE LAWS OF TAUTOLOGY. Repetition—by addition or multiplication—does not alter the truth value of an element.	$a \cup a = a$ $a \cap a = a$	$a \vee a = a$ $a \wedge a = a$	$a + a = a$ $a \cdot a = a$
2. THE LAWS OF COMMUTATION. Disjunction or conjunction is not affected by sequential change. (Disjunction—OR: if either input **a** or input **b**, or both inputs **a** and **b**, are conducting, then the output (a + b) is conducting. Conjunction—AND: If, and only if, both inputs **a** and **b** are conducting, then the output (a·b) is conducting.)	$a \cup b = b \cup a$ $a \cap b = b \cap a$	$a \vee b = b \vee a$ $a \wedge b = b \wedge a$	$a + b = b + a$ $a \cdot b = b \cdot a$
3. THE LAWS OF ASSOCIATION. Disjunction or conjunction is unaffected by grouping.	$(a \cup b) \cup c =$ $a \cup (b \cup c)$ $(a \cap b) \cap c =$ $a \cap (b \cap c)$	$(a \vee b) \vee c =$ $a \vee (b \vee c)$ $(a \wedge b) \wedge c =$ $a \wedge (b \wedge c)$	$(a + b) + c =$ $a + (b + c)$ $(a \cdot b) \cdot c =$ $a \cdot (b \cdot c)$
4. THE LAWS OF DISTRIBUTION. An element is added to a product by adding the element to each member of the product. A sum is multiplied by an element by multiplying every member of the sum by the element.	$a \cup (b \cap c) =$ $(a \cup b) \cap (a \cup c)$ $a \cap (b \cup c) =$ $(a \cap b) \cup (a \cap c)$	$a \vee (b \wedge c) =$ $(a \vee b) \wedge (a \vee c)$ $a \wedge (b \vee c) =$ $(a \wedge b) \vee (a \wedge c)$	$a + (b \cdot c) =$ $(a + b) \cdot (a + c)$ $a \cdot (b + c) =$ $(a \cdot b) + (a \cdot c)$
5. THE LAWS OF ABSORPTION. The disjunction of a product by one of its members is equivalent to this member. The conjunction of a sum by one of its members is equivalent to this member.	$a \cup (a \cap b) = a$ $a \cap (a \cup b) = a$	$a \vee (a \wedge b) = a$ $a \wedge (a \vee b) = a$	$a + (a \cdot b) = a$ $a \cdot (a + b) = a$

THE LAWS OF COMBINATION

THE LAWS OF THE UNIQUE ELEMENTS

1. THE LAWS OF THE UNIVERSE CLASS. The sum consisting of an element and the universe class is equivalent to the universe class. The product consisting of an element and the universe class is equivalent to the element.

$a \cup 1 = 1$ | $a \vee 1 = 1$ | $a + 1 = 1$

$a \cap 1 = a$ | $a \wedge 1 = a$ | $a \cdot 1 = a$

2. THE LAWS OF THE NULL CLASS. The sum consisting of an element and the null class is equivalent to the element. The product consisting of an element and the null class is equivalent to the null class.

$a \cup 0 = a$ | $a \vee 0 = a$ | $a + 0 = a$

$a \cap 0 = 0$ | $a \wedge 0 = 0$ | $a \cdot 0 = 0$

THE LAWS OF NEGATION (COMPLEMENT)

1. THE LAWS OF COMPLEMENTATION. The sum consisting of an element and its complement is equivalent to the universe class. The product consisting of an element and its complement is equivalent to the null class.

$a \cup a' = 1$ | $a \vee {\sim}a = 1$ | $a + \bar{a} = 1$

$a \cap a' = 0$ | $a \wedge {\sim}a = 0$ | $a \cdot \bar{a} = 0$

2. THE LAW OF CONTRAPOSITION. If an element **a** is equivalent to the complement of an element **b**, it is implied that the element **b** is equivalent to the complement of the element **a**.

$a = b' .\supset. b = a'$ | $a \equiv {\sim}b .\supset. b \equiv {\sim}a$ | $a = \bar{b} .\supset. b = \bar{a}$

3. THE LAW OF DOUBLE NEGATION. The complement of the negation of an element is equivalent to the element.

$a = a''$ | $a = {\sim}a'$ | $a = \bar{\bar{a}}$

4. THE LAWS OF EXPANSION. The disjunction of a product composed of the elements **a** and **b** and a product composed of the complement of element **a** and the complement of element **b** is equivalent to the element **a**. The conjunction of a sum composed of the elements **a** and **b** and a sum composed of the element **a** and the complement of element **b** is equivalent to the element **a**.

$(a \cap b) \cup (a \cap b') = a$ | $(a \wedge b) \vee (a \wedge {\sim}b) = a$ | $(a \cdot b) + (a \cdot \bar{b}) = a$

$(a \cup b) \cap (a \cup b') = a$ | $(a \vee b) \wedge (a \vee {\sim}b) = a$ | $(a + b) \cdot (a + \bar{b}) = a$

5. THE LAWS OF DUALITY. The complement of a sum composed of the elements **a** and **b** is equivalent to the conjunction of the complement of element **a** and the complement of element **b**. The complement of a product composed of the elements **a** and **b** is equivalent to the disjunction of the complement of the elements **a** and the complement of element **b**.

$(a \cup b)' = a' \cap b'$ | ${\sim}(a \vee b) = {\sim}a \wedge {\sim}b$ | $(a + b)' = \bar{a} \cdot \bar{b}$

$(a \cap b)' = a' \cup b'$ | ${\sim}(a \wedge b) = {\sim}a \vee {\sim}b$ | $(a \cdot b)' = \bar{a} + \bar{b}$

THE POSTULATES OF BOOLEAN ALGEBRA (Continued)

Boolean Relationships

Idempoint: $a + 0 = a$ $a0 = 0$ where
$$a + 1 = 1 \quad a1 = a \qquad 0 \equiv \bar{a}$$
$$a + a = a \quad aa = a$$

Commutative: $a + b = b + a$
$$ab = ba$$

Associative: $(a + b) + c = a + (b + c)$
$$(ab)c = a(bc)$$

Distributive: $ab + ac = a(b + c)$
$$a + bc = (a + b)(a + c)$$

Absorption: $a(a + b) \equiv a + ab \equiv a$

DeMorgan Theorem: $\bar{\bar{a}} = a$

$$\overline{(ab)} = \bar{a} + \bar{b} \qquad \overline{\overline{(ab)}} = a + b$$

$$\overline{a + b} = \overline{\bar{a}\bar{b}} \qquad \overline{\overline{a + b}} = ab$$

Legend:

NOT: The line over a term indicates a false or not true state.

AND: Two terms directly adjacent to each other are called an "AND" function.

OR: Two terms separated by "+" are called an "OR" function.

Examples: $a\bar{b}$ reads as "a and not b"

$\bar{a}b$ reads as "Not a and b"

$\bar{a}\bar{b}$ reads as "Not a and Not b"

$\overline{ab}$ reads as "Not a or Not b" (See De-Morgan)

CONVERSION CHART OF
STANDARD METRIC PREFIXES

This chart shows, in their relative positions, symbols, multiples (10^n), and abbreviations for all the international multiples and submultiples as recommended by the International Committee on Weights and Measures (1962) and adapted by the National Bureau of Standards.

This chart provides a fast and easy method of conversion from any metric notation to any other. "Unity" represents the basic unit of measurement such as volts, ohms, watts, amperes, grams, hertz, etc. The number of steps up or down between the two prefixes which are being compared is equal to the direction *and* the number of places in which the decimal point has to be moved to convert from one to the other.

FOR EXAMPLE:
To convert 0.0032 milliampere to nanoampere—move six places down. Answer: 3200 nA.
To convert 43,280 kilohertz to megahertz—move three places up. Answer: 43.28 MHz.
To convert 10.74 microns to millimeters—move three places up. Answer: 0.01074 mm.

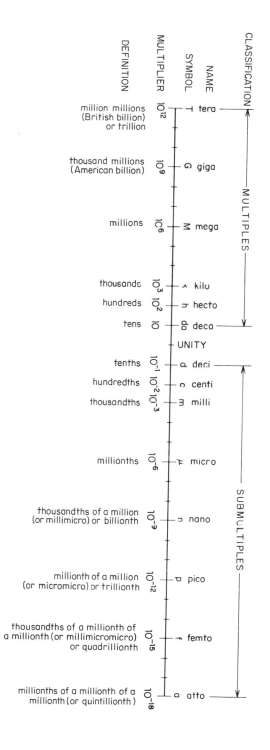

HARMONIC REJECTION
NOMOGRAM

This scale relates the magnitude of harmonic distortion, expressed as a rejection ratio in decibels, to percentage of distortion.

FOR EXAMPLE:

1. A design specifies that a given audio sine-wave oscillator should have its closest harmonic at least 28 dB below the fundamental. The chart indicates that the closest harmonic must be less than 3.9% of the magnitude of the fundamental.

2. Find the harmonic content of a signal made up of the following:

    Fundamental frequency 100 V rms
    Second harmonic   5 V rms
    Third harmonic   2 V rms

Adding harmonics vectorially gives

$$\sqrt{5^2 + 2^2} = 5.39$$

$$\% \text{ distortion} = \frac{\text{harmonic voltage}}{\text{fundamental voltage}} \times 100 =$$

$$\frac{5.39}{100} \times 100$$

Thus the distortion is 5.39%, which means that the harmonic content of the signal is 25.2 dB below the fundamental.

2^n	n	2^{-n}
1	0	1.0
2	1	0.5
4	2	0.25
8	3	0.125
16	4	0.062 5
32	5	0.031 25
64	6	0.015 625
128	7	0.007 812 5
256	8	0.003 906 25
512	9	0.001 953 125
1 024	10	0.000 976 562 5
2 048	11	0.000 488 281 25
4 096	12	0.000 244 140 625
8 192	13	0.000 122 070 312 5
16 384	14	0.000 061 035 156 25
32 768	15	0.000 030 517 578 125
65 536	16	0.000 015 258 789 062 5
131 072	17	0.000 007 629 394 531 25
262 144	18	0.000 003 814 697 265 625
524 288	19	0.000 001 907 348 632 812 5
1 048 576	20	0.000 000 953 674 316 406 25
2 097 152	21	0.000 000 476 837 158 203 125
4 194 304	22	0.000 000 238 418 579 101 562 5
8 388 608	23	0.000 000 119 209 289 550 781 25
16 777 216	24	0.000 000 059 604 644 775 390 625
33 554 432	25	0.000 000 029 802 322 387 695 312 5
67 108 864	26	0.000 000 014 901 161 193 847 656 25
134 217 728	27	0.000 000 007 450 580 596 923 828 125
268 435 456	28	0.000 000 003 725 290 298 461 914 062 5
536 870 912	29	0.000 000 001 862 645 149 230 957 031 25
1 073 741 824	30	0.000 000 000 931 322 574 615 478 515 625
2 147 483 648	31	0.000 000 000 465 661 287 307 739 257 812 5
4 294 967 296	32	0.000 000 000 232 830 643 653 869 628 906 25
8 589 934 592	33	0.000 000 000 116 415 321 826 934 814 453 125
17 179 869 184	34	0.000 000 000 058 207 660 913 467 407 226 562 5
34 359 738 368	35	0.000 000 000 029 103 830 456 733 703 613 281 25
68 719 476 736	36	0.000 000 000 014 551 915 228 366 851 806 640 625
137 438 953 472	37	0.000 000 000 007 275 957 614 183 425 903 320 312 5
274 877 906 944	38	0.000 000 000 003 637 978 807 091 712 951 660 156 25
549 755 813 888	39	0.000 000 000 001 818 989 403 545 856 475 830 078 125
1 099 511 627 776	40	0.000 000 000 000 909 494 701 772 928 237 915 039 062 5
2 199 023 255 552	41	0.000 000 000 000 454 747 350 886 464 118 957 519 531 25
4 398 046 511 104	42	0.000 000 000 000 227 373 675 443 232 059 478 759 765 625
8 796 093 022 208	43	0.000 000 000 000 113 686 837 721 616 029 739 379 882 812 5
17 592 186 044 416	44	0.000 000 000 000 056 843 418 860 808 014 869 689 941 406 25
35 184 372 088 832	45	0.000 000 000 000 028 421 709 430 404 007 434 844 970 703 125
70 368 744 177 664	46	0.000 000 000 000 014 210 854 715 202 003 717 422 485 351 562 5
140 737 488 355 328	47	0.000 000 000 000 007 105 427 357 601 001 858 711 242 675 781 25
281 474 976 710 656	48	0.000 000 000 000 003 552 713 678 800 500 929 355 621 337 890 625
562 949 953 421 312	49	0.000 000 000 000 001 776 356 839 400 250 464 677 810 668 945 312 5
1 125 899 906 842 624	50	0.000 000 000 000 000 888 178 419 700 125 232 338 905 334 472 656 25
2 251 799 813 685 248	51	0.000 000 000 000 000 444 089 209 850 062 616 169 452 667 236 328 125
4 503 599 627 370 496	52	0.000 000 000 000 000 222 044 604 925 031 308 084 726 333 618 164 062 5
9 007 199 254 740 992	53	0.000 000 000 000 000 111 022 302 462 515 654 042 363 166 809 082 031 25
18 014 398 509 481 984	54	0.000 000 000 000 000 055 511 151 231 257 827 021 181 583 404 541 015 625
36 028 797 018 963 968	55	0.000 000 000 000 000 027 755 575 615 628 913 510 590 791 702 270 507 812 5
72 057 594 037 927 936	56	0.000 000 000 000 000 013 877 787 807 814 456 755 295 395 851 135 253 906 25
144 115 188 075 855 872	57	0.000 000 000 000 000 006 938 893 903 907 228 377 647 697 925 567 626 953 125
288 230 376 151 711 744	58	0.000 000 000 000 000 003 469 446 951 953 614 188 823 848 962 783 813 476 562 5
576 460 752 303 423 488	59	0.000 000 000 000 000 001 734 723 475 976 807 094 411 924 481 391 906 738 281 25
1 152 921 504 606 846 976	60	0.000 000 000 000 000 000 867 361 737 988 403 547 205 962 240 695 953 369 140 625
2 305 843 009 213 693 952	61	0.000 000 000 000 000 000 433 680 868 994 201 773 602 981 120 347 976 684 570 312 5
4 611 686 018 427 387 904	62	0.000 000 000 000 000 000 216 840 434 497 100 886 801 490 560 173 988 342 285 156 25
9 223 372 036 854 775 808	63	0.000 000 000 000 000 000 108 420 217 248 550 443 400 745 280 086 994 171 142 578 125
18 446 744 073 709 551 616	64	0.000 000 000 000 000 000 054 210 108 624 275 221 700 372 640 043 497 085 571 289 062 5
36 893 488 147 419 103 232	65	0.000 000 000 000 000 000 027 105 054 312 137 610 850 186 320 021 748 542 785 644 531 25
73 786 976 294 838 206 464	66	0.000 000 000 000 000 000 013 552 527 156 068 805 425 093 160 010 874 271 392 822 265 625
147 573 952 589 676 412 928	67	0.000 000 000 000 000 000 006 776 263 578 034 402 712 546 580 005 437 135 696 411 132 812 5
295 147 905 179 352 825 856	68	0.000 000 000 000 000 000 003 388 131 789 017 201 356 273 290 002 718 567 848 205 566 406 25
590 295 810 358 705 651 712	69	0.000 000 000 000 000 000 001 694 065 894 508 600 678 136 645 001 359 283 924 102 783 203 125
1 180 591 620 717 411 303 424	70	0.000 000 000 000 000 000 000 847 032 947 254 300 339 068 322 500 679 641 962 051 391 601 562 5
2 361 183 241 434 822 606 848	71	0.000 000 000 000 000 000 000 423 516 473 627 150 169 534 161 250 339 820 981 025 695 800 781 25
4 722 366 482 869 645 213 696	72	0.000 000 000 000 000 000 000 211 758 236 813 575 084 767 080 625 169 910 490 512 847 900 390 625

n	2^n
73	94447 32965 73929 04273 92
74	18889 46593 14785 80854 784
75	37778 93186 29571 61709 568
76	75557 86372 59143 23419 136
77	15111 57274 51828 64683 8272
78	30223 14549 03657 29367 6544
79	60446 29098 07314 58735 3088
80	12089 25819 61462 91747 06176
81	24178 51639 22925 83494 12352
82	48357 03278 45851 66988 24704
83	96714 06556 91703 33976 49408
84	19342 81311 38340 66795 29881 6
85	38685 62622 76681 33590 59763 2
86	77371 25245 53362 67181 19526 4
87	15474 25049 10672 53436 23905 28
88	30948 50098 21345 06872 47810 56
89	61897 00196 42690 13744 95621 12
90	12379 40039 28538 02748 99124 224
91	24758 80078 57076 05497 98248 448
92	49517 60157 14152 10995 96496 896
93	99035 20314 28304 21991 92993 792
94	19807 04062 85660 84398 38598 7584
95	39614 08125 71321 68796 77197 5168
96	79228 16251 42643 37593 54395 0336
97	15845 63250 28528 67518 70879 00672
98	31691 26500 57057 35037 41758 01344
99	63382 53001 14114 70074 83516 02688
100	12676 50600 22822 94014 96703 20537 6

SQUARES. CUBES, AND ROOTS

n	n^2	$\sqrt{n}$	$\sqrt{10n}$	n^3		n	$\sqrt[3]{n}$	$\sqrt[3]{10n}$	$\sqrt[3]{100n}$
1	1	1.000000	3.162278	1		1	1.000000	2.154435	4.641589
2	4	1.414214	4.472136	8		2	1.259921	2.714418	5.848035
3	9	1.732051	5.477226	27		3	1.442250	3.107233	6.694330
4	16	2.000000	6.324555	64		4	1.587401	3.419952	7.368063
5	25	2.236068	7.071068	125		5	1.709976	3.684031	7.937005
6	36	2.449490	7.745967	216		6	1.817121	3.914868	8.434327
7	49	2.645751	8.366600	343		7	1.912931	4.121285	8.879040
8	64	2.828427	8.944272	512		8	2.000000	4.308869	9.283178
9	81	3.000000	9.486833	729		9	2.080084	4.481405	9.654894
10	100	3.162278	10.00000	1,000		10	2.154435	4.641589	10.00000
11	121	3.316625	10.48809	1,331		11	2.223980	4.791420	10.32280
12	144	3.464102	10.95445	1,728		12	2.289428	4.932424	10.62659
13	169	3.605551	11.40175	2,197		13	2.351335	5.065797	10.91393
14	196	3.741657	11.83216	2,744		14	2.410142	5.192494	11.18689
15	225	3.872983	12.24745	3,375		15	2.466212	5.313293	11.44714
16	256	4.000000	12.64911	4,096		16	2.519842	5.428835	11.69607
17	289	4.123106	13.03840	4,913		17	2.571282	5.539658	11.93483
18	324	4.242641	13.41641	5,832		18	2.620741	5.646216	12.16440
19	361	4.358899	13.78405	6,859		19	2.668402	5.748897	12.38562
20	400	4.472136	14.14214	8,000		20	2.714418	5.848035	12.59921
21	441	4.582576	14.49138	9,261		21	2.758924	5.943922	12.80579
22	484	4.690416	14.83240	10,648		22	2.802039	6.036811	13.00591
23	529	4.795832	15.16575	12,167		23	2.843867	6.126926	13.20006
24	576	4.898979	15.49193	13,824		24	2.884499	6.214465	13.38866
25	625	5.000000	15.81139	15,625		25	2.924018	6.299605	13.57209
26	676	5.099020	16.12452	17,576		26	2.962496	6.382504	13.75069
27	729	5.196152	16.43168	19,683		27	3.000000	6.463304	13.92477
28	784	5.291503	16.73320	21,952		28	3.036589	6.542133	14.09460
29	841	5.385165	17.02939	24,389		29	3.072317	6.619106	14.26043
30	900	5.477226	17.32051	27,000		30	3.107233	6.694330	14.42250
31	961	5.567764	17.60682	29,791		31	3.141381	6.767899	14.58100
32	1,024	5.656854	17.88854	32,768		32	3.174802	6.839904	14.73613
33	1,089	5.744563	18.16590	35,937		33	3.207534	6.910423	14.88806
34	1,156	5.830952	18.43909	39,304		34	3.239612	6.979532	15.03695
35	1,225	5.916080	18.70829	42,875		35	3.271066	7.047299	15.18294
36	1,296	6.000000	18.97367	46,656		36	3.301927	7.113787	15.32619
37	1,369	6.082763	19.23538	50,653		37	3.332222	7.179054	15.46680
38	1,444	6.164414	19.49359	54,872		38	3.361975	7.243156	15.60491
39	1,521	6.244998	19.74842	59,319		39	3.391211	7.306144	15.74061
40	1,600	6.324555	20.00000	64,000		40	3.419952	7.368063	15.87401
41	1,681	6.403124	20.24846	68,921		41	3.448217	7.428959	16.00521
42	1,764	6.480741	20.49390	74,088		42	3.476027	7.488872	16.13429
43	1,849	6.557439	20.73644	79,507		43	3.503398	7.547842	16.26133
44	1,936	6.633250	20.97618	85,184		44	3.530348	7.605905	16.38643
45	2,025	6.708204	21.21320	91,125		45	3.556893	7.663094	16.50964
46	2,116	6.782330	21.44761	97,336		46	3.583048	7.719443	16.63103
47	2,209	6.855655	21.67948	103,823		47	3.608826	7.774980	16.75069
48	2,304	6.928203	21.90890	110,592		48	3.634241	7.829735	16.86865
49	2,401	7.000000	22.13594	117,649		49	3.659306	7.883735	16.98499
50	2,500	7.071068	22.36068	125,000		50	3.684031	7.937005	17.09976

n	n²	√n	√10n	n³
50	2,500	7.071068	22.36068	125,000
51	2,601	7.141428	22.58318	132,651
52	2,704	7.211103	22.80351	140,608
53	2,809	7.280110	23.02173	148,877
54	2,916	7.348469	23.23790	157,464
55	3,025	7.416198	23.45208	166,375
56	3,136	7.483315	23.66432	175,616
57	3,249	7.549834	23.87467	185,193
58	3,364	7.615773	24.06319	195,112
59	3,481	7.681146	24.28992	205,379
60	3,600	7.745967	24.49490	216,000
61	3,721	7.810250	24.69818	226,981
62	3,844	7.874008	24.89980	238,328
63	3,969	7.937254	25.09980	250,047
64	4,096	8.000000	25.29822	262,144
65	4,225	8.062258	25.49510	274,625
66	4,356	8.124038	25.69047	287,496
67	4,489	8.185353	25.88436	300,763
68	4,624	8.246211	26.07681	314,432
69	4,761	8.306624	26.26785	328,509
70	4,900	8.366600	26.45751	343,000
71	5,041	8.426150	26.64583	357,911
72	5,184	8.485281	26.83282	373,248
73	5,329	8.544004	27.01851	389,017
74	5,476	8.602325	27.20294	405,224
75	5,625	8.660254	27.38613	421,875
76	5,776	8.717798	27.56810	438,976
77	5,929	8.774964	27.74887	456,533
78	6,084	8.831761	27.92848	474,552
79	6,241	8.888194	28.10694	493,039
80	6,400	8.944272	28.28427	512,000
81	6,561	9.000000	28.46050	531,441
82	6,724	9.055385	28.63564	551,368
83	6,889	9.110434	28.80972	571,787
84	7,056	9.165151	28.98275	592,704
85	7,225	9.219544	29.15476	614,125
86	7,396	9.273618	29.32576	636,056
87	7,569	9.327379	29.49576	658,503
88	7,744	9.380832	29.66479	681,472
89	7,921	9.433981	29.83287	704,969
90	8,100	9.486833	30.00000	729,000
91	8,281	9.539392	30.16621	753,571
92	8,464	9.591663	30.33150	778,688
93	8,649	9.643651	30.49590	804,357
94	8,836	9.695360	30.65942	830,584
95	9,025	9.746794	30.82207	857,375
96	9,216	9.797959	30.98387	884,736
97	9,409	9.848858	31.14482	912,673
98	9,604	9.899495	31.30495	941,192
99	9,801	9.949874	31.46427	970,299
100	10,000	10.00000	31.62278	1,000,000

n	∛n	∛10n	∛100n
50	3.684031	7.937005	17.09976
51	3.708430	7.989570	17.21301
52	3.732511	8.041452	17.32478
53	3.756286	8.092672	17.43513
54	3.779763	8.143253	17.54411
55	3.802952	8.193213	17.65174
56	3.825862	8.242571	17.75808
57	3.848501	8.291344	17.86316
58	3.870877	8.339551	17.96702
59	3.892996	8.387207	18.06969
60	3.914868	8.434327	18.17121
61	3.936497	8.480926	18.27160
62	3.957892	8.527019	18.37091
63	3.979057	8.572619	18.46915
64	4.000000	8.617739	18.56636
65	4.020726	8.662391	18.66256
66	4.041240	8.706588	18.75777
67	4.061548	8.750340	18.85204
68	4.081655	8.793659	18.94536
69	4.101566	8.836556	19.03778
70	4.121285	8.879040	19.12931
71	4.140818	8.921121	19.21997
72	4.160168	8.962809	19.30979
73	4.179339	9.004113	19.39877
74	4.198336	9.045042	19.48695
75	4.217163	9.085603	19.57434
76	4.235824	9.125805	19.66095
77	4.254321	9.165656	19.74681
78	4.272659	9.205164	19.83192
79	4.290840	9.244335	19.91632
80	4.308869	9.283178	20.00000
81	4.326749	9.321698	20.08299
82	4.344481	9.359902	20.16530
83	4.362071	9.397796	20.24694
84	4.379519	9.435388	20.32793
85	4.396830	9.472682	20.40828
86	4.414005	9.509685	20.48800
87	4.431048	9.546403	20.56710
88	4.447960	9.582840	20.64560
89	4.464745	9.619002	20.72351
90	4.481405	9.654894	20.80084
91	4.497941	9.690521	20.87759
92	4.514357	9.725888	20.95379
93	4.530655	9.761000	21.02944
94	4.546836	9.795861	21.10454
95	4.562903	9.830476	21.17912
96	4.578857	9.864848	21.25317
97	4.594701	9.898983	21.32671
98	4.610436	9.932884	21.39975
99	4.626065	9.966555	21.47229
100	4.641589	10.00000	21.54435

n	n^4	n^5	n^6	n^7	n^8
1	1	1	1	1	1
2	16	32	64	128	256
3	81	243	729	2187	6561
4	256	1024	4096	16384	65536
5	625	3125	15625	78125	390625
6	1296	7776	46656	279936	1679616
7	2401	16807	117649	823543	5764801
8	4096	32768	262144	2097152	16777216
9	6561	59049	531441	4782969	43046721
					$\times 10^8$
10	10000	100000	1000000	10000000	1.000000
11	14641	161051	1771561	19487171	2.143589
12	20736	248832	2985984	35831808	4.299817
13	28561	371293	4826809	62748517	8.157307
14	38416	537824	7529536	105413504	14.757891
15	50625	759375	11390625	170859375	25.628906
16	65536	1048576	16777216	268435456	42.949673
17	83521	1419857	24137569	410338673	69.757574
18	104976	1889568	34012224	612220032	110.199606
19	130321	2476099	47045881	893871739	169.835630
				$\times 10^9$	$\times 10^{10}$
20	160000	3200000	64000000	1.280000	2.560000
21	194481	4084101	85766121	1.801089	3.782286
22	234256	5153632	113379904	2.494358	5.487587
23	279841	6436343	148035889	3.404825	7.831099
24	331776	7962624	191102976	4.586471	11.007531
25	390625	9765625	244140625	6.103516	15.258789
26	456976	11881376	308915776	8.031810	20.882706
27	531441	14348907	387420489	10.460353	28.242954
28	614656	17210368	481890304	13.492929	37.780200
29	707281	20511149	594823321	17.249876	50.024641
			$\times 10^8$	$\times 10^{10}$	$\times 10^{11}$
30	810000	24300000	7.290000	2.187000	6.561000
31	923521	28629151	8.875037	2.751261	8.528910
32	1048576	33554432	10.737418	3.435974	10.995116
33	1185921	39135393	12.914680	4.261844	14.064086
34	1336336	45435424	15.448044	5.252335	17.857939
35	1500625	52521875	18.382656	6.433930	22.518754
36	1679616	60466176	21.767823	7.836416	28.211099
37	1874161	69343957	25.657264	9.493188	35.124795
38	2085136	79235168	30.109364	11.441558	43.477921
39	2313441	90224199	35.187438	13.723101	53.520093
			$\times 10^9$	$\times 10^{10}$	$\times 10^{12}$
40	2560000	102400000	4.096000	16.384000	6.553600
41	2825761	115856201	4.750104	19.475427	7.984925
42	3111696	130691232	5.489032	23.053933	9.682652
43	3418801	147008443	6.321363	27.181861	11.688200
44	3748096	164916224	7.256314	31.927781	14.048224
45	4100625	184528125	8.303766	37.366945	16.815125
46	4477456	205962976	9.474297	43.581766	20.047612
47	4879681	229345007	10.779215	50.662312	23.811287
48	5308416	254803968	12.230590	58.706834	28.179280
49	5764801	282475249	13.841287	67.822307	33.232931
50	6250000	312500000	15.625000	78.125000	39.062500

n	n^4	n^5	n^6	n^7	n^8
			$\times 10^9$	$\times 10^{11}$	$\times 10^{13}$
50	6250000	312500000	15.625000	7.812500	3.906250
51	6765201	345025251	17.596288	8.974107	4.576794
52	7311616	380204032	19.770610	10.280717	5.345973
53	7890481	418195493	22.164361	11.747111	6.225969
54	8503056	459165024	24.794911	13.389252	7.230196
55	9150625	503284375	27.680641	15.224352	8.373394
56	9834496	550731776	30.840979	17.270948	9.671731
57	10556001	601692057	34.296447	19.548975	11.142916
58	11316496	656356768	38.068693	22.079842	12.806308
59	12117361	714924299	42.180534	24.886515	14.683044
		$\times 10^8$	$\times 10^{10}$	$\times 10^{11}$	$\times 10^{13}$
60	12960000	7.776000	4.665600	27.993600	16.796160
61	13845841	8.445963	5.152037	31.427428	19.170731
62	14776336	9.161328	5.680024	35.216146	21.834011
63	15752961	9.924365	6.252350	39.389806	24.815578
64	16777216	10.737418	6.871948	43.980465	28.147498
65	17850625	11.602906	7.541889	49.022279	31.864481
66	18974736	12.523326	8.265395	54.551607	36.004061
67	20151121	13.501251	9.045838	60.607116	40.606768
68	21381376	14.539336	9.886748	67.229888	45.716324
69	22667121	15.640313	10.791816	74.463533	51.379837
		$\times 10^8$	$\times 10^{10}$	$\times 10^{12}$	$\times 10^{14}$
70	24010000	16.807000	11.764900	8.235430	5.764801
71	25411681	18.042294	12.810028	9.095120	6.457535
72	26873856	19.349176	13.931407	10.030613	7.222041
73	28398241	20.730716	15.133423	11.047399	8.064601
74	29986576	22.190066	16.420649	12.151280	8.991947
75	31640625	23.730469	17.797852	13.348389	10.011292
76	33362176	25.355254	19.269993	14.645195	11.130348
77	35153041	27.067842	20.842238	16.048523	12.357363
78	37015056	28.871744	22.519960	17.565569	13.701144
79	38950081	30.770564	24.308746	19.203909	15.171088
		$\times 10^8$	$\times 10^{10}$	$\times 10^{12}$	$\times 10^{14}$
80	40960000	32.768000	26.214400	20.971520	16.777216
81	43046721	34.867844	28.242954	22.876792	18.530202
82	45212176	37.073984	30.400667	24.928547	20.441409
83	47458321	39.390406	32.694037	27.136051	22.522922
84	49787136	41.821744	35.129803	29.509035	24.787589
85	52200625	44.370531	37.714952	32.057709	27.249053
86	54700816	47.042702	40.456724	34.792782	29.921793
87	57289761	49.842092	43.362620	37.725479	32.821167
88	59969536	52.773192	46.440409	40.867560	35.963452
89	62742241	55.840594	49.698129	44.231335	39.365888
		$\times 10^9$	$\times 10^{11}$	$\times 10^{13}$	$\times 10^{15}$
90	65610000	5.904900	5.314410	4.782969	4.304672
91	68574961	6.240321	5.678693	5.167610	4.702525
92	71639296	6.590815	6.063550	5.578466	5.132189
93	74805201	6.956884	6.469902	6.017009	5.595818
94	78074896	7.339040	6.898698	6.484776	6.095689
95	81450625	7.737809	7.350919	6.983373	6.634204
96	84934656	8.153727	7.827578	7.514475	7.213896
97	88529281	8.587340	8.329720	8.079828	7.837434
98	92236816	9.039208	8.858424	8.681255	8.507630
99	96059601	9.509900	9.414801	9.320653	9.227447
100	100000000	10.000000	10.000000	10.000000	10.000000

FACTORIALS

Numerical

n	$\frac{1}{n!}$					n!			n
1	1.							1	1
2	0.5							2	2
3	.16666	66666	66666	66666	66667			6	3
4	.04166	66666	66666	66666	66667			24	4
5	.00833	33333	33333	33333	33333			120	5
6	0.00138	88888	88888	88888	88889			720	6
7	.00019	84126	98412	69841	26984			5040	7
8	.00002	48015	87301	58730	15873			40320	8
9	.00000	27557	31922	39858	90653		3	62880	9
10	.00000	02755	73192	23985	89065		36	28800	10
11	0.00000	00250	52108	38544	17188		399	16800	11
12	.00000	00020	87675	69878	68099		4790	01600	12
13	.00000	00001	60590	43836	82161		62270	20800	13
14	.00000	00000	11470	74559	77297	8	71782	91200	14
15	.00000	00000	00764	71637	31820	130	76743	68000	15
16	0.00000	00000	00047	79477	33239	2092	27898	88000	16
17	.00000	00000	00002	81145	72543	35568	74280	96000	17
18	.00000	00000	00000	15619	20697	G 40237	37057	28000	18
19	.00000	00000	00000	00822	06352	121 64510	04088	32000	19
20	.00000	00000	00000	00041	10318	2432 90200	81766	40000	20

$$n! = 1 \times 2 \times 3 \times 4 \times 5 \ldots n$$

FOR EXAMPLE:

For $n = 7$, $n! = 5040$,

$1/n! = 0.00019841269841269841269984$,

and $\log (n!) = 3.702431$.

Logarithmic

Logarithms of the products $1 \times 2 \times 3 \ldots n$, n from 1 to 100.

n	log (n!)	n	log (n!)	n	log (n!)	n	log (n!)
1	0.000000	26	26.605619	51	66.190645	76	111.275425
2	0.301030	27	28.036983	52	67.906648	77	113.161916
3	0.778151	28	29.484141	53	69.630924	78	115.054011
4	1.380211	29	30.946539	54	71.363318	79	116.951638
5	2.079181	30	32.423660	55	73.103681	80	118.854728
6	2.857332	31	33.915022	56	74.851869	81	120.763213
7	3.702431	32	35.420172	57	76.607744	82	122.677027
8	4.605521	33	36.938686	58	78.371172	83	124.596105
9	5.559763	34	38.470165	59	80.142024	84	126.520384
10	6.559763	35	40.014233	60	81.920175	85	128.449803
11	7.601156	36	41.570535	61	83.705505	86	130.384301
12	8.680337	37	43.138737	62	85.497896	87	132.323821
13	9.794280	38	44.718520	63	87.297237	88	134.268303
14	10.940408	39	46.309585	64	89.103417	89	136.217693
15	12.116500	40	47.911645	65	90.916330	90	138.171936
16	13.320620	41	49.524429	66	92.735874	91	140.130977
17	14.551069	42	51.147678	67	94.561949	92	142.094765
18	15.806341	43	52.781147	68	96.394458	93	144.063248
19	17.085095	44	54.424599	69	98.233307	94	146.036376
20	18.386125	45	56.077812	70	100.078405	95	148.014099
21	19.708344	46	57.740570	71	101.929663	96	149.996371
22	21.050767	47	59.412668	72	103.786996	97	151.983142
23	22.412494	48	61.093909	73	105.650319	98	153.974368
24	23.792706	49	62.784105	74	107.519550	99	155.970004
25	25.190646	50	64.483075	75	109.394612	100	157.970004

RECTANGULAR-POLAR
CONVERSION CHART

This chart quickly converts between cartesian (rectangular) and polar forms of notation. The horizontal (real) and the vertical (imaginary) coordinates are used for rectangular notation, and the angular (magnitude) and circular (angle) coordinates are used for polar notation. The same units of measurement are used for both systems. This makes conversion from one system to the other readily possible. The range of the chart can be extended by multiplying the horizontal and vertical axes by the same power of ten.

FOR EXAMPLE:
1. $2 + j3$ is equivalent to $3.6 \, \underline{/56°}$
2. $70 \, \underline{/55°}$ is equivalent to $40 + j57$
3. $6 - j3$ is equivalent to $6.7 \, \underline{/333°}$

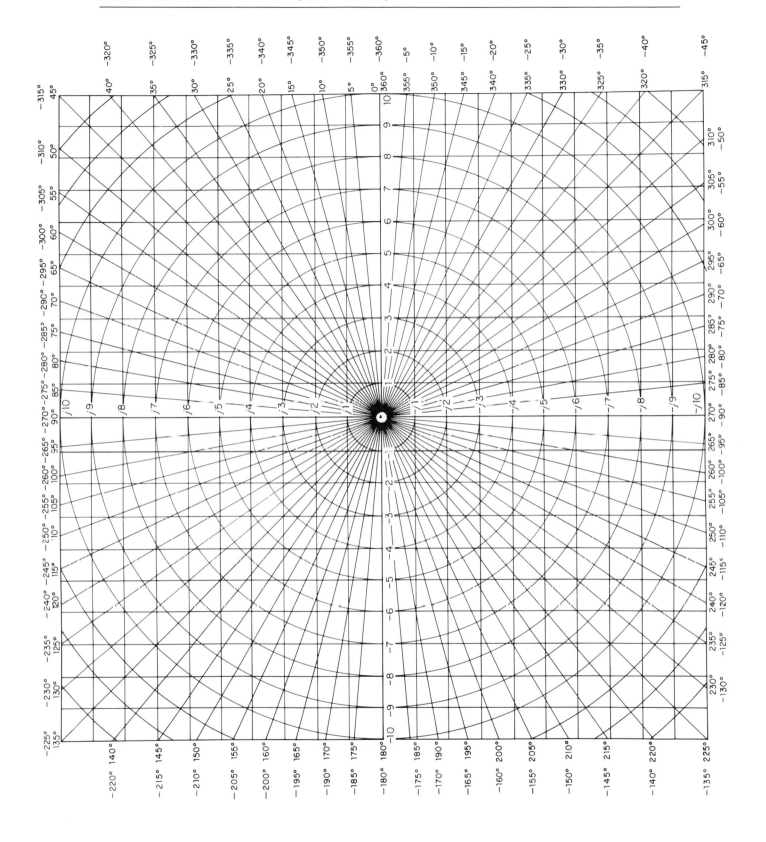

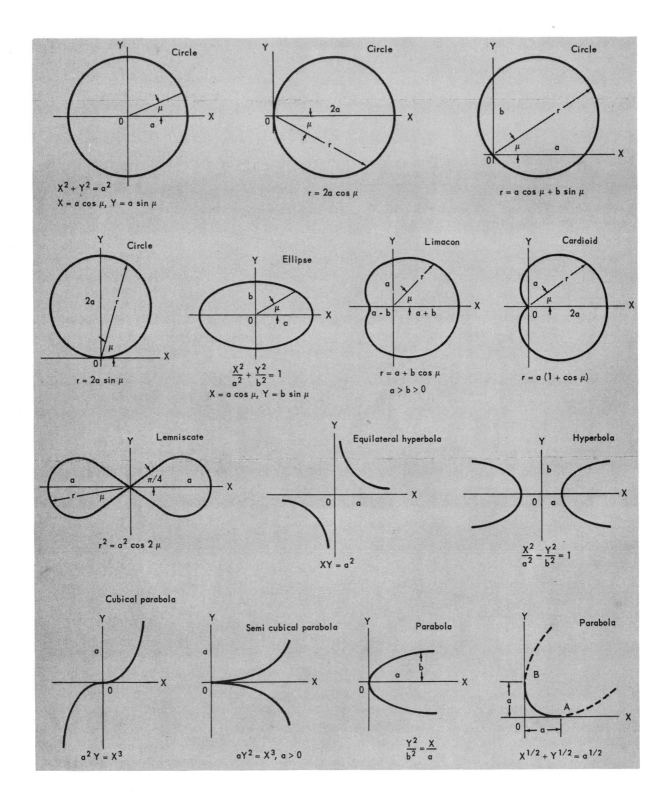

Circle
$$X^2 + Y^2 = a^2$$
$$X = a \cos \mu, \ Y = a \sin \mu$$

Circle
$$r = 2a \cos \mu$$

Circle
$$r = a \cos \mu + b \sin \mu$$

Circle
$$r = 2a \sin \mu$$

Ellipse
$$\frac{X^2}{a^2} + \frac{Y^2}{b^2} = 1$$
$$X = a \cos \mu, \ Y = b \sin \mu$$

Limacon
$$r = a + b \cos \mu$$
$$a > b > 0$$

Cardioid
$$r = a \left(1 + \cos \mu\right)$$

Lemniscate
$$r^2 = a^2 \cos 2\mu$$

Equilateral hyperbola
$$XY = a^2$$

Hyperbola
$$\frac{X^2}{a^2} - \frac{Y^2}{b^2} = 1$$

Cubical parabola
$$a^2 Y = X^3$$

Semi cubical parabola
$$aY^2 = X^3, \ a > 0$$

Parabola
$$\frac{Y^2}{b^2} = \frac{X}{a}$$

Parabola
$$X^{1/2} + Y^{1/2} = a^{1/2}$$

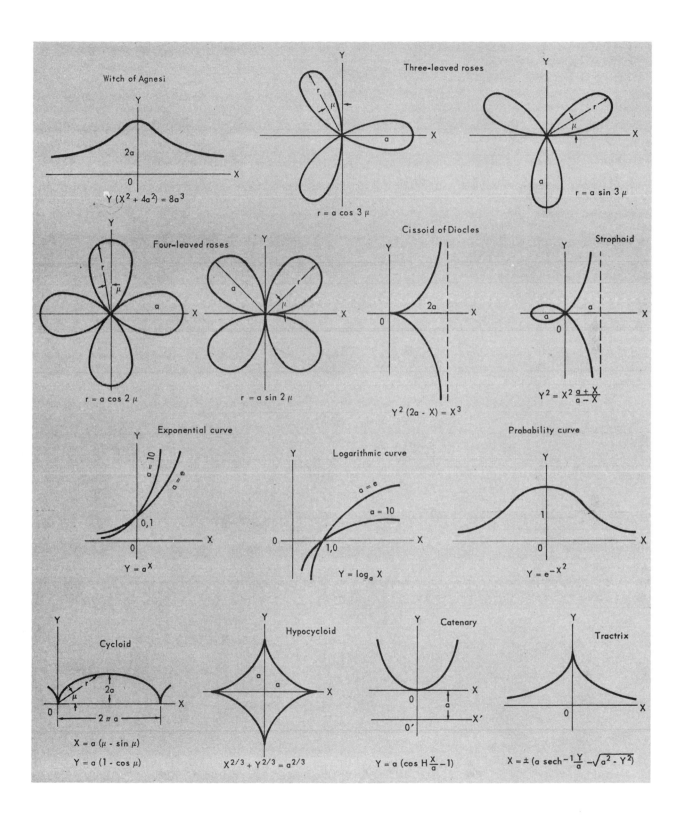

Witch of Agnesi

$$Y(X^2 + 4a^2) = 8a^3$$

Three-leaved roses

$$r = a \cos 3\mu$$

$$r = a \sin 3\mu$$

Four-leaved roses

$$r = a \cos 2\mu$$

$$r = a \sin 2\mu$$

Cissoid of Diocles

$$Y^2(2a - X) = X^3$$

Strophoid

$$Y^2 = X^2 \frac{a + X}{a - X}$$

Exponential curve

$$Y = a^X$$

Logarithmic curve

$$Y = \log_a X$$

Probability curve

$$Y = e^{-X^2}$$

Cycloid

$$X = a(\mu - \sin \mu)$$
$$Y = a(1 - \cos \mu)$$

Hypocycloid

$$X^{2/3} + Y^{2/3} = a^{2/3}$$

Catenary

$$Y = a\left(\cos H \frac{X}{a} - 1\right)$$

Tractrix

$$X = \pm \left(a \, \text{sech}^{-1} \frac{Y}{a} - \sqrt{a^2 - Y^2}\right)$$

FORMULAS FOR SOLIDS

Cube

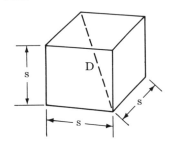

Surface Area
$$A = 6s^2$$

Volume
$$V = s^3$$

Diagonal
$$D = 1.7321s$$

Parallelopiped

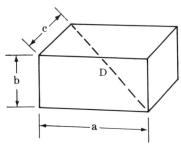

Surface Area
$$A = 2(ab + bc + ac)$$

Volume
$$V = abc$$

Diagonal
$$D = \sqrt{a^2 + b^2 + c^2}$$

Right Circular Cylinder

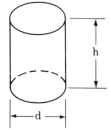

Surface Area
$$A = 1.5708\, d\, (2h + d)$$

Volume
$$V = .7854d^2h$$

Frustrum of Right Circular Cylinder

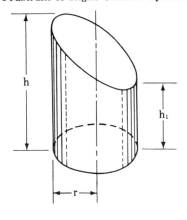

Lateral Area
$$A = 3.1416\, r\, (h + h_1)$$

Area of Top Section
$$A = 3.1416r\sqrt{r^2 + \left(\frac{h_1 - h}{2}\right)^2}$$

Area of Base
$$A = 3.1416r^2$$

Volume
$$V = 1.5708r^2(h + h_1)$$

Right Regular Pyramid

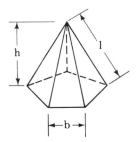

Surface Area
$$A = \tfrac{1}{2}nbl + A_B(\text{area of base})$$

Volume
$$V = \tfrac{1}{3}A_Bh$$

Right Regular Cone

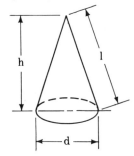

Surface Area
$$A = 1.5708d\ (.5d + l)$$

Volume
$$V = .2618d^2h$$

Frustrum of Right Regular Pyramid

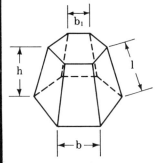

Surface Area
$$A = \tfrac{1}{2}[n\ (b + b_1) + A_B + A_T]$$

Volume
$$V = \tfrac{1}{3}h\ (A_B + A_T + \sqrt{A_BA_T})$$

Frustrum of Right Regular Cone

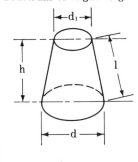

Surface Area
$$A = .3927[d^2 + d^2_1 + 4l(d + d_1)]$$

Volume
$$V = .2618h\ (d^2 + dd_1 + d_1^2)$$

FORMULAS FOR SOLIDS
(Continued)

Sphere

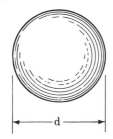

Surface Area
$$A = 3.1416d^2$$

Volume
$$V = .5236d^3$$

Sector of Sphere

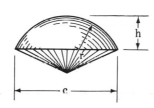

Surface Area
$$A = 1.5708\, r\ (4h + c)$$

Volume
$$V = 2.0944r^2h$$

Segment of Sphere

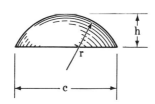

Surface Area of Top Section
$$A = 6.2832rh \text{ or}$$
$$A = .7854\ (4h^2 + c^2)$$

Total Surface Area
$$A = 1.5708\ (2h^2 + c^2)$$

Volume
$$V = 1.0472h^2\ (3r - h) \text{ or}$$
$$V = .1318h\ (3c^2 + 4h^2)$$

Zone of Sphere

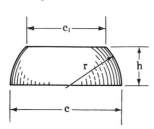

Area of Spherical Surface
$$A = 6.2832rh$$

Total Surface Area
$$A = .7854\ (8rh + c^2 + c_1^2)$$

Volume
$$V = .1318h\ (3c^2 + 3c_1^2 + 4h^2)$$

Torus

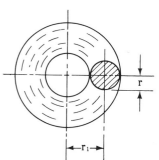

Surface Area
$$A = 39.478rr_1$$

Volume
$$V = 19.739r^2r_1$$

Ellipsoid

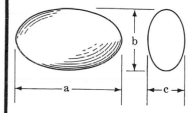

Volume
$$V = .5236abc$$

Paraboloid

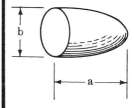

Volume
$$V = .3927ab^2$$

AREAS OF A FEW COMMON SHAPES

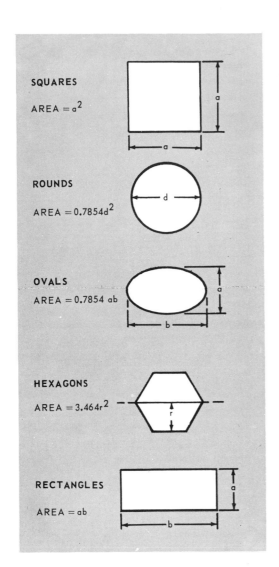

SQUARES

AREA $= a^2$

ROUNDS

AREA $= 0.7854d^2$

OVALS

AREA $= 0.7854\ ab$

HEXAGONS

AREA $= 3.464r^2$

RECTANGLES

AREA $= ab$

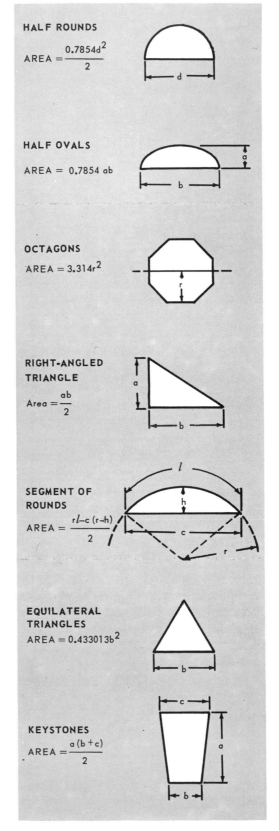

HALF ROUNDS

AREA $= \dfrac{0.7854d^2}{2}$

HALF OVALS

AREA $= 0.7854\ ab$

OCTAGONS

AREA $= 3.314r^2$

RIGHT-ANGLED TRIANGLE

Area $= \dfrac{ab}{2}$

SEGMENT OF ROUNDS

AREA $= \dfrac{rl - c\,(r-h)}{2}$

EQUILATERAL TRIANGLES

AREA $= 0.433013b^2$

KEYSTONES

AREA $= \dfrac{a\,(b+c)}{2}$

TRIANGLES

RIGHT-ANGLED

Known	Find	FORMULAS
a, c	A, B, b	$\sin A = \dfrac{a}{c}$, $\cos B = \dfrac{a}{c}$, $b = \sqrt{c^2 - a^2}$
	Area	$\dfrac{a}{2}\sqrt{c^2 - a^2}$
a, b	A, B, c	$\tan A = \dfrac{a}{b}$, $\tan B = \dfrac{b}{a}$, $c = \sqrt{a^2 + b^2}$
	Area	$\dfrac{ab}{2}$
A, a	B, b, c	$B = 90° - A$, $b = a \cot A$, $c = \dfrac{a}{\sin A}$
	Area	$\dfrac{a^2 \cot A}{2}$
A, b	B, a, c	$B = 90° - A$, $a = b \tan A$, $c = \dfrac{b}{\cos A}$
	Area	$\dfrac{b^2 \tan A}{2}$
A, c	B, a, b	$B = 90° - A$, $a = c \sin A$, $b = c \cos A$
	Area	$\dfrac{c^2 \sin A \cos A}{2} = \dfrac{c^2 \sin 2A}{4}$

$s = \dfrac{a + b + c}{2}$

OBLIQUE-ANGLED

Known	Find	FORMULAS
a, b, c	A	$\sin \frac{1}{2} A = \sqrt{\dfrac{(s\text{-}b)(s\text{-}c)}{bc}}$, $\cos \frac{1}{2} A = \sqrt{\dfrac{s(s\text{-}a)}{bc}}$, $\tan \frac{1}{2} A = \sqrt{\dfrac{(s\text{-}b)(s\text{-}c)}{s(s\text{-}a)}}$
	B	$\sin \frac{1}{2} B = \sqrt{\dfrac{(s\text{-}a)(s\text{-}c)}{ac}}$, $\cos \frac{1}{2} B = \sqrt{\dfrac{s(s\text{-}b)}{ac}}$, $\tan \frac{1}{2} B = \sqrt{\dfrac{(s\text{-}a)(s\text{-}c)}{s(s\text{-}b)}}$
	C	$\sin \frac{1}{2} C = \sqrt{\dfrac{(s\text{-}a)(s\text{-}b)}{ab}}$, $\cos \frac{1}{2} C = \sqrt{\dfrac{s(s\text{-}c)}{ab}}$, $\tan \frac{1}{2} C = \sqrt{\dfrac{(s\text{-}a)(s\text{-}b)}{s(s\text{-}c)}}$
	Area	$\sqrt{s(s\text{-}a)(s\text{-}b)(s\text{-}c)}$
a, A, B,	b, c	$b = \dfrac{a \sin B}{\sin A}$, $c = \dfrac{a \sin C}{\sin A} = \dfrac{a \sin (A + B)}{\sin A}$
	C	$C = 180° - (A + B)$
	Area	$\frac{1}{2} a b \sin C = \dfrac{a^2 \sin B \sin C}{2 \sin A}$
a, b, A	B	$\sin B = \dfrac{b \sin A}{a}$
	C	$C = 180° - (A + B)$
	c	$c = \dfrac{a \sin C}{\sin A} = \dfrac{b \sin C}{\sin B} = \sqrt{a^2 + b^2 - 2 a b \cos C}$
	Area	$\frac{1}{2} a b \sin C = \frac{1}{2} a c \sin B = \frac{1}{2} b c \sin A$
a, b, C	A	$\tan A = \dfrac{a \sin c}{b - a \cos c}$
	B	$B = 180° - (A + C)$, $\tan \frac{1}{2} (A - B) = \dfrac{a - b}{a + b} \cot \frac{1}{2} C$
	C	$c = \dfrac{a \sin c}{\sin A} = \sqrt{a^2 + b^2 - 2 a b \cos C}$
	Area	$\frac{1}{2} a b \sin C$

$a^2 = b^2 + c^2 - 2 bc \cos A$, $b^2 = a^2 + c^2 - 2 a c \cos B$,

$c^2 = a^2 + b^2 - 2 a b \cos C$

$\dfrac{a}{\sin A} = \dfrac{b}{\sin B} = \dfrac{c}{\sin C}$

VALUES OF FUNCTIONS FOR CERTAIN ANGLES

Angle deg.	Arc	Sin	Cos	Tan	Cot	Sec	Csc	Chord.
0	0	0	+1	0	∞	+1	∞	0
30	$1/6\,\pi$	$1/2$	$1/2\sqrt{3}$	$1/3\sqrt{3}$	$\sqrt{3}$	$2/3\sqrt{3}$	2	$\sqrt{2-\sqrt{3}}$
45	$1/4\,\pi$	$1/2\sqrt{2}$	$1/2\sqrt{2}$	+1	+1	$\sqrt{2}$	$\sqrt{2}$	$\sqrt{2-\sqrt{2}}$
60	$1/3\,\pi$	$1/2\sqrt{3}$	$1/2$	$\sqrt{3}$	$1/3\sqrt{3}$	2	$2/3\sqrt{3}$	1
90	$1/2\,\pi$	+1	0	∞	0	∞	+1	$\sqrt{2}$
120	$2/3\,\pi$	$1/2\sqrt{3}$	$-1/2$	$-\sqrt{3}$	$-1/3\sqrt{3}$	-2	$2/3\sqrt{3}$	$\sqrt{3}$
135	$3/4\,\pi$	$1/2\sqrt{2}$	$-1/2\sqrt{2}$	-1	-1	$-\sqrt{2}$	$\sqrt{2}$	$\sqrt{2+\sqrt{2}}$
150	$5/6\,\pi$	$1/2$	$-1/2\sqrt{3}$	$-1/3\sqrt{3}$	$-\sqrt{3}$	$-2/3\sqrt{3}$	2	$\sqrt{2+\sqrt{3}}$
180	π	0	-1	0	∞	-1	∞	2
210	$7/6\,\pi$	$-1/2$	$-1/2\sqrt{3}$	$1/3\sqrt{3}$	$\sqrt{3}$	$-2/3\sqrt{3}$	-2	$\sqrt{2+\sqrt{3}}$
225	$5/4\,\pi$	$-1/2\sqrt{2}$	$-1/2\sqrt{2}$	+1	+1	$-\sqrt{2}$	$-\sqrt{2}$	$\sqrt{2+\sqrt{2}}$
240	$4/3\,\pi$	$-1/2\sqrt{3}$	$-1/2$	$\sqrt{3}$	$1/3\sqrt{3}$	-2	$-2/3\sqrt{3}$	$\sqrt{3}$
270	$3/2\,\pi$	-1	0	∞	0	∞	-1	$\sqrt{2}$
300	$5/3\,\pi$	$-1/2\sqrt{3}$	$1/2$	$-\sqrt{3}$	$-1/3\sqrt{3}$	2	$-2/3\sqrt{3}$	1
315	$7/4\,\pi$	$-1/2\sqrt{2}$	$1/2\sqrt{2}$	-1	-1	$\sqrt{2}$	$-\sqrt{2}$	$\sqrt{2-\sqrt{2}}$
330	$11/6\,\pi$	$-1/2$	$1/2\sqrt{3}$	$-1/3\sqrt{3}$	$-\sqrt{3}$	$2/3\sqrt{3}$	-2	$\sqrt{2-\sqrt{3}}$
360	$2\,\pi$	0	+1	0	∞	+1	∞	0

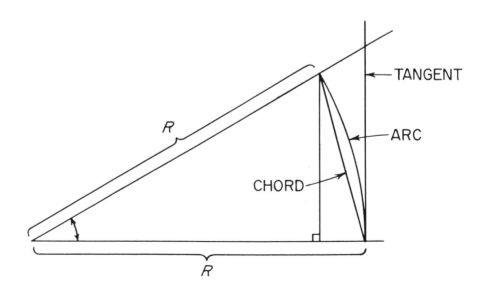

TRIGONOMETRIC FUNCTIONS

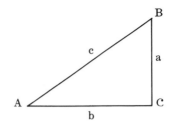

Fundamental Trigonometric Functions

$$\sin A = \frac{a}{c} \qquad \csc A = \frac{c}{a}$$

$$\cos A = \frac{b}{c} \qquad \sec A = \frac{c}{b}$$

$$\tan A = \frac{a}{b} \qquad \cot A = \frac{b}{a}$$

Functions of one angle

$$\sin^2 A + \cos^2 A = 1$$
$$\sec^2 A - \tan^2 A = 1$$
$$\csc^2 A - \cot^2 A = 1$$

Functions of the sum of two angles

$$\sin (A+B) = \sin A \cos B + \cos A \sin B$$
$$\cos (A+B) = \cos A \cos B - \sin A \sin B$$
$$\tan (A+B) = \frac{\tan A + \tan B}{1 - \tan A \tan B}$$
$$\cot (A+B) = \frac{\cot A \cot B - 1}{\cot B + \cot A}$$

Functions of the difference of two angles

$$\sin (A-B) = \sin A \cos B - \cos A \sin B$$
$$\cos (A-B) = \cos A \cos B + \sin A \sin B$$
$$\tan (A-B) = \frac{\tan A - \tan B}{1 + \tan A \tan B}$$
$$\cot (A-B) = \frac{\cot A \cot B + 1}{\cot B - \cot A}$$

Functions of one-half an angle

$$\sin \tfrac{1}{2}A = \frac{\sin A}{2\cos \tfrac{1}{2}A} = \pm \sqrt{\frac{1-\cos A}{2}}$$
$$\cos \tfrac{1}{2}A = \frac{\sin A}{2\sin \tfrac{1}{2}A} = \pm \sqrt{\frac{1+\cos A}{2}}$$
$$\tan \tfrac{1}{2}A = \frac{1-\cos A}{\sin A} = \pm \sqrt{\frac{1-\cos A}{1+\cos A}}$$
$$\cot \tfrac{1}{2}A = \pm \sqrt{\frac{1-\cos A}{1+\cos A}}$$

Functions of twice an angle

$$\sin 2A = 2 \sin A \cos A = \frac{2\tan A}{1+\tan^2 A}$$
$$\cos 2A = \cos^2 A - \sin^2 A = 1 - 2\sin^2 A$$
$$= 2\cos^2 A - 1 = \frac{1-\tan^2 A}{1+\tan^2 A}$$
$$\tan 2A = \frac{2\tan A}{1-\tan^2 A} = \frac{\sin 3A - \sin A}{\cos 3A + \cos A}$$
$$\cot 2A = \frac{\cot^2 A - 1}{2\cot A}$$

Functions of three times an angle

$$\sin 3A = 3\sin A - 4\sin^3 A$$
$$\cos 3A = 4\cos^3 A - 3\cos A$$
$$\tan 3A = \frac{3\tan A - \tan^3 A}{1 - 3\tan^2 A}$$
$$\cot 3A = \frac{\cot^3 A - 3\cot A}{3\cot^2 - 1}$$

Functions of angles squared

$$\sin^2 A = \frac{1-\cos 2A}{2}$$
$$\cos^2 A = \frac{1+\cos 2A}{2}$$
$$\tan^2 A = \frac{1-\cos 2A}{1+\cos 2A}$$
$$\cot^2 A = \frac{1+\cos 2A}{1-\cos 2A}$$
$$\sin^2 A - \sin^2 B = \sin (A+B) \sin (A-B)$$
$$\cos^2 A - \sin^2 B = \cos (A+B) \cos (A-B)$$

Functions — Relationships

$$\sin A = \frac{\cos A}{\cot A} - \frac{1}{\csc A} = \cos A \tan A = \sqrt{1-\cos^2 A}$$
$$\cos A = \frac{\sin A}{\tan A} = \frac{1}{\sec A} = \sin A \cot A = \sqrt{1-\sin^2 A}$$
$$\tan A = \frac{\sin A}{\cos A} = \frac{1}{\cot A} = \sin A \sec A$$
$$\cot A = \frac{\cos A}{\sin A} = \frac{1}{\tan A} = \cos A \csc A$$
$$\sec A = \frac{\tan A}{\sin A} = \frac{1}{\cos A}$$
$$\csc A = \frac{\cot A}{\sin A} = \frac{1}{\sin A}$$
$$\sin A + \sin B = 2\sin \tfrac{1}{2} (A+B) \cos \tfrac{1}{2} (A-B)$$
$$\sin A - \sin B = 2\cos \tfrac{1}{2} (A+B) \sin \tfrac{1}{2} (A-B)$$
$$\cos A + \cos B = 2\cos \tfrac{1}{2} (A+B) \cos \tfrac{1}{2} (A-B)$$
$$\cos A - \cos B = -2\sin \tfrac{1}{2} (A+B) \sin \tfrac{1}{2} (A-B)$$
$$\tan A + \tan B = \frac{\sin (A+B)}{\cos A \cos B}$$
$$\tan A - \tan B = \frac{\sin (A-B)}{\cos A \cos B}$$
$$\cot A + \cot B = \frac{\sin (A+B)}{\sin A \sin B}$$
$$\cot A - \cot B = \frac{\sin (B-A)}{\sin A \sin B}$$

For two signals having the same frequency, the phase can be determined by measuring the major and minor axes of the ellipse. The phase angle is equal to twice the angle whose tangent is the ratio of the major axis to the minor axis. The absolute accuracy of this method is dependent upon the phase in the horizontal and vertical amplifiers of the oscilloscope being equal and the care that is taken to make the horizontal and vertical amplitudes equal.

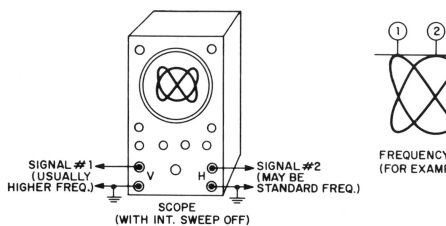

SIGNAL #1 (USUALLY HIGHER FREQ.)

V   H

SIGNAL #2 (MAY BE STANDARD FREQ.)

SCOPE (WITH INT. SWEEP OFF)

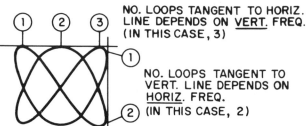

NO. LOOPS TANGENT TO HORIZ. LINE DEPENDS ON VERT. FREQ. (IN THIS CASE, 3)

NO. LOOPS TANGENT TO VERT. LINE DEPENDS ON HORIZ. FREQ. (IN THIS CASE, 2)

FREQUENCY RATIO = VERT. FREQ. : HORIZ. FREQ. :: 3:2
(FOR EXAMPLE; IF HORIZ. FREQ. IS 600Hz
THEN VERT. FREQ. IS 900Hz

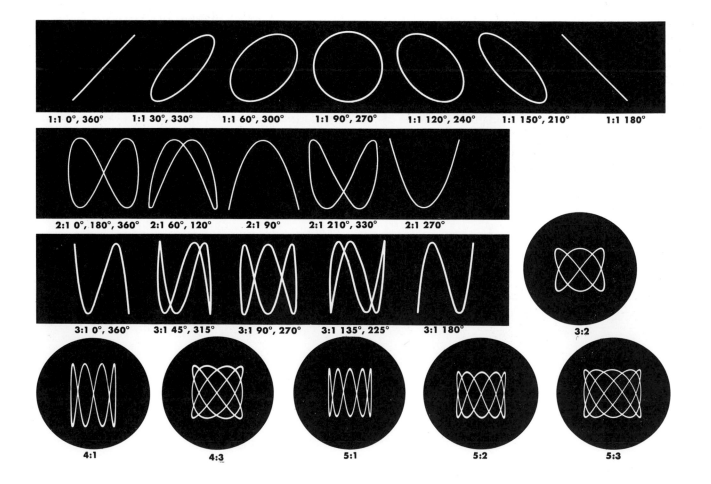

1:1 0°, 360°   1:1 30°, 330°   1:1 60°, 300°   1:1 90°, 270°   1:1 120°, 240°   1:1 150°, 210°   1:1 180°

2:1 0°, 180°, 360°   2:1 60°, 120°   2:1 90°   2:1 210°, 330°   2:1 270°

3:1 0°, 360°   3:1 45°, 315°   3:1 90°, 270°   3:1 135°, 225°   3:1 180°   3:2

4:1   4:3   5:1   5:2   5:3

PULSE PARAMETER NOMOGRAM

This normalized nomogram relates pulse rise time, repetition frequency, and pulse width to data channel bandwidth. To use the nomogram, connect a horizontal line through the selected bandwidth. The intersection with the other columns gives maximum pulse repetition frequency, minimum pulse width, and minimum risetime. For a given bandwidth, any combination of factors below the line can be used.

FOR EXAMPLE:

For a bandwidth of 10 MHz (10×10^6 Hz) the fastest risetime is 0.035×10^{-6} sec, the maximum pulse repetition frequency is 3.34×10^6 pulses per second, and the minimum pulse width is 0.15×10^{-6} sec.

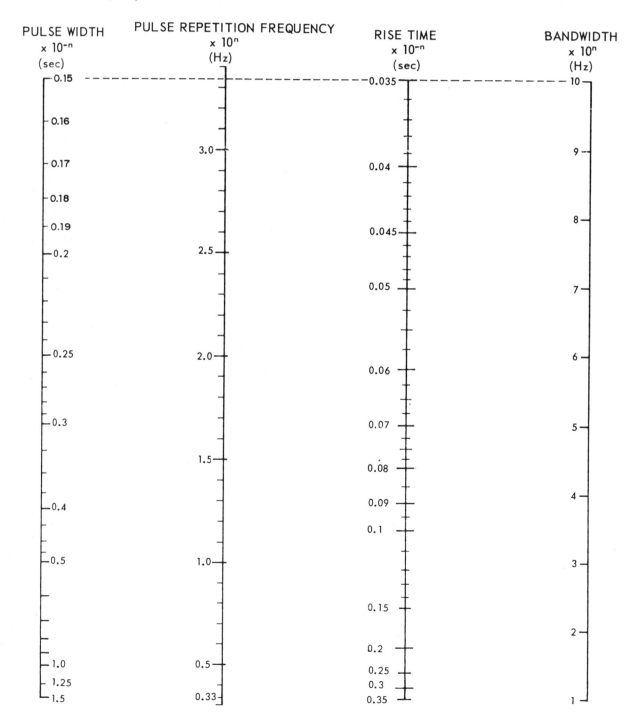

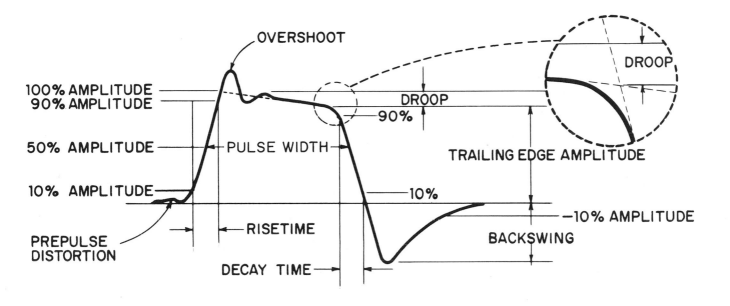

This scale is based on the formula $f = 1/T$. It converts between the frequency (f) and the period (T) of any recurrent waveform between 1 Hz and 10,000 GHz. It is useful where a large number of conversions are required as is the case when an oscilloscope with a time-calibrated sweep is used for frequency measurements.

FOR EXAMPLE:

The period of a 40-MHz signal is 25 nsec. The frequency of a signal with a period of 12.5 μsec is 80 kHz.

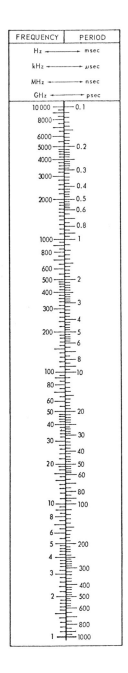

PHASE ANGLE, TIME INTERVAL, AND FREQUENCY NOMOGRAM

Time delay, phase angle, and frequency are related by the following formula:

$$t = \frac{10^2 \theta}{36f}$$

where

t is in milliseconds
θ is in degrees
f is in hertz

FOR EXAMPLE:

A phase angle of 90° between two 60-Hz wave shapes has a time interval of 4.16 msec.

Note: Corresponding right-hand frequency and time scales are used together as are left-hand frequency and time scales. The range of the nomogram can be extended by multiplying the frequency scale by any power of 10 and dividing the time scale by the same power of 10.

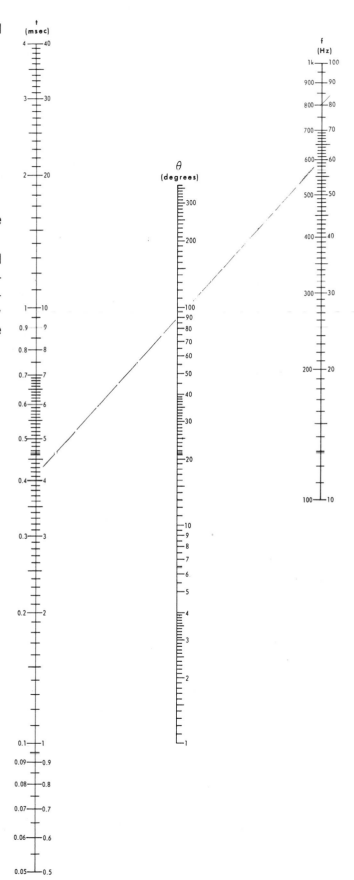

CHARACTERISTICS OF RECURRENT WAVEFORMS— RELATIONSHIP BETWEEN PEAK, RMS, AND AVERAGE VALUES

Description	Waveform	E_{rms}	E_{ave}
Alternating sine wave		$\dfrac{E_{peak}}{\sqrt{2}}$	$\dfrac{2E_{peak}}{\pi}$
Sawtooth wave		$\dfrac{E_{peak}}{\sqrt{3}}$	$\dfrac{E_{peak}}{2}$
Clipped sawtooth wave		$E_{peak}\sqrt{\dfrac{T_o}{3T}}$	$\dfrac{E_{peak}\,T_o}{2T}$
Square wave		$E_{peak}\sqrt{\dfrac{1}{2}}$	$\dfrac{E_{peak}}{2}$
Rectified sine wave		$\dfrac{E_{peak}}{\sqrt{2}}$	$\dfrac{2E_{peak}}{\pi}$
Clipped sine wave		$E_{peak}\sqrt{\dfrac{T_o}{2T}}\quad$ or if $T = T_o$ $\dfrac{E_{peak}}{2}$	$\dfrac{E_{peak}}{\pi}$
Alternating square wave		E_{peak}	E_{peak}
Rectangular wave		$E_{peak}\sqrt{\dfrac{T_o}{T}}$	$\dfrac{E_{peak}\,T_o}{T}$
Triangular wave		$\dfrac{E_{peak}}{\sqrt{3}}$	$\dfrac{E_{peak}}{2}$

FOURIER CONTENT OF COMMON PERIODIC WAVEFORMS

The Fourier content of five common periodic waveforms, out to the seventh harmonic, is given in this table. Magnitudes only are tabulated—not phase relationships. The magnitudes are those of the voltage waveform, followed by the corresponding percentage values in parentheses. If energy content is desired, these values must be squared. Note that there are no even harmonics present in any of the symmetrical waveforms.

Waveform	Name	Harmonic Composition (magnitude)						
		Fund.	2nd	3rd	4th	5th	6th	7th
Square Wave	Square Wave	$\frac{4}{\pi}E$ (127%)	0 (0%)	$\frac{4}{3\pi}E$ (42.5%)	0 (0%)	$\frac{4}{5\pi}E$ (25.5%)	0 (0%)	$\frac{4}{7\pi}E$ (18.2%)
Triangular Wave	Triangular Wave	$\frac{8}{\pi^2}E$ (81%)	0 (0%)	$\frac{8}{9\pi^2}E$ (9%)	0 (0%)	$\frac{8}{25\pi^2}E$ (3.2%)	0 (0%)	$\frac{8}{49\pi^2}E$ (1.6%)
Sawtooth Wave	Sawtooth Wave	$\frac{2}{\pi}E$ (63.6%)	$\frac{1}{\pi}E$ (31.8%)	$\frac{2}{3\pi}E$ (21.2%)	$\frac{1}{2\pi}E$ (15.9%)	$\frac{2}{5\pi}E$ (12.7%)	$\frac{1}{3\pi}E$ (10.6%)	$\frac{2}{7\pi}E$ (9.1%)
Half-Wave Rectifier Output	Half-Wave Rectifier Output	$\frac{1}{\pi}E$ (31.8%)	$\frac{2}{3\pi}E$ (21.2%)	0 (0%)	$\frac{2}{15\pi}E$ (4.2%)	0 (0%)	$\frac{2}{35\pi}E$ (1.8%)	0 (0%)
Full-Wave Rectifier Output	Full-Wave Rectifier Output	$\frac{2}{\pi}E$ (63.6%)	$\frac{4}{3\pi}E$ (42.3%)	0 (0%)	$\frac{4}{15\pi}E$ (8.5%)	0 (0%)	$\frac{4}{35\pi}E$ (3.6%)	0 (0%)

SYNTHESIS OF A SQUARE WAVE

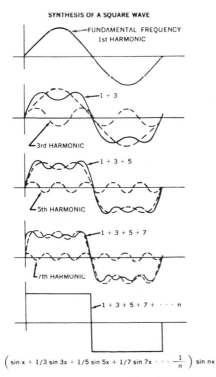

$$\left(\sin x + 1/3 \sin 3x + 1/5 \sin 5x + 1/7 \sin 7x \cdots \frac{1}{n} \right) \sin nx$$

SYNTHESIS OF A TRIANGULAR WAVE

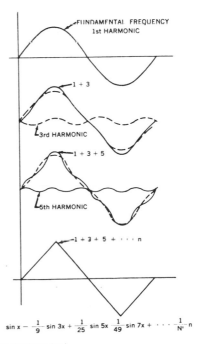

$$\sin x - \frac{1}{9} \sin 3x + \frac{1}{25} \sin 5x - \frac{1}{49} \sin 7x + \cdots \frac{1}{N} n$$

(From May 1968 *Electronic Instrument Digest*. © Milton S. Kiver Publications, Inc.)

CONVERSIONS FROM dB AND dBm TO VOLTAGE AND POWER RATIOS, AND FROM dBm TO POWER AND VOLTAGE LEVELS

Ratio in db or dBm	Relationships for Either dB or dBm				Relationships for dBm Only			
	Voltage Ratio (per unit)	Power Ratio (per unit)	Voltage Ratio (per cent)	Power Ratio (per cent)	Power (Referred to 1 mW)	Voltage Across 50 ohms	Voltage Across 70 ohms	Voltage Across 600 ohms
+120.0	10^6	10^{12}	——	——	10 GW	224 kV	265 kV	775 kV
+80.0	10^4	10^8	——	——	100 kW	2.24 kV	2.65 kV	7.75 kV
+60.0	10^3	10^6	——	——	1 kW	224 V	265 V	775 V
+50.0	316	10^5	——	——	100 W	70.7 V	83.7 V	245 V
+40.0	100	10^4	——	——	10.0 W	22.4 V	26.5 V	77.5 V
+30.0	31.6	10^3	3160	——	1.00 W	7.07 V	8.37 V	24.5 V
+20.0	10.00	100.0	1000	——	100 mW	2.24 V	2.65 V	7.75 V
+17.0	7.08	50.1	708	5010	50 mW	1.59 V	1.88 V	5.49 V
+13.98	5.00	25.0	500	2500	25 mW	1.12 V	1.325 V	3.875 V
+12.04	4.00	16.0	400	1600	16 mW	895 mV	1.060 V	3.100 V
+9.54	3.00	9.00	300	900	9 mW	672 mV	795 mV	2.325 V
+6.02	2.00	4.00	200	400	4 mW	448 mV	530 mV	1.550 V
+3.01	1.41	2.00	141	200	2 mW	316 mV	374 mV	1.092 V
+2.00	1.26	1.58	126	158	1.26 mW	282 mV	334 mV	976 mV
+1.00	1.12	1.26	112	126	1.12 mW	251 mV	297 mV	868 mV
0.00	1.00	1.000	100	100	1.00 mW	224 mV	265 mV	775 mV
—1.00	0.893	0.793	89.3	79.3	790 μW	201 mV	237 mV	693 mV
—2.00	0.793	0.633	79.3	63.3	630 μW	178 mV	215 mV	615 mV
—3.01	0.707	0.500	70.7	50.0	500 μW	158 mV	187 mV	548 mV
—6.02	0.500	0.250	50.0	25.0	250 μW	114 mV	133 mV	388 mV
—9.54	0.333	0.111	33.3	11.1	110 μW	74.5 mV	88.3 mV	258 mV
—12.04	0.250	0.063	25.0	6.3	62.5 μW	56.0 mV	66.2 mV	194 mV
—13.98	0.200	0.040	20.0	4.0	40.0 μW	44.8 mV	53.0 mV	155 mV
—17.0	0.141	0.020	14.1	2.0	20.0 μW	31.6 mV	37.4 mV	109 mV
—20.0	0.100	0.010	10.0	1.0	10.0 μW	22.4 mV	26.5 mV	77.5 mV
—30.0	0.032	0.001	3.16	0.1	1.0 μW	7.07 mV	8.37 mV	24.5 mV
—40.0	0.010	10^{-4}	1.000	0.01	100 nW	2.24 mV	2.65 mV	7.75 mV
—50.0	0.0032	10^{-5}	0.316	0.001	10 nW	707 μV	837 μV	2.45 mV
—60.0	0.001	10^{-6}	0.100	10^{-4}	1 nW	224 μV	265 μV	775 μV
—80.0	10^{-4}	10^{-8}	0.010	10^{-6}	10 pW	22.4 μV	26.5 μV	77.5 μV
—120.0	10^{-6}	10^{-12}	——	——	1 fW	224. nV	265 nV	775 nV

(From May 1968 *Electronic Instrument Digest.* © Milton S. Kiver Publications, Inc.)

CONVERSIONS FROM dB AND dBm TO VOLTAGE AND POWER RATIOS, AND FROM dBm TO POWER AND VOLTAGE LEVELS (Continued)

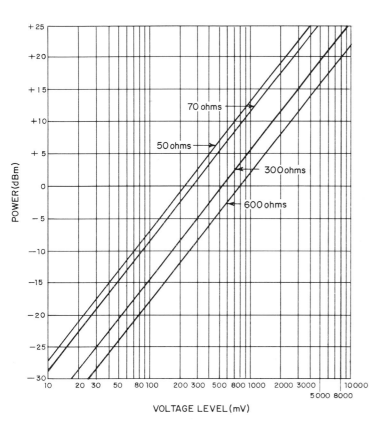

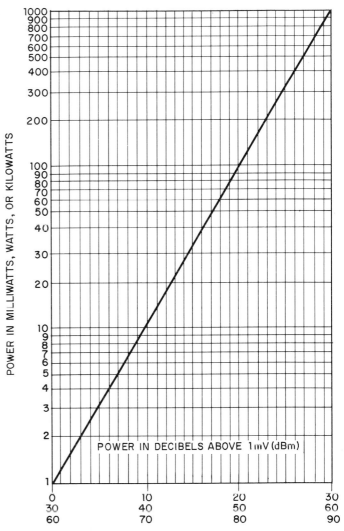

(From May 1968 *Electronic Instrument Digest*. © Milton S. Kiver Publications, Inc.)

With these two nomograms dB gain or loss of any equipment can be determined (even if input and output impedances differ) if input and output voltages and resistances can be measured. The nomograms cover a power range of 10,000 to 1, a voltage range of 100 to 1, and a decibel range from +40 to -40 dB. Voltage and resistance scales of nomogram I bearing the same suffix are used together.

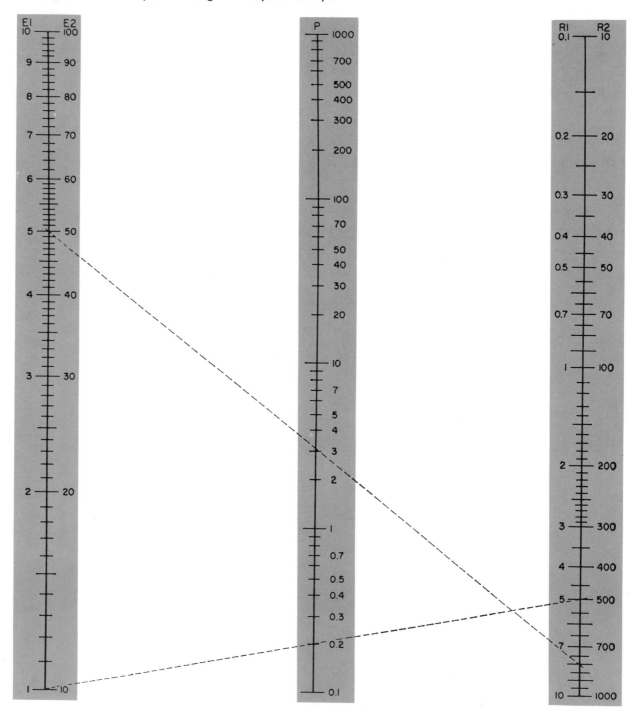

DECIBEL NOMOGRAMS (Continued)

FOR EXAMPLE:
Determine the gain of an amplifier that produces an output of 5 V across 8 ohms with a 10-V signal applied to its 500-ohm input. From nomogram I, the input power is 0.2 W and the output power is 3.1 W. Connecting input and output power on nomogram II shows the amplifier gain to be slightly less than 12 dB.

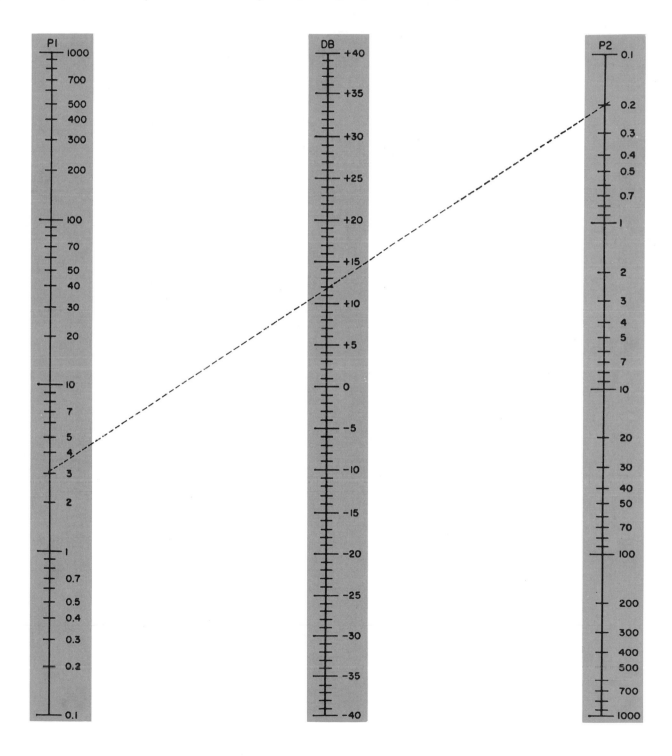

LETTER SYMBOLS FOR QUANTITIES USED IN ELECTRICAL SCIENCE AND ELECTRICAL ENGINEERING

Extracted from IEEE Standard No. 280

These tables list quantities grouped in several categories, and give quantity symbols, units based on the International System,* and unit symbols.

Those quantity symbols that are separated by a comma are alternatives on equal standing. Where two symbols for a quantity are separated by three dots (...), the second is a reserve symbol, which is to be used only where there is specific need to avoid a conflict. As a rule the tables do not indicate the vectorial or tensorial character that some of the quantities may have.

The International System of Units (Système International d'Unités) is the coherent system of units based on the following units and quantities:

Unit	Quantity
meter	length
kilogram	mass
second	time
ampere	electric current
kelvin	temperature
candela	luminous intensity
radian	plane angle
steradian	solid angle

This system was named (and given the international designation SI) in 1960 by the Conférence Générale des Poids et Mesures (CGPM). The SI units include as subsystems the MKS system of units, which covers mechanics, and the MKSA or Giorgi system, which covers mechanics, electricity, and magnetism.

*The name of the unit is given as a further guide to the definition of the symbol. A quantity shall be represented by the standard letter symbol appearing in the table regardless of the system of units in which the quantity is expressed.

214

LETTER SYMBOLS FOR QUANTITIES USED IN ELECTRICAL SCIENCE AND ELECTRICAL ENGINEERING (Continued)

Item Quantity	Quantity Symbol[a]	Unit Based on International System	Unit Symbol	Remarks
1. Space and Time angle, plane	$\alpha, \beta, \gamma, \theta, \Phi, \psi$	radian	rad	Other Greek letters are permitted where no conflict results.
angle, solid	$\Omega \ldots \omega$	steradian	sr	
length	l	meter	m	
breadth, width	b	meter	m	
height	h	meter	m	
thickness	d, δ	meter	m	
radius	r	meter	m	
diameter	d	meter	m	
length of path line segment	s	meter	m	
wavelength	λ	meter	m	
wave number	$\sigma \ldots \bar{\nu}$	reciprocal meter	m^{-1}	$\sigma = 1/\lambda$ The symbol $\bar{\nu}$ is used in spectroscopy.
circular wave number angular wave number	k	radian per meter	rad/m	$k = 2\pi/\lambda$
area	$A \ldots S$	square meter	m^2	
volume	V, v	cubic meter	m^3	
time	t	second	s	
period time of one cycle	T	second	s	
time constant	$\tau \ldots T$	second	s	
frequency	$f \ldots \nu$	hertz	Hz	The name *cycle per second* is also used for this unit. The symbol for the unit *cycle per second* is c/s; the use of cps as a symbol is deprecated. The symbol f is used in circuit theory, sound, and mechanics; ν is used in optics and quantum theory.
speed of rotation rotational frequency	n	revolution per second	r/s	
angular frequency	ω	radian per second	rad/s	$\omega = 2\pi f$
angular velocity	ω	radian per second	rad/s	
complex (angular) frequency oscillation constant	$p \ldots s$	reciprocal second	s^{-1}	$p = -\delta + j\omega$
angular acceleration	α	radian per second squared	rad/s^2	
velocity	ν	meter per second	m/s	
speed of propagation of electromagnetic waves	c	meter per second	m/s	In vacuum, c_0; *see* 8.1.
acceleration (linear)	a	meter per second squared	m/s^2	
acceleration of free fall gravitational acceleration	g	meter per second squared	m/s^2	Standard value, g_n; *see* 8.10.
damping coefficient	δ	neper per second	Np/s	If F is a function of time given by $F = Ae^{-\delta t} \sin(2\pi t/T)$,

[a]Commas separate symbols on equal standing. Where two symbols are separated by three dots the second is a reserve symbol and is to be used only when there is specific need to avoid a conflict. See Introduction to the Tables.

Item	Quantity	Quantity Symbol[a]	Unit Based on International System	Unit Symbol	Remarks
	logarithmic decrement	Λ	(numeric)		then δ is the damping coefficient. $\Lambda = T\delta$, where T and δ are as given in the equation of 1.28.
	attenuation coefficient	α	neper per meter	Np/m	
	phase coefficient	β	radian per meter	rad/m	
	propagation coefficient	γ	reciprocal meter	m^{-1}	$\gamma = \alpha + j\beta$.
2. Mechanics[b]					
	mass	m	kilogram	kg	
	(mass) density	ρ	kilogram per cubic meter	kg/m^3	Mass divided by volume.
	momentum	ρ	kilogram meter per second	$kg \cdot m/s$	
	moment of inertia	I, J	kilogram meter squared	$kg \cdot m^2$	
	second (axial) moment of area	I, I_a	meter to the fourth power	m^4	Quantities 2.4a and 2.4b should be distinguished from 2.4. They have often been given the name "moment of inertia."
	second (polar) moment of area	J, I_p	meter to the fourth power	m^4	
	force	F	newton	N	
	weight	W	newton	N	Varies with acceleration of free fall.
	weight density	γ	newton per cubic meter	N/m^3	Weight divided by volume.
	moment of force	M	newton meter	$N \cdot m$	
	torque	$T \ldots M$	newton meter	$N \cdot m$	
	pressure	p	newton per square meter	N/m^2	The name *pascal* has been suggested for this unit.
	normal stress	σ	newton per square meter	N/m^2	
	shear stress	τ	newton per square meter	N/m^2	
	stress tensor	σ	newton per square meter	N/m^2	
	linear strain	ϵ	(numeric)		
	shear strain	γ	(numeric)		
	strain tensor	ϵ	(numeric)		
	volume strain	θ	(numeric)		
	Poisson's ratio	μ, ν	(numeric)		Lateral contraction divided by elongation.
	Young's modulus modulus of elasticity	E	newton per square meter	N/m^2	$E = \sigma/\epsilon$
	shear modulus modulus of rigidity	G	newton per square meter	N/m^2	$G = \tau/\gamma$
	bulk modulus	K	newton per square meter	N/m^2	$K = -p/\theta$
	work	W	joulé	J	
	energy	E, W	joule	J	U is recommended in thermodynamics for internal energy and for blackbody radiation.

[b]The units and corresponding unit symbols are included for use in electrical science and electrical engineering. In mechanics and mechanical engineering other units and corresponding unit symbols are also used. (USAS Y10.3 now being revised.)

LETTER SYMBOLS FOR QUANTITIES USED IN ELECTRICAL SCIENCE AND ELECTRICAL ENGINEERING (Continued)

Item	Quantity	Quantity Symbol[a]	Unit Based on International System	Unit Symbol	Remarks
	energy (volume) density	w	joule per cubic meter	J/m^3	
	power	P	watt	W	Rate of energy transfer. $W = J/s$
	efficiency	η	(numeric)		
3. Heat[c]					
	absolute temperature thermodynamic temperature	$T \ldots \Theta$	kelvin	K	In 1967 the CGPM voted to give the name *kelvin* to the SI unit of temperature, which was formerly called *degree Kelvin*, and to assign it the symbol K (without the symbol $^\circ$).
	temperature customary temperature	$t \ldots \theta$	degree Celsius	°C	The symbol °C is printed without space between ° and the letter that follows. The word *centigrade* has been abandoned as the name of a temperature scale. The units of temperature interval or difference are identical on the Kelvin and Celsius scales. The name kelvin and symbol K were adopted by the CGPM. The name *degree* and the symbol *deg* are also used. When it is necessary to distinguish between the Fahrenheit degree and the Celsius degree, the symbols deg F and deg C may be used.
	heat	Q	joule	J	
	internal energy	U	joule	J	
	heat flow rate	$\Phi \ldots q$	watt	W	Heat crossing a surface divided by time.
	temperature coefficient	α	reciprocal kelvin	K^{-1}	A temperature coefficient is not completely defined unless the quantity that changes is specified (e.g., resistance, length, pressure). The pressure (temperature) coefficient is designated by β; the cubic expansion (temperature) coefficient, by α, β, or γ.
	thermal diffusivity	α	square meter per second	m^2/s	

[c]The units and corresponding unit symbols are included for use in electrical science and engineering. In mechanical engineering other units and corresponding unit symbols are also used. (Cf. USAS Y10.4.)

LETTER SYMBOLS FOR QUANTITIES USED IN ELECTRICAL SCIENCE AND ELECTRICAL ENGINEERING (Continued)

Item	Quantity	Quantity Symbol[a]	Unit Based on International System	Unit Symbol	Remarks
	thermal conductivity	$\lambda \ldots k$	watt per meter kelvin	$W/(m \cdot K)$	
	thermal conductance	G_θ	watt per kelvin	W/K	
	thermal resistivity	ρ_θ	meter kelvin per watt	$m \cdot K/W$	
	thermal resistance	R_θ	kelvin per watt	K/W	
	thermal capacitance heat capacity	C_θ	joule per kelvin	J/K	
	thermal impedance	Z_θ	kelvin per watt	K/W	
	specific heat capacity	c	joule per kelvin kilogram	$J/(K \cdot kg)$	Heat capacity divided by mass.
	entropy	S	joule per kelvin	J/K	
	specific entropy	s	joule per kelvin kilogram	$J/(K \cdot kg)$	Entropy divided by mass.
	enthalpy	H	joule	J	
4. Radiation and Light					
	radiant intensity	$I \ldots I_e$	watt per steradian	W/sr	
	radiant power radiant flux	$P, \Phi \ldots \Phi_e$	watt	W	
	radiant energy	$W, Q \ldots Q_e$	joule	J	The symbol U is used for the special case of blackbody radiant energy.
	radiance	$L \ldots L_e$	watt per steradian square meter	$W/(sr \cdot m^2)$	
	radiant exitance	$M \ldots M_e$	watt per square meter	W/m^2	
	irradiance	$E \ldots E_e$	watt per square meter	W/m^2	
	luminous intensity	$I \ldots I_v$	candela	cd	
	luminous flux	$\Phi \ldots \Phi_v$	lumen	lm	
	quantity of light	$Q \ldots Q_v$	lumen second	$lm \cdot s$	
	luminance	$L \ldots L_v$	candela per square meter	cd/m^2	The name nit is sometimes used for this unit.
	luminous exitance	$M \ldots M_v$	lumen per square meter	lm/m^2	
	illuminance illumination	$E \ldots E_v$	lux	lx	$lx = lm/m^2$
	luminous efficacy	$K(\lambda)$	lumen per watt	lm/W	(λ) is not part of the basic symbol but indicates that luminous efficacy is a function of wavelength.
	total luminous efficacy	K, K_t	lumen per watt	lm/W	$K = \Phi/P$
	refractive index index of refraction	n	(numeric)		
	emissivity	$e(\lambda)$	(numeric)		(λ) is not part of the basic symbol but indicates that emissivity is a function of wavelength.
	total emissivity	e, ϵ_t	(numeric)		
	absorptance	$\alpha(\lambda)$	(numeric)		(λ) is not part of the basic symbol but indicates that the absorptance is a function of wavelength.
	transmittance	$\tau(\lambda)$	(numeric)		(λ) is not part of the basic symbol but indicates that the transmittance is a

LETTER SYMBOLS FOR QUANTITIES USED IN ELECTRICAL SCIENCE AND ELECTRICAL ENGINEERING (Continued)

Item	Quantity	Quantity Symbol[a]	Unit Based on International System	Unit Symbol	Remarks
	reflectance	$\rho(\lambda)$	(numeric)		function of wavelength. (λ) is not part of the basic symbol but indicates that the reflectance is a function of wavelength.
5. Fields and Circuits					
	electric charge quantity of electricity	Q	coulomb	C	
	linear density of charge	λ	coulomb per meter	C/m	
	surface density of charge	σ	coulomb per square meter	C/m^2	
	volume density of charge	ρ	coulomb per cubic meter	C/m^3	
	electric field strength	$E \ldots K$	volt per meter	V/m	
	electrostatic potential potential difference	$V \ldots \phi$	volt	V	
	retarded scalar potential	V_r	volt	V	
	voltage electromotive force	$V, E \ldots U$	volt	V	
	electric flux	Ψ	coulomb	C	
	electric flux density (electric) displacement	D	coulomb per square meter	C/m^2	
	capacitivity permittivity absolute permittivity	ϵ	farad per meter	F/m	Of vacuum, e_v.
	relative capacitivity relative permittivity dielectric constant	ϵ_r, κ	(numeric)		
	complex relative capacitivity complex relative permittivity complex dielectric constant	$\epsilon_r{}^*, \kappa^*$	(numeric)		$e_r{}^* = e_r' - je_r''$ e_r'' is positive for lossy materials. The complex absolute permittivity ϵ^* is defined in analogous fashion.
	electric susceptibility	$\chi_e \ldots \epsilon_i$	(numeric)		$\chi_e = \epsilon_r - 1$
	electrization	$E_i \ldots K_i$	volt per meter	V/m	$E_i = (D/\Gamma_e) - E$
	electric polarization	P	coulomb per square meter	C/m^2	$P = D - \Gamma_e E$
	electric dipole moment	p	coulomb meter	C·m	
	(electric) current	I	ampere	A	
	current density	$J \ldots S$	ampere per square meter	A/m^2	
	linear current density	$A \ldots \alpha$	ampere per meter	A/m	Current divided by the breadth of the conducting sheet.
	magnetic field strength	H	ampere per meter	A/m	

Item	Quantity	Quantity Symbol[d]	Unit Based on International System	Unit Symbol	Remarks
	magnetic (scalar) potential magnetic potential difference	U, U_m	ampere	A	
	magnetomotive force	$F, F_m \ldots \mathfrak{F}$	ampere	A	
	magnetic flux	Φ	weber	Wb	
	magnetic flux density magnetic induction	B	tesla	T	$T = Wb/m^2$
	magnetic flux linkage	Λ	weber	Wb	
	(magnetic) vector potential	A	weber per meter	Wb/m	
	retarded (magnetic) vector potential	A_r	weber per meter	Wb/m	
	(magnetic) permeability absolute permeability	μ	henry per meter	H/m	Of vacuum, μ_v.
	relative (magnetic) permeability	μ_t	(numeric)		
	initial (relative) permeability	μ_o	(numeric)		
	complex relative permeability	μ_r^*	(numeric)		$\mu_r^* = \mu_t' - j\mu_r''$ μ_r'' is positive for lossy materials. The complex absolute permeability μ^* is defined in analogous fashion.
	magnetic susceptibility	$\chi_m \ldots \mu_i$	(numeric)		$\chi_m = \mu_r - 1$
	reluctivity	ν	meter per henry	m/H	$\nu = 1/\mu$
	magnetization	H_i, M	ampere per meter	A/m	$H_i = (B/\Gamma_m) - H$
	magnetic polarization intrinsic magnetic flux density	J, B_i	tesla	T	$J = B - \Gamma_m H$
	magnetic (area) moment	m	ampere meter squared	A·m²	The vector product $m \times B$ is equal to the torque.
	capacitance	C	farad	F	
	elastance	S	reciprocal farad	F⁻¹	$S = 1/C$
	(self) inductance	L	henry	H	
	reciprocal inductance	Γ	reciprocal henry	H⁻¹	
	mutual inductance	L_{ij}, M_{ij}	henry	H	If only a single mutual inductance is involved, M may be used without subscripts.
	coupling coefficient	$k \ldots \kappa$	(numeric)		$k = L_{ij}(L_i L_j)^{-1/2}$
	leakage coefficient	σ	(numeric)		$\sigma = 1 - k^2$
	number of turns (in a winding)	N, n	(numeric)		
	number of phases	m	(numeric)		
	turns ratio	$n \ldots n_*$	(numeric)		
	transformer ratio	a	(numeric)		Square root of the ratio of secondary to primary self inductance. Where

Item	Quantity	Quantity Symbol[d]	Unit Based on International System	Unit Symbol	Remarks
					the coefficient of coupling is high, $a \approx n_*$.
	resistance	R	ohm	Ω	
	resistivity	ρ	ohm meter	$\Omega \cdot m$	
	volume resistivity				
	conductance	G	mho	mho	$G = \mathrm{Re}\, Y$ The IEC has adopted the name *siemens* (S) for this unit. The CGPM has not yet adopted a name.
	conductivity	γ, σ	mho per meter	mho/m	$\gamma = 1/\rho$ The symbol σ is used in field theory, as γ is there used for the propagation coefficient. See remark for 5.50.
	reluctance	$R, R_m \ldots \mathcal{R}$	reciprocal henry	H^{-1}	Magnetic potential difference divided by magnetic flux.
	permeance	$P, P_m \ldots \mathcal{P}$	henry	H	$P_m = 1/R_m$
	impedance	Z	ohm	Ω	$Z = R + jX$
	reactance	X	ohm	Ω	
	capacitive reactance	X_C	ohm	Ω	For a pure capacitance, $X_C = -1/\omega C$
	inductive reactance	X_L	ohm	Ω	For a pure inductance, $X_L = \omega L$
	quality factor	Q	(numeric)		$Q = \dfrac{2\pi \,(\text{peak energy stored})}{(\text{energy dissipated per cycle})}$ For a simple reactor, $Q = \lvert X \rvert / R$
	admittance	Y	mho	mho	$Y = 1/Z = G + jB$ See remark for 5.50.
	susceptance	B	mho	mho	$B = \mathrm{Im}\, Y$ See remark for conductance.
	loss angle	δ	radian	radian	$\delta = \arctan(R/\lvert X \rvert)$
	active power	P	watt	W	
	reactive power	$Q \ldots P_q$	var	var	
	apparent power	$S \ldots P_s$	voltampere	VA	
	power factor	$\cos\phi \ldots F_p$	(numeric)		
	reactive factor	$\sin\phi \ldots F_q$	(numeric)		
	input power	P_i	watt	W	
	output power	P_o	watt	W	
	Poynting vector	S	watt per square meter	W/m^2	
	characteristic impedance surge impedance	Z_0	ohm	Ω	
	intrinsic impedance of a medium	η	ohm	Ω	
	voltage standing-wave ratio	S	(numeric)		
	resonance frequency	f_r	hertz	Hz	The name *cycle per second* (c/s) is also used for this unit.
	critical frequency cutoff frequency	f_c	hertz	Hz	
	resonance angular frequency	ω_r	radian per second	rad/s	
	critical angular frequency cutoff angular frequency	ω_c	radian per second	rad/s	

Item	Quantity	Quantity Symbol[d]	Unit Based on International System	Unit Symbol	Remarks
	resonance wavelength	λ_r	meter	m	
	critical wavelength	λ_c	meter	m	
	cutoff wavelength				
	wavelength in a guide	λ_g	meter	m	
	hysteresis coefficient	k_h	(numeric)		
	eddy-current coefficient	k_e	(numeric)		
	phase angle	ϕ, θ	radian	rad	
	phase difference				
6.	Electronics and Telecommunication carrier frequency	f_c	hertz	Hz	The name *cycle per second* (c/s) is also used for this unit.
	instantaneous frequency	f, f_i	hertz	Hz	
	intermediate frequency	f_i, f_{if}	hertz	Hz	
	modulation frequency	f_m	hertz	Hz	
	pulse repetition frequency	f_p	hertz	Hz	
	frequency deviation	f_d	hertz	Hz	
	Doppler frequency shift	f_D	hertz	Hz	
	pulse duration	t_p	second	s	
	rise time (of a pulse)	t_r	second	s	
	fall time (of a pulse)	t_f	second	s	
	decay time (of a pulse)				
	duty factor	D	(numeric)		$D = t_p f_p$
	pulse duty factor				
	phase propagation time	t_ϕ	second	s	
	group propagation time	t_g	second	s	
	duration of a signal element	τ	second	s	
	signaling speed	$1/\tau$	baud	Bd	
	cathode-heating time	t_k	second	s	
	deionization time	t_d	second	s	
	ionization time	t_i	second	s	
	form factor	k_f	(numeric)		
	peak factor	k_{pk}	(numeric)		
	distortion factor	d	(numeric)		
	modulation factor (AM)	m	(numeric)		
	modulation index (FM)	η	(numeric)		
	signal power	P_s, S	watt	W	
	noise power	P_n, N	watt	W	
	noise-power density	N_o	watt per hertz	W/Hz	
	energy of a signal element	E	joule	J	
	signal-to-noise power, ratio[e]	$R, S/N$	(numeric)		$R = P_s/P_n$
	elementary signal-to-noise ratio[e]	R, R_e	(numeric)		$R_e = E/N_o$

LETTER SYMBOLS FOR QUANTITIES USED IN ELECTRICAL SCIENCE AND ELECTRICAL ENGINEERING (Continued)

Item Quantity	Quantity Symbol[d]	Unit Based on International System	Unit Symbol	Remarks
gain (power)[e]	G	(numeric)		
amplification (current or voltage)[e]	A	(numeric)		
noise factor[e]	F	(numeric)		
noise figure				
bandwidth	B	hertz	Hz	See remark for carrier frequency.
feedback transfer ratio	β	(numeric)		
critical frequency of an ionized layer	f_c	hertz	Hz	See remark for carrier frequency.
plasma frequency	f_n	hertz	Hz	See remark for carrier frequency.
ion (number) density	$n^+; n^-$	ion per cubic meter	m^{-3}	
mobility (of a charge carrier in a medium)	μ	square meter per volt second	$m^2/(V \cdot s)$	
rate of production of electrons per unit volume	q	electron per cubic meter second	$m^{-3} s^{-1}$	
recombination coefficient	α	cubic meter per second	m^3/s	
effective attachment coefficient	β	reciprocal second	s^{-1}	
μ-factor	μ_{ij}	(numeric)		$\mu_{ij} = \|\partial v_i/\partial v_j\|$ where v_i and v_j are the voltages of the ith and jth electrodes, and the current to the ith electrode and all electrode voltages other than v_i and v_j are held constant.
amplification factor	μ	(numeric)		The amplification factor is the μ-factor for the anode and control-grid electrodes.
interelectrode transadmittance	y_{ij}	mho	mho	See remark for conductance.
interelectrode transconductance	g_{ij}	mho	mho	The real part of the interelectrode transmittance. See conductance.
mutual conductance transconductance	g_m, g_{ag}	mho	mho	The mutual conductance is the control-grid-to-anode transconductance. See conductance
conversion transconductance	g_c	mho	mho	Transconductance defined for a heterodyne conversion transducer. See conductance.
plate resistance anode resistance	r_a	ohm	Ω	
anode dissipation power	P_a	watt	W	
grid dissipation power	P_g	watt	W	
saturation current of a cathode	I_s	ampere	A	

Item	Quantity	Quantity Symbol[d]	Unit Based on International System	Unit Symbol	Remarks
	secondary-emission ratio	δ	(numeric)		
	temperature of mercury condensate	T_{Hg}	kelvin	K	
	radiant sensitivity of a phototube, dynamic	s	ampere per watt	A/W	
	radiant sensitivity of a phototube, static	S	ampere per watt	A/W	
	luminous sensitivity of a phototube, dynamic	s_v	ampere per lumen	A/lm	
	luminous sensitivity of a phototube, static	S_v	ampere per lumen	A/lm	
	Subscripts, electronic tubes				
	anode	a			
	cathode	k			
	grid	g			
	heater	h			
	filament (emitting)	f			
	fluorescent screen or target	t			
	external conducting coating	M			
	internal conducting coating	m			
	deflector electrode	x or y			
	internal shield	s			
	wave-retardation electrode	wr			
	beam-forming plate	bp			
	switch, moving contact	cm			
	switch, fixed contact	cf			
	Subscripts, semiconductor devices				
	emitter terminal	E, e			
	base terminal	B, b			
	collector terminal	C, c			
	anode	A, a			
	cathode	K, k			
	control terminal (gate)	G, g			
	junction (general)	J, j			
7. Machines and Power Engineering					
	synchronous speed (of rotation)	n_1	revolution per second	r/s	
	synchronous angular frequency	ω_1	radian per second	rad/s	
	slip	s	(numeric)		
	number of poles	$p, 2p$	(numeric)		The IEC gives p for the number of pairs of poles, although p has been widely used in the U.S. for the number of poles. Where ambiguity may occur, the intended meaning should be indicated.
	pole strength	$p \ldots m$	weber	Wb	

LETTER SYMBOLS FOR UNITS USED IN ELECTRICAL SCIENCE AND ELECTRICAL ENGINEERING

Extracted from IEEE Standard No. 260

The use of unit symbols, instead of the spelled-out names of the units, is frequently desirable where space is restricted. Their use presupposes that the reader will find them intelligible. If there is any doubt that the reader will understand a symbol, the name of the unit should be written in full. When an unfamiliar unit symbol is first used in text, it should be followed by its name in parentheses; only the symbol need be used thereafter. Explanatory notes or keys should be included where appropriate on drawings and in tabular matter.

The use of unit symbols is never mandatory, but when unit symbols are employed they must conform to those given in this Standard.

List of Symbols

Symbols for units are listed alphabetically by name of unit below. The list is intended to be reasonably complete, but could not possibly include all units that might conceivably be used in modern electrical technology. Many compound symbols and many illustrations of the use of the metric prefixes are included. Other combined forms may easily be constructed.

Every effort should be made to maintain the distinction between upper- and lower-case letters shown in the list, wherever the symbols for units are used, even if the surrounding text uses upper-case style.

In the notes accompanying the symbols, some units are identified as SI units. These units belong to the International System of Units (Système International d'Unités), which is the name given in 1960 by the Conférence Générale des Poids et Mesures to the coherent system of units based on the following basic units and quantities:

Unit	Quantity
meter	length
kilogram	mass
second	time
ampere	electric current
kelvin	temperature
candela	luminous intensity

The SI units include as subsystems the MKS system of units, which covers mechanics, and the MKSA or Giorgi system, which covers mechanics, electricity, and magnetism.

Unit	Symbol	Remarks
ampere	A	
ampere-hour	Ah	
ampere-turn	At	
angstrom	Å	
atmosphere		
normal atmosphere	atm	1 atm = 101 325 N/m^2
technical atmosphere	at	1 at = 1 kgf/cm^2
atomic mass unit (unified)	u	The (unified) atomic mass unit is defined as one-twelfth of the mass of an atom of the ^{12}C nuclide. Use of the old atomic mass unit (amu), defined by reference to oxygen, is deprecated.
bar	bar	1 bar = 100 000 N/m^2
barn	b	1 b = 10^{-28} m^2
bel	B	
billion electronvolts	GeV	The name *billion electronvolts* is deprecated; see *gigaelectronvolt*.
British thermal unit	Btu	

Unit	Symbol	Remarks
calorie (International Table calorie)	cal_{IT}	$1 \ cal_{IT} = 4.1868 \ J$ The 9th Conférénce Générale des Poids et Mesures has adopted the joule as the unit of heat, avoiding the use of the calorie as far as possible
calorie (thermochemical calorie)	cal_{th}	$1 \ cal_{th} = 4.1840 \ J$ (See note for International Table calorie.)
candela	cd	
candela per square foot	cd/ft^2	
candela per square meter	cd/m^2	The name *nit* is sometimes used for this unit.
candle	cd	The unit of luminous intensity has been given the name *candela*; use of the name *candle* for this unit is deprecated.
centimeter	cm	
circular mil	cmil	$1 \ cmil = (\pi/4) \cdot 10^{-6} \ in^2$
coulomb	C	
cubic centimeter	cm^3	
cubic foot	ft^3	
cubic foot per minute	ft^3/min	
cubic foot per second	ft^3/s	
cubic inch	in^3	
cubic meter	m^3	
cubic meter per second	m^3/s	
cubic yard	yd^3	
curie	Ci	Unit of activity in the field of radiation dosimetry
cycle per second	c/s	The name *hertz* (Hz) is internationally accepted for this unit.
decibel	dB	
decibel referred to one milliwatt	dBm	
degree (plane angle)	$\ldots^0$	
degree (temperature)		Note that there is no space between the symbol ° and the letter. The use of the word *centigrade* for the Celsius temperature scale was abandoned by the Conférence Générale des Poids et Mesures in 1948. In 1967 the CGPM gave the name *kelvin* to the SI unit of temperature, which was formerly called *degree Kelvin*, and assigned it the symbol K (without the symbol °).
degree Celsius	°C	
degree Fahrenheit	°F	
kelvin	K	
dyne	dyn	
electronvolt	eV	
erg	erg	
farad	F	
foot	ft	
footcandle	fc	The name *lumen per square foot* (lm/ft^2) is preferred for this unit.

LETTER SYMBOLS FOR UNITS
USED IN ELECTRICAL SCIENCE
AND ELECTRICAL ENGINEERING
(Continued)

Unit	Symbol	Remarks
footlambert	fL	If luminance is to be measured in English units, the candela per square foot (cd/ft^2) is preferred.
foot per minute	ft/min	
foot per second	ft/s	
foot per second squared	ft/s^2	
foot poundal	ft · pdl	
foot pound-force	ft · lbf	
gal	Gal	$1 \text{ Gal} = 1 \text{ cm/s}^2$
gallon	gal	The gallon, quart, and pint differ in the U.S. and the U.K. and their use is deprecated.
gallon per minute	gal/min	
gauss	G	The gauss is the electromagnetic CGS unit of magnetic flux density. Use of SI unit, the tesla, is preferred.
gigacycle per second	Gc/s	See note for cycle per second.
gigaelectronvolt	GeV	
gigahertz	GHz	
gilbert	Gb	The gilbert is the electromagnetic CGS unit of magnetomotive force. Use of the SI unit, the ampere (or ampere turn), is preferred.
gram	g	
henry	H	
hertz	Hz	
horsepower	hp	
hour	h	Time may be designated as in the following example: $9^h 46^m 30^s$.
inch	in	
inch per second	in/s	
joule	J	
joule per kelvin	J/K	
kelvin	K	In 1967 the CGPM gave the name *kelvin* to the SI unit of temperature which had formerly been called *degree Kelvin* and assigned it the symbol K (without the symbol °).
kilocycle per second	kc/s	See note for cycle per second.
kiloelectronvolt	keV	
kilogauss	kG	
kilogram	kg	
kilogram-force	kgf	In some countries the name *kilopond* (kp) has been adopted for this unit.
kilohertz	kHz	
kilojoule	kJ	
kilohm	kΩ	
kilometer	km	
kilometer per hour	km/h	
kilovar	kvar	

Unit	Symbol	Remarks
kilovolt	kV	
kilovoltampere	kVA	
kilowatt	kW	
kilowatthour	kWh	
knot	knot	
lambert	L	The lambert is the CGS unit of luminance. Use of the SI unit, the candela per square meter, is preferred.
liter	l	
liter per second	l/s	
lumen	lm	
lumen per square foot	lm/ft^2	
lumen per square meter	lm/m^2	
lumen per watt	lm/W	
lumen second	lm · s	
lux	lx	1 lx = 1 lm/m^2
maxwell	Mx	The maxwell is the electromagnetic CGS unit of magnetic flux. Use of the SI unit, the weber, is preferred.
megacycle per second	Mc/s	See note for cycle per second.
megaelectronvolt	MeV	
megahertz	MHz	
megavolt	MV	
megawatt	MW	
megohm	MΩ	
meter	m	
mho	mho	The IEC has adopted the name *siemens* (S) for this unit.
microampere	μA	
microbar	μbar	
microfarad	μF	
microgram	μg	
microhenry	μH	
micrometer	μm	
micromho	μmho	See note for mho.
micron	μm	The name *micrometer* is preferred.
microsecond	μs	
microsiemens	μS	
microwatt	μW	
mil	mil	1 mil = 0.001 in
mile (statute)	mi	
nautical mile	nmi	
mile per hour	mi/h	
milliampere	mA	
millibar	mbar	
millibarn	mb	
milligal	mGal	
milligram	mg	
millihenry	mH	

Unit	Symbol	Remarks
milliliter	ml	
millimeter	mm	
conventional millimeter of mercury	mmHg	1 mmHg = 133.322 N/m^2
millimicron	nm	The name *nanometer* is preferred.
millisecond	ms	
millisiemens	mS	
millivolt	mV	
milliwatt	mW	
minute (plane angle)	. . .	
minute (time)	min	Time may be designated as in the following example: $9^h 46^m 30^s$
nanoampere	nA	
nanofarad	nF	
nanometer	nm	
nanosecond	ns	
nanowatt	nW	
nautical mile	nmi	
neper	Np	
newton	N	
newton meter	N · m	
newton per square meter	N/m^2	
oersted	Oe	The oersted is the electromagnetic CGS unit of magnetic field strength. Use of the SI unit, the ampere per meter, is preferred.
ohm	Ω	
ounce (avoirdupois)	oz	
picoampere	pA	
picofarad	pF	
picosecond	ps	
picowatt	pW	
pint	pt	The gallon, quart, and pint differ in the U.S. and the U.K., and their use is deprecated.
pound	lb	
poundal	pdl	
pound-force	lbf	
pound-force foot	lbf · ft	
pound-force per square inch	lbf/in^2	
pound per square inch		Although use of the abbreviation psi is common, it is not recommended. See pound-force per square inch.
quart	qt	The gallon, quart, and pint differ in the U.S. and the U.K., and their use is deprecated.
rad	rd	Unit of absorbed dose in the field of radiation dosimetry.
radian	rad	
rem	rem	Unit of dose equivalent in the field of radiation dosimetry.

Unit	Symbol	Remarks
revolution per minute	r/min	Although use of the abbreviation rpm is common, it is not recommended.
revolution per second	r/s	
roentgen	R	Unit of exposure in the field of radiation dosimetry.
second (plane angle)	$\ldots''$	
second (time)	s	Time may be designated as in the following example: $9^h 46^m 30^s$.
siemens	S	$1\ S = 1\ \Omega^{-1}$
square foot	ft^2	
square inch	in^2	
square meter	m^2	
square yard	yd^2	
steradian	sr	
tesla	T	$1\ T = 1\ Wb/m^2$
tonne	t	$1\ t = 1000\ kg$
(unified) atomic mass unit	u	The (unified) atomic mass unit is defined as one-twelfth of the mass of an atom of the ^{12}C nuclide. Use of the old atomic mass unit (amu), defined by reference to oxygen, is deprecated.
var	var	Unit of reactive power
volt	V	
voltampere	VA	Unit of apparent power
watt	W	
watthour	Wh	
watt per steradian	W/sr	
watt per steradian square meter	$W/sr \cdot m^2$)	
weber	Wb	$1\ Wb = 1\ V \cdot s$
yard	yd	

CONVERSION OF ELECTROMAGNETIC UNITS

Three common systems of electromagnetic units are in universal employ. They are:

1. The absolute system or CGS electromagnetic system.
2. The practical CGS electromagnetic system.
3. The MKS system (Gaussian or Giorgi depending upon the choice of constants).

The chart allows rapid conversion from one system to another. In any one row, any quantity divided by any other quantity produces unity.

These Quantities Are Those Effected By Rationalization

Quantity	Rationalized			Unrationalized		
	MKS	CGS EM	CGS ES	MKS	CGS EM	CGS ES
Dielectric displacement	1	10^{-5}	3×10^5	4π	$4\pi \times 10^{-5}$	$12\pi \times 10^5$
	10^5	1	3×10^{10}	$4\pi \times 10^5$	4π	$12\pi \times 10^{10}$
	$1/3 \times 10^{-5}$	$1/3 \times 10^{-10}$	1	$4\pi/3 \times 10^{-5}$	$4\pi/3 \times 10^{-10}$	4π
	$1/4\pi$	$1/4\pi \times 10^{-5}$	$3/4\pi \times 10^5$	1	10^{-5}	3×10^{-5}
	$1/4\pi \times 10^5$	$1/4\pi$	$3/4\pi \times 10^{10}$	10^5	1	3×10^{10}
	$1/12\pi \times 10^{-5}$	$1/12\pi \times 10^{-10}$	$1/4\pi$	$1/3 \times 10^{-5}$	$1/3 \times 10^{-10}$	1
Units	Coulomb/m^2	Abcoulomb/m^2	Statcoulomb/cm^2	Coulomb/m^2	Abcoulomb/cm^2	Statcoulomb/cm^2
Magnetic field intensity	1	10^{-3}	3×10^7	4π	$4\pi \times 10^{-3}$	$12\pi \times 10^7$
	10^3	1	3×10^{10}	$4\pi \times 10^3$	4π	$12\pi \times 10^{10}$
	$1/3 \times 10^{-7}$	$1/3 \times 10^{-10}$	1	$4\pi/3 \times 10^{-7}$	$4\pi/3 \times 10^{-10}$	4π
	$1/4\pi$	$1/4\pi \times 10^{-3}$	$3/4\pi \times 10^7$	1	10^{-3}	3×10^7
	$1/4\pi \times 10^3$	$1/4\pi$	$3/4\pi \times 10^{10}$	10^3	1	3×10^{10}
	$1/12\pi \times 10^{-7}$	$1/12\pi \times 10^{-10}$	$1/4\pi$	$1/3 \times 10^{-7}$	$1/3 \times 10^{-10}$	1
Units	Amp-turn/m	Oersted	ESU	Amp-turn/m	Oersted	ESU
Magnetomotive force	1	10^{-1}	3×10^9	4π	$4\pi \times 10^{-1}$	$12\pi \times 10^9$
	10	1	3×10^{10}	40π	4π	$12\pi \times 10^{10}$
	$1/3 \times 10^{-9}$	$1/3 \times 10^{-10}$	1	$4\pi/3 \times 10^{-9}$	$4\pi/3 \times 10^{-10}$	4π
	$1/4\pi$	$1/4\pi \times 10^{-1}$	$3/4\pi \times 10^9$	1	10^{-1}	3×10^9
	$10/4\pi$	$1/4\pi$	$3/4\pi \times 10^{10}$	10	1	3×10^{10}
	$1/12\pi \times 10^{-9}$	$1/12\pi \times 10^{-10}$	$1/4\pi$	$1/3 \times 10^{-9}$	$1/3 \times 10^{-10}$	1
Units	Amp-turn	Gilbert	ESU	Amp-turn	Gilbert	ESU

CONVERSION OF ELECTROMAGNETIC UNITS
(Continued)

	Practical Unit	Electromagnetic Unit	Electrostatic Unit
Quantity	MKS	CGS EM	CGS ES
1. Capacitance	1 Farad	10^{-9} Abfarad	9×10^{11} Statfarad
	10^9 Farad	1 Abfarad	9×10^{20} Statfarad
	$1/9 \times 10^{-11}$ Farad	$1/9 \times 10^{-20}$ Abfarad	1 Statfarad
2. Charge	1 Coulomb	10^{-1} Abcoulomb	3×10^9 Statcoulomb
	10 Coulomb	1 Abcoulomb	3×10^{10} Statcoulomb
	$1/3 \times 10^{-9}$ Coulomb	$1/3 \times 10^{-10}$ Abcoulomb	1 Statcoulomb
3. Charge density	1 Coulomb/m^3	10^{-7} Abcoulomb/cm^3	3×10^3 Statcoulomb/cm^3
	10^7 Coulomb/m^3	1 Abcoulomb/cm^3	3×10^{10} Statcoulomb/cm^3
	$1/3 \times 10^{-3}$ Coulomb/m^3	$1/3 \times 10^{-10}$ Abcoulomb/cm^3	1 Statcoulomb/cm^3
4. Conductivity	1 Mho/m	10^{-11} Abmho/cm	9×10^9 Statmho/cm
	10^{11} Mho/m	1 Abmho/cm	9×10^{20} Statmho/cm
	$1/9 \times 10^{-9}$ Mho/m	$1/9 \times 10^{-20}$ Abmho/cm	1 Statmho/cm
5. Current	1 Ampere	10^{-1} Abampere	3×10^9 Statampere
	10 Ampere	1 Abampere	3×10^{10} Statampere
	$1/3 \times 10^{-9}$ Ampere	$1/3 \times 10^{-10}$ Abampere	1 Statampere
6. Current density	1 Ampere/m^2	10^{-5} Abampere/cm^2	3×10^5 Statampere/cm^2
	10^5 Ampere/m^2	1 Abampere/cm^2	3×10^{10} Statampere/cm^2
	$1/3 \times 10^{-5}$ Ampere/m^2	$1/3 \times 10^{-10}$ Abampere/cm^2	1 Statampere/cm^2
7. Electric field intensity	1 Volt/meter	10^6 Abvolt/cm	$1/3 \times 10^{-4}$ Statvolt/cm
	10^{-6} Volt/meter	1 Abvolt/cm	$1/3 \times 10^{-10}$ Statvolt/cm
	3×10^4 Volt/meter	3×10^{10} Abvolt/cm	1 Statvolt/cm
8. Electric potential	1 Volt	10^8 Abvolts	$1/3 \times 10^{-2}$ Statvolts
	10^{-8} Volt	1 Abvolt	$1/3 \times 10^{-10}$ Statvolts
	3×10^2 Volt	3×10^{10} Abvolts	1 Statvolt
9. Electric dipole moment	1 Coulomb–meter	10 Abcoulomb–cm	3×10^{11} Statcoulomb–cm
	10^{-1} Coulomb–meter	1 Abcoulomb–cm	3×10^{10} Statcoulomb–cm
	$1/3 \times 10^{-11}$ Coulomb–meter	$1/3 \times 10^{-10}$ Abcoulomb–cm	1 Statcoulomb–cm
10. Energy	1 Joule	10^7 Erg	10^7 Erg
	10^{-7} Joule	1 Erg	1 Erg
	10^{-7} Joule	1 Erg	1 Erg
11. Force	1 Newton	10^5 Dyne	10^5 Dyne
	10^{-5} Newton	1 Dyne	1 Dyne
	10^{-5} Newton	1 Dyne	1 Dyne
12. Flux density	1 Weber/m^2	10^4 Gauss	$1/3 \times 10^{-6}$ esu
	10^{-4} Weber/m^2	1 Gauss	$1/3 \times 10^{-10}$ esu
	3×10^6 Weber/m^2	3×10^{10} Gauss	1 esu
13. Inductance	1 Henry	10^9 Abhenry	$1/9 \times 10^{-11}$ Stathenry
	10^{-9} Henry	1 Abhenry	$1/9 \times 10^{-20}$ Stathenry
	9×10^{11} Henry	9×10^{20} Abhenry	1 Stathenry
14. Inductive capacity	1 Farad/meter	10^{-11} Abfarad/cm	9×10^9 Statfarad/cm
	10^{11} Farad/meter	1 Abfarad/cm	9×10^{20} Statfarad/cm
	$1/9 \times 10^{-9}$ Farad/meter	$1/9 \times 10^{-20}$ Abfarad/cm	1 Statfarad/cm
15. Magnetic flux	1 Weber	10^8 Maxwell	$1/3 \times 10^{-2}$ esu
	10^{-8} Weber	1 Maxwell	$1/3 \times 10^{-10}$ esu
	3×10^2 Weber	3×10^{10} Maxwell	1 esu
16. Magnetic dipole moment	1 Ampere–$meter^2$	10^3 Abamp–cm^2	3×10^{13} Statamp–cm^2
	10^{-3} Ampere–$meter^2$	1 Abamp–cm^2	3×10^{10} Statamp–cm^2
	$1/3 \times 10^{-13}$ Ampere–$meter^2$	$1/3 \times 10^{-10}$ Abamp–cm^2	1 Statamp–cm^2
17. Permeability	1 Henry/meter	10^7 Abhenry/cm	$1/9 \times 10^{-13}$ Stathenry/cm
	10^{-7} Henry/meter	1 Abhenry/cm	$1/9 \times 10^{-20}$ Stathenry/cm
	9×10^{13} Henry/meter	9×10^{20} Abhenry/cm	1 Stathenry/cm
18. Power	1 Watt	10^7 erg/sec	10^7 erg/sec
	10^{-7} Watt	1 erg/sec	1 erg/sec
	10^{-7} Watt	1 erg/sec	1 erg/sec
19. Resistance	1 Ohm	10^9 Abohm	$1/9 \times 10^{-11}$ Statohm
	10^{-9} Ohm	1 Abohm	$1/9 \times 10^{-20}$ Statohm
	9×10^{11} Ohm	9×10^{20} Abohm	1 Statohm

SPACE-TIME-VELOCITY AND ACCELERATION FORMULAS

This tabulation presents all basic linear motion formulas with all their variations. Terms are defined and units of measurement are specified.

A = Acceleration or deceleration—ft/sec/sec (32.2 for gravity)

D = Distance—ft (may be used in lieu of "H" in vertical free fall)

E = Energy—ft—lbs

F = Force—lbs

H = Height—ft (may be used lieu of "D" with A—32.2)

$M = \text{Mass}—\dfrac{W}{32.2} = \dfrac{\text{lb—sec}^2}{\text{ft}}$

T = Time—sec

V_a = Average velocity—ft/sec

V_f = Final velocity—ft/sec

V_i = Initial velocity—ft/sec

W = Weight—lbs

To Find	Formulae					
A	$\dfrac{V_f - V_i}{T}$	$\left(\begin{smallmatrix}\text{When}\\ V_i=0\end{smallmatrix}\right)\dfrac{V_f}{T}$	$\left(\begin{smallmatrix}\text{When}\\ V_i=0\end{smallmatrix}\right)\dfrac{V_f^2}{2D}$	$\dfrac{2D}{T^2}$	$\dfrac{WV_a}{FT}$	$\dfrac{F}{M}$
D	V_aT	$\dfrac{T(V_i + V_f)}{2}$	$\left(\begin{smallmatrix}\text{When}\\ V_i=0\end{smallmatrix}\right)\dfrac{V_f T}{2}$ ⟶ $\dfrac{V_a^2}{2A}$ ⟶ $\dfrac{AT^2}{2}$		$\dfrac{E}{F}$	
E	FD	WH				
F	MA	$\dfrac{M(V_f - V_i)}{T}$	$\dfrac{E}{D}$	$\dfrac{WV_a}{AT}$		
H	$\dfrac{E}{W}$	$16.1\,T^2$				
M	$\dfrac{W}{32.2}$	$\dfrac{F}{A}$	$\dfrac{FT}{V_f - V_i}$			
T	$\dfrac{D}{V_a}$	$\dfrac{2D}{V_f + V_i}$	$\dfrac{V_f - V_i}{A}$	$\left(\begin{smallmatrix}\text{When}\\ V_i=0\end{smallmatrix}\right)\dfrac{V_f}{A}$	$\left(\begin{smallmatrix}\text{When}\\ V=0\end{smallmatrix}\right)\dfrac{2D}{V_f}$	
	$\sqrt{\dfrac{2D}{A}}$	$\sqrt{\dfrac{H}{4}}$	$\dfrac{WV_a}{FA}$	$\dfrac{M(V_f - V_i)}{F}$		
V_f	$2V_a - V_i$	$\left(\begin{smallmatrix}\text{When}\\ V_i=0\end{smallmatrix}\right)2V_a$	$\dfrac{2D}{T} - V_i$	$\left(\begin{smallmatrix}\text{When}\\ V_i=0\end{smallmatrix}\right)\dfrac{2D}{T}$	$AT + V_i$	$\left(\begin{smallmatrix}\text{When}\\ V_i=0\end{smallmatrix}\right)AT$
V_i	$2V_a - V_f$	$\dfrac{2D}{T} - V_f$	$V_f - AT$	$V_f - \dfrac{FT}{M}$		
W	$\dfrac{AFT}{V_a}$	$32.2\,M,$	$\dfrac{E}{H}$			

CONVERSION FACTORS

To Convert	Into	Multiply By
A		
Abcoulomb	Statcoulombs	2.998×10^{10}
Acre	Sq. chain (Gunters)	10
Acre	Rods	160
Acre	Square links (Gunters)	1×10^5
Acre	Hectare or sq. hectometer	.4047
acres	sq feet	43,560.0
acres	sq meters	4,047.
acres	sq miles	1.562×10^{-3}
acres	sq yards	4,840.
acre-feet	cu feet	43,560.0
acre-feet	gallons	3.259×10^5
amperes/sq cm	amps/sq in.	6.452
amperes/sq cm	amps/sq meter	10^4
amperes/sq in.	amps/sq cm	0.1550
amperes/sq in.	amps/sq meter	1,550.0
amperes/sq meter	amps/sq cm	10^{-4}
amperes/sq meter	amps/sq in.	6.452×10^{-4}
ampere-hours	coulombs	3,600.0
ampere-hours	faradays	0.03731
ampere-turns	gilberts	1.257
ampere-turns/cm	amp-turns/in.	2.540
ampere-turns/cm	amp-turns/meter	100.0
ampere-turns/cm	gilberts/cm	1.257
ampere-turns/in.	amp-turns/cm	0.3937
ampere-turns/in.	amp-turns/meter	39.37
ampere-turns/in.	gilberts/cm	0.4950
ampere-turns/meter	amp-turns/cm	0.01
ampere-turns/meter	amp-turns/in.	0.0254
ampere-turns/meter	gilberts/cm	0.01257
Angstrom unit	Inch	3937×10^{-9}
Angstrom unit	Meter	1×10^{-10}
Angstrom unit	Micron or (Mu)	1×10^{-4}
Are	Acre (US)	.02471
Ares	sq. yards	119.60
ares	acres	0.02471
ares	sq meters	100.0
Astronomical Unit	Kilometers	1.495×10^8
Atmospheres	Ton/sq. inch	.007348
atmospheres	cms of mercury	76.0
atmospheres	ft of water (at 4°C)	33.90
atmospheres	in. of mercury (at 0°C)	29.92
atmospheres	kgs/sq cm	1.0333
atmospheres	kgs/sq meter	10,332.
atmospheres	pounds/sq.in.	14.70
atmospheres	tons/sq ft	1.058
B		
Barrels (U.S., dry)	cu. inches	7056.
Barrels (U.S., dry)	quarts (dry)	105.0
Barrels (U.S., liquid)	gallons	31.5
barrels (oil)	gallons (oil)	42.0
bars	atmospheres	0.9869
bars	dynes/sq cm	10^6
bars	kgs/sq meter	1.020×10^4
bars	pounds/sq ft	2,089.
bars	pounds/sq in.	14.50
Baryl	Dyne/sq. cm.	1.000
Bolt (US Cloth)	Meters	36.576
BTU	Liter—Atmosphere	10.409

To Convert	Into	Multiply By
Btu	ergs	1.0550×10^{10}
Btu	foot-lbs	778.3
Btu	gram-calories	252.0
Btu	horsepower-hrs	3.931×10^{-4}
Btu	joules	1,054.8
Btu	kilogram-calories	0.2520
Btu	kilogram-meters	107.5
Btu	kilowatt-hrs	2.928×10^{-4}
Btu/hr	foot-pound/sec	0.2162
Btu/hr	gram-cal/sec	0.0700
Btu/hr	horsepower-hrs	3.929×10^{-4}
Btu/hr	watts	0.2931
Btu/min	foot-lbs/sec	12.96
Btu/min	horsepower	0.02356
Btu/min	kilowatts	0.01757
Btu/min	watts	17.57
Btu/sq ft/min	watts/sq in.	0.1221
Bucket (Br. dry)	Cubic Cm.	1.818×10^4
bushels	cu ft	1.2445
bushels	cu in.	2,150.4
bushels	cu meters	0.03524
bushels	liters	35.24
bushels	pecks	4.0
bushels	pints (dry)	64.0
bushels	quarts (dry)	32.0
C		
Calories, gram (mean)	B.T.U. (mean)	3.9685×10^{-3}
centares (centiares)	sq meters	1.0
Centigrade (Celsius)	Fahrenheit	$(C° \times 9/5) + 32$
Centigrams	grams	0.01
Centiliter	Ounce fluid (US)	.3382
Centiliter	Cubic inch	.6103
Centiliter	drams	2.705
centiliters	liters	0.01
centimeters	feet	3.281×10^{-2}
centimeters	inches	0.3937
centimeters	kilometers	10^{-5}
centimeters	meters	0.01
centimeters	miles	6.214×10^{-6}
centimeters	millimeters	10.0
centimeters	mils	393.7
centimeters	yards	1.094×10^{-2}
centimeter-dynes	cm-grams	1.020×10^{-3}
centimeter-dynes	meter-kgs.	1.020×10^{-8}
centimeter-dynes	pound-feet	7.376×10^{-8}
centimeter-grams	cm-dynes	980.7
centimeter-grams	meter-kgs	10^{-5}
centimeter-grams	pound-feet	7.233×10^{-5}
centimeters of mercury	atmospheres	0.01316
centimeters of mercury	feet of water	0.4461
centimeters of mercury	kgs/sq meter	136.0
centimeters of mercury	pounds/sq ft	27.85
centimeters of mercury	pounds/sq in.	0.1934
centimeters/sec	feet/min	1.9685
centimeters/sec	feet/sec	0.03281
centimeters/sec	kilometers/hr	0.036
centimeters/sec	knots	0.1943
centimeters/sec	meters/min	0.6
centimeters/sec	miles/hr	0.02237
centimeters/sec	miles/min	3.728×10^{-4}

CONVERSION FACTORS (Continued)

To Convert	Into	Multiply By
centimeters/sec/sec	feet/sec/sec	0.03281
centimeters/sec/sec	kms/hr/sec	0.036
centimeters/sec/sec	meters/sec/sec	0.01
centimeters/sec/sec	miles/hr/sec	0.02237
Chain	Inches	792.00
Chain	meters	20.12
Chains (surveyors' or Gunter's)	yards	22.00
circular mils	sq cms	5.067×10^{-6}
circular mils	sq mils	0.7854
Circumference	Radians	6.283
circular mils	sq inches	7.854×10^{-7}
Cords	cord feet	8
Cord feet	cu feet	16
Coulomb	Statcoulombs	2.998×10^{9}
coulombs	faradays	1.036×10^{-5}
coulombs/sq cm	coulombs/sq in.	64.52
coulombs/sq cm.	coulombs/sq meter	10^{4}
coulombs/sq in.	coulombs/sq cm	0.1550
coulombs/sq in.	coulombs/sq meter	1,550.
coulombs/sq meter	coulombs/sq cm	10^{-4}
coulombs/sq meter	coulombs/sq in.	6.452×10^{-4}
cubic centimeters	cu feet	3.531×10^{-5}
cubic centimeters	cu inches	0.06102
cubic centimeters	cu meters	10^{-6}
cubic centimeters	cu yards	1.308×10^{-6}
cubic centimeters	gallons (U.S. liq.)	2.642×10^{-4}
cubic centimeters	liters	0.001
cubic centimeters	pints (U.S. liq.)	2.113×10^{-3}
cubic centimeters	quarts (U.S. liq.)	1.057×10^{-3}
cubic feet	bushels (dry)	0.8036
cubic feet	cu cms	28,320.0
cubic feet	cu inches	1,728.0
cubic feet	cu meters	0.02832
cubic feet	cu yards	0.03704
cubic feet	gallons (U.S. liq.)	7.48052
cubic feet	liters	28.32
cubic feet	pints (U.S. liq.)	59.84
cubic feet	quarts (U.S. liq.)	29.92
cubic feet/min	cu cms/sec	472.0
cubic feet/min	gallons/sec	0.1247
cubic feet/min	liters/sec	0.4720
cubic feet/min	pounds of water/min	62.43
cubic feet/sec	million gals/day	0.646317
cubic feet/sec	gallons/min	448.831
cubic inches	cu cms	16.39
cubic inches	cu feet	5.787×10^{-4}
cubic inches	cu meters	1.639×10^{-5}
cubic inches	cu yards	2.143×10^{-5}
cubic inches	gallons (U.S. liquid)	4.329×10^{-3}
cubic inches	liters	0.01639
cubic inches	mil-feet	1.061×10^{5}
cubic inches	pints (U.S. liq.)	0.03463
cubic inches	quarts (U.S. liq.)	0.01732
cubic meters	bushels (dry)	28.38
cubic meters	cu cms	10^{6}
cubic meters	cu feet	35.31
cubic meters	cu inches	61,023.0
cubic meters	cu yards	1.308
cubic meters	gallons (U.S. liq.)	264.2
cubic meters	liters	1,000.0
cubic meters	pints (U.S. liq.)	2,113.0

To Convert	Into	Multiply By
cubic meters	quarts (U.S. liq.)	1,057.
cubic yards	cu cms	7.646×10^{5}
cubic yards	cu feet	27.0
cubic yards	cu inches	46,656.0
cubic yards	cu meters	0.7646
cubic yards	gallons (U.S. liq.)	202.0
cubic yards	liters	764.6
cubic yards	pints (U.S. liq.)	1,615.9
cubic yards	quarts (U.S. liq.)	807.9
cubic yards/min	cubic ft/sec	0.45
cubic yards/min	gallons/sec	3.367
cubic yards/min	liters/sec	12.74

D

To Convert	Into	Multiply By
Dalton	Gram	1.650×10^{-24}
days	seconds	86,400.0
decigrams	grams	0.1
deciliters	liters	0.1
decimeters	meters	0.1
degrees (angle)	quadrants	0.01111
degrees (angle)	radians	0.01745
degrees (angle)	seconds	3,600.0
degrees/sec	radians/sec	0.01745
degrees/sec	revolutions/min	0.1667
degrees/sec	revolutions/sec	2.778×10^{-3}
dekagrams	grams	10.0
dekaliters	liters	10.0
dekameters	meters	10.0
Drams (apothecaries' or troy)	ounces (avoirdupois)	0.1371429
Drams (apothecaries' or troy)	ounces (troy)	0.125
Drams (U.S., fluid or apoth.)	cubic cm.	3.6967
drams	grams	1.7718
drams	grains	27.3437
drams	ounces	0.0625
Dyne/cm	Erg/sq. millimeter	.01
Dyne/sq. cm.	Atmospheres	9.869×10^{-7}
Dyne/sq. cm.	Inch of Mercury at 0°C	2.953×10^{-5}
Dyne/sq. cm	Inch of Water at 4°C	4.015×10^{-4}
dynes	grams	1.020×10^{-3}
dynes	joules/cm	10^{-7}
dynes	joules/meter (newtons)	10^{-5}
dynes	kilograms	1.020×10^{-6}
dynes	poundals	7.233×10^{-5}
dynes	pounds	2.248×10^{-6}
dynes/sq cm	bars	10^{-6}

E

To Convert	Into	Multiply By
Ell	Cm.	114.30
Ell	Inches	45
Em, Pica	Inch	.167
Em, Pica	Cm.	.4233
Erg/sec	Dyne—cm/sec	1.000
ergs	Btu	9.480×10^{-11}
ergs	dyne-centimeters	1.0
ergs	foot-pounds	7.367×10^{-8}
ergs	gram-calories	0.2389×10^{-7}
ergs	gram-cms	1.020×10^{-3}
ergs	horsepower-hrs	3.7250×10^{-14}

To Convert	Into	Multiply By
ergs	joules	10^{-7}
ergs	kg-calories	2.389×10^{-11}
ergs	kg-meters	1.020×10^{-8}
ergs	kilowatt-hrs	0.2778×10^{-13}
ergs	watt-hours	0.2778×10^{-10}
ergs/sec	Btu/min	$5,688 \times 10^{-9}$
ergs/sec	ft-lbs/min	4.427×10^{-6}
ergs/sec	ft-lbs/sec	7.3756×10^{-8}
ergs/sec	horsepower	1.341×10^{-10}
ergs/sec	kg-calories/min	1.433×10^{-9}
ergs/sec	kilowatts	10^{-10}

F

To Convert	Into	Multiply By
farads	microfarads	10^{6}
Faraday/sec	Ampere (absolute)	9.6500×10^{4}
faradays	ampere-hours	26.80
faradays	coulombs	9.649×10^{4}
Fathom	Meter	1.828804
fathoms	feet	6.0
feet	centimeters	30.48
feet	kilometers	3.048×10^{-4}
feet	meters	0.3048
feet	miles (naut.)	1.645×10^{-4}
feet	miles (stat.)	1.894×10^{-4}
feet	millimeters	304.8
feet	mils	1.2×10^{4}
feet of water	atmospheres	0.02950
feet of water	in. of mercury	0.8826
feet of water	kgs/sq cm	0.03048
feet of water	kgs/sq meter	304.8
feet of water	pounds/sq ft	62.43
feet of water	pounds/sq in.	0.4335
feet/min	cm/sec	0.5080
feet/min	feet/sec	0.01667
feet/min	kms/hr	0.01829
feet/min	meters/min	0.3048
feet/min	miles/hr	0.01136
feet/sec	cms/sec	30.48
feet/sec	kms/hr	1.097
feet/sec	knots	0.5921
feet/sec	meters/min	18.29
feet/sec	miles/hr	0.6818
feet/sec	miles/min	0.01136
feet/sec/sec	cms/sec/sec	30.48
feet/sec/sec	kms/hr/sec	1.097
feet/sec/sec	meters/sec/sec	0.3048
feet/sec/sec	miles/hr/sec	0.6818
feet/100 feet	per cent grade	1.0
foot-candle	lumen/sq.,meter	10.764
foot-pounds	Btu	1.286×10^{-3}
foot-pounds	ergs	1.356×10^{7}
foot-pounds	gram-calories	0.3238
foot-pounds	hp-hrs	5.050×10^{-7}
foot-pounds	joules	1.356
foot-pounds	kg-calories	3.24×10^{-4}
foot-pounds	kg-meters	0.1383
foot-pounds	kilowatt-hrs	3.766×10^{-7}
foot-pounds/min	Btu/min	1.286×10^{-3}
foot-pounds/min	foot-pounds/sec	0.01667
foot-pounds/min	horsepower	3.030×10^{-5}
foot-pounds/min	kg-calories/min	3.24×10^{-4}

To Convert	Into	Multiply By
foot-pounds/min	kilowatts	2.260×10^{-5}
foot-pounds/sec	Btu/hr	4.6263
foot-pounds/sec	Btu/min	0.07717
foot-pounds/sec	horsepower	1.818×10^{-3}
foot-pounds/sec	kg-calories/min	0.01945
foot-pounds/sec	kilowatts	1.356×10^{-3}
Furlongs	miles (U.S.)	0.125
furlongs	rods	40.0
furlongs	feet	660.0

G

To Convert	Into	Multiply By
gallons	cu cms	3,785.0
gallons	cu feet	0.1337
gallons	cu inches	231.0
gallons	cu meters	3.785×10^{-3}
gallons	cu yards	4.951×10^{-3}
gallons	liters	3.785
gallons (liq. Br. Imp.)	gallons (U.S. liq.)	1.20095
gallons (U.S.)	gallons (Imp.)	0.83267
gallons of water	pounds of water	8.3453
gallons/min	cu ft/sec	2.228×10^{-3}
gallons/min	liters/sec	0.06308
gallons/min	cu ft/hr	8.0208
gausses	lines/sq in.	6.452
gausses	webers/sq cm	10^{-8}
gausses	webers/sq in.	6.452×10^{-8}
gausses	webers/sq meter	10^{-4}
gilberts	ampere-turns	0.7958
gilberts/cm	amp-turns/cm	0.7958
gilberts/cm	amp-turns/in	2.021
gilberts/cm	amp-turns/meter	79.58
Gills (British)	cubic cm.	142.07
gills	liters	0.1183
gills	pints (liq.)	0.25
Grade	Radian	.01571
Grains	drams (avoirdupois)	0.03657143
grains (troy)	grains (avdp)	1.0
grains (troy)	grams	0.06480
grains (troy)	ounces (avdp)	2.0833×10^{-3}
grains (troy)	pennyweight (troy)	0.04167
grains/U.S. gal	parts/million	17.118
grains/U.S. gal	pound/million gal	142.86
grains/Imp. gal	parts/million	14.286
grams	dynes	980.7
grams	grains	15.43
grams	joules/cm	9.807×10^{-5}
grams	joules/meter (newtons)	9.807×10^{-3}
grams	kilograms	0.001
grams	milligrams	1,000.
grams	ounces (avdp)	0.03527
grams	ounces (troy)	0.03215
grams	poundals	0.07093
grams	pounds	2.205×10^{-3}
grams/cm	pounds/inch	5.600×10^{-3}
grams/cu cm	pounds/cu ft	62.43
grams/cu cm	pounds/cu in	0.03613
grams/cu cm	pounds/mil-foot	3.405×10^{-7}
grams/liter	grains/gal	58.417
grams/liter	pounds/1,000 gal	8.345
grams/liter	pounds/cu ft	0.062427

CONVERSION FACTORS (Continued)

To Convert	Into	Multiply By
grams/liter	parts/million	1,000.0
grams/sq cm	pounds/sq ft	2.0481
gram-calories	Btu	3.9683×10^{-3}
gram-calories	ergs	4.1868×10^{7}
gram-calories	foot-pounds	3.0880
gram-calories	horsepower-hrs	1.5596×10^{-6}
gram-calories	kilowatt-hrs	1.1630×10^{-6}
gram-calories	watt-hrs	1.1630×10^{-3}
gram-calories/sec	Btu/hr	14.286
gram-centimeters	Btu	9.297×10^{-8}
gram-centimeters	ergs	980.7
gram-centimeters	joules	9.807×10^{-5}
gram-centimeters	kg-cal	2.343×10^{-8}
gram-centimeters	kg-meters	10^{-5}

H

To Convert	Into	Multiply By
Hand	Cm.	10.16
hectares	acres	2.471
hectares	sq feet	1.076×10^{5}
hectograms	grams	100.0
hectoliters	liters	100.0
hectometers	meters	100.0
hectowatts	watts	100.0
henries	millihenries	1,000.0
Hogsheads (British)	cubic ft.	10.114
Hogsheads (U.S.)	cubic ft.	8.42184
Hogsheads (U.S.)	gallons (U.S.)	63
horsepower	Btu/min	42.44
horsepower	foot-lbs/min	33,000.
horsepower	foot-lbs/sec	550.0
horsepower (metric) (542.5 ft lb/sec)	horsepower (550 ft lb/sec)	0.9863
horsepower (550 ft lb/sec)	horsepower (metric) (542.5 ft lb/sec)	1.014
horsepower	kg-calories/min	10.68
horsepower	kilowatts	0.7457
horsepower	watts	745.7
horsepower (boiler)	Btu/hr	33.479
horsepower (boiler)	kilowatts	9.803
horsepower-hrs	Btu	2,547.
horsepower-hrs	ergs	2.6845×10^{13}
horsepower hrs	foot lbs	1.98×10^{6}
horsepower-hrs	gram-calories	641,190.
horsepower-hrs	joules	2.684×10^{6}
horsepower-hrs	kg-calories	641.1
horsepower-hrs	kg-meters	2.737×10^{5}
horsepower-hrs	kilowatt-hrs	0.7457
hours	days	4.167×10^{-2}
hours	weeks	5.952×10^{-3}
Hundredweights (long)	pounds	112
Hundredweights (long)	tons (long)	0.05
Hundredweights (short)	ounces (avoirdupois)	1600
Hundredweights (short)	pounds	100
Hundredweights (short)	tons (metric)	0.0453592
Hundredweights (short)	tons (long)	0.0446429

I

To Convert	Into	Multiply By
inches	centimeters	2.540
inches	meters	2.540×10^{-2}
inches	miles	1.578×10^{-5}
inches	millimeters	25.40
inches	mils	1,000.0
inches	yards	2.778×10^{-2}
inches of mercury	atmospheres	0.03342
inches of mercury	feet of water	1.133
inches of mercury	kgs/sq cm	0.03453
inches of mercury	kgs/sq meter	345.3
inches of mercury	pounds/sq ft	70.73
inches of mercury	pounds/sq in.	0.4912
inches of water (at 4°C)	atmospheres	2.458×10^{-3}
inches of water (at 4°C)	inches of mercury	0.07355
inches of water (at 4°C)	kgs/sq cm	2.540×10^{-3}
inches of water (at 4°C)	ounces/sq in.	0.5781
inches of water (at 4°C)	pounds/sq ft	5.204
inches of water (at 4°C)	pounds/sq in.	0.03613
International Ampere	Ampere (absolute)	.9998
International Volt	Volts (absolute)	1.0003
International volt	Joules (absolute)	1.593×10^{-19}
International volt	Joules	9.654×10^{4}

J

To Convert	Into	Multiply By
joules	Btu	9.480×10^{-4}
joules	ergs	10^{7}
joules	foot-pounds	0.7376
joules	kg-calories	2.389×10^{-4}
joules	kg-meters	0.1020
joules	watt-hrs	2.778×10^{-4}
joules/cm	grams	1.020×10^{4}
joules/cm	dynes	10^{7}
joules/cm	joules/meter (newtons)	100.0
joules/cm	poundals	723.3
joules/cm	pounds	22.48

K

To Convert	Into	Multiply By
kilograms	dynes	980,665.
kilograms	grams	1,000.0
kilograms	joules/cm	0.09807
kilograms	joules/meter (newtons)	9.807
kilograms	poundals	70.93
kilograms	pounds	2.205
kilograms	tons (long)	9.842×10^{-4}
kilograms	tons (short)	1.102×10^{-3}
kilograms/cu meter	grams/cu cm	0.001
kilograms/cu meter	pounds/cu ft	0.06243
kilograms/cu meter	pounds/cu in.	3.613×10^{-5}
kilograms/cu meter	pounds/mil-foot	3.405×10^{-10}
kilograms/meter	pounds/ft	0.6720
Kilogram/sq. cm.	Dynes	980,665
kilograms/sq cm	atmospheres	0.9678
kilograms/sq cm	feet of water	32.81
kilograms/sq cm	.inches of mercury	28.96
kilograms/sq cm	pounds/sq ft	2,048.
kilograms/sq cm	pounds/sq in.	14.22
kilograms/sq meter	atmospheres	9.678×10^{-5}
kilograms/sq meter	bars	98.07×10^{-6}
kilograms/sq meter	feet of water	3.281×10^{-3}
kilograms/sq meter	inches of mercury	2.896×10^{-3}
kilograms/sq meter	pounds/sq ft	0.2048
kilograms/sq meter	pounds/sq in.	1.422×10^{-3}
kilograms/sq mm	kgs/sq meter	10^{6}
kilogram-calories	Btu	3.968
kilogram-calories	foot-pounds	3,088.
kilogram-calories	hp-hrs	1.560×10^{-3}
kilogram-calories	joules	4,186.

To Convert	Into	Multiply By
kilogram-calories	kg-meters	426.9
kilogram-calories	kilojoules	4.186
kilogram-calories	kilowatt-hrs	1.163×10^{-3}
kilogram meters	Btu	9.294×10^{-3}
kilogram meters	ergs	9.804×10^{7}
kilogram meters	foot-pounds	7.233
kilogram meters	joules	9.804
kilogram meters	kg-calories	2.342×10^{-3}
kilogram meters	kilowatt-hrs	2.723×10^{-6}
kilolines	maxwells	1,000.0
kiloliters	liters	1,000.0
kilometers	centimeters	10^{5}
kilometers	feet	3,281.
kilometers	inches	3.937×10^{4}
kilometers	meters	1,000.0
kilometers	miles	0.6214
kilometers	millimeters	10^{6}
kilometers	yards	1,094.
kilometers/hr	cms/sec	27.78
kilometers/hr	feet/min	54.68
kilometers/hr	feet/sec	0.9113
kilometers/hr	knots	0.5396
kilometers/hr	meters/min	16.67
kilometers/hr	miles/hr	0.6214
kilometers/hr/sec	cm/sec/sec	27.78
kilometers/hr/sec	ft/sec/sec	0.9113
kilometers/hr/sec	meters/sec/sec	0.2778
kilometers/hr/sec	miles/hr/sec	0.6214
kilowatts	Btu/min	56.92
kilowatts	foot-lbs/min	4.426×10^{4}
kilowatts	foot-lbs/sec	737.6
kilowatts	horsepower	1.341
kilowatts	kg-calories/min	14.34
kilowatts	watts	1,000.0
kilowatt-hrs	Btu	3,413.
kilowatt-hrs	ergs	3.600×10^{13}
kilowatt-hrs	foot-lbs	2.655×10^{6}
kilowatt-hrs	gram-calories	859,850.
kilowatt-hrs	horsepower-hrs	1.341
kilowatt-hrs	joules	3.6×10^{6}
kilowatt-hrs	kg-calories	859.85
kilowatt-hrs	kg-meters	3.671×10^{5}
kilowatt-hrs	pounds of water evaporated from and at 212°F.	3.53
kilowatt-hrs	pounds of water raised from 62° to 212°F.	22.75
knots	feet/hr	6,080.
knots	kilometers/hr	1.8532
knots	nautical miles/hr	1.0
knots	statute miles/hr	1.151
knots	yards/hr	2,027.
knots	feet/sec	1.689

L

To Convert	Into	Multiply By
league	miles (approx.)	3.0
Light year	Miles	5.9×10^{12}
Light year	Kilometers	9.46091×10^{12}
lines/sq cm	gausses	1.0
lines/sq in.	gausses	0.1550
lines/sq in.	webers/sq cm	1.550×10^{-9}
lines/sq in.	webers/sq in.	10^{-8}
lines/sq in.	webers/sq meter	1.550×10^{-5}

To Convert	Into	Multiply By
links (engineer's)	inches	12.0
links (surveyor's)	inches	7.92
liters	bushels (U.S. dry)	0.02838
liters	cu cm	1,000.0
liters	cu feet	0.03531
liters	cu inches	61.02
liters	cu meters	0.001
liters	cu yards	1.308×10^{-3}
liters	gallons (U.S. liq.)	0.2642
liters	pints (U.S. liq.)	2.113
liters	quarts (U.S. liq.)	1.057
liters/min	cu ft/sec	5.886×10^{-4}
liters/min	gals/sec	4.403×10^{-3}
lumens/sq ft	foot-candles	1.0
Lumen	Spherical candle power	.07958
Lumen	Watt	.001496
Lumen/sq. ft.	Lumen/sq. meter	10.76
lux	foot-candles	0.0929

M

To Convert	Into	Multiply By
maxwells	kilolines	0.001
maxwells	webers	10^{-8}
megalines	maxwells	10^{6}
megohms	microhms	10^{12}
megohms	ohms	10^{6}
meters	centimeters	100.0
meters	feet	3.281
meters	inches	39.37
meters	kilometers	0.001
meters	miles (naut.)	5.396×10^{-4}
meters	miles (stat.)	6.214×10^{-4}
meters	millimeters	1,000.0
meters	yards	1.094
meters	varas	1.179
meters/min	cms/sec	1.667
meters/min	feet/min	3.281
meters/min	feet/sec	0.05468
meters/min	kms/hr	0.06
meters/min	knots	0.03238
meters/min	miles/hr	0.03728
meters/sec	feet/min	196.8
meters/sec	feet/sec	3.281
meters/sec	kilometers/hr	3.6
meters/sec	kilometers/min	0.06
meters/sec	miles/hr	2.237
meters/sec	miles/min	0.03728
meters/sec/sec	cms/sec/sec	100.0
meters/sec/sec	ft/sec/sec	3.281
meters/sec/sec	kms/hr/sec	3.6
meters/sec/sec	miles/hr/sec	2.237
meter-kilograms	cm-dynes	9.807×10^{7}
meter-kilograms	cm-grams	10^{5}
microfarad	farads	10^{-6}
micrograms	grams	10^{-6}
microhms	megohms	10^{-12}
microhms	ohms	10^{-6}
microliters	liters	10^{-6}
Microns	meters	1×10^{-6}
miles (naut.)	feet	6,080.27
miles (naut.)	kilometers	1.853
miles (naut.)	meters	1,853.
miles (naut.)	miles (statute)	1.1516

To Convert	Into	Multiply By
miles (naut.)	yards	2,027.
miles (statute)	centimeters	1.609×10^5
miles (statute)	feet	5,280.
miles (statute)	inches	6.336×10^4
miles (statute)	kilometers	1.609
miles (statute)	meters	1,609.
miles (statute)	miles (naut.)	0.8684
miles (statute)	yards	1,760.
miles/hr	cms/sec	44.70
miles/hr	feet/min	88.
miles/hr	feet/sec	1.467
miles/hr	kms/hr	1.609
miles/hr	kms/min	0.02682
miles/hr	knots	0.8684
miles/hr	meters/min	26.82
miles/hr	miles/min	0.1667
miles/hr/sec	cms/sec/sec	44.70
miles/hr/sec	feet/sec/sec	1.467
miles/hr/sec	kms/hr/sec	1.609
miles/hr/sec	meters/sec/sec	0.4470
miles/min	cms/sec	2,682.
miles/min	feet/sec	88.
miles/min	kms/min	1.609
miles/min	knots/min	0.8684
miles/min	miles/hr	60.0
mil feet	cu inches	9.425×10^{-6}
milliers	kilograms	1,000.
Millimicrons	meters	1×10^{-9}
Milligrams	grains	0.01543236
milligrams	grams	0.001
milligrams/liter	parts/million	1.0
millihenries	henries	0.001
milliliters	liters	0.001
millimeters	centimeters	0.1
millimeters	feet	3.281×10^{-3}
millimeters	inches	0.03937
millimeters	kilometers	10^{-6}
millimeters	meters	0.001
millimeters	miles	6.214×10^{-7}
millimeters	mils	39.37
millimeters	yards	1.094×10^{-3}
million gals/day	cu ft/sec	1.54723
mils	centimeters	2.540×10^{-3}
mils	feet	8.333×10^{-5}
mils	inches	0.001
mils	kilometers	2.540×10^{-8}
mils	yards	2.778×10^{-5}
miner's inches	cu ft/min	1.5
Minims (British)	cubic cm.	0.059192
Minims (U.S., fluid)	cubic cm.	0.061612
Minutes (angles)	degrees	0.01667
minutes (angles)	quadrants	1.852×10^{-4}
minutes (angles)	radians	2.909×10^{-4}
minutes (angles)	seconds	60.0
myriagrams	kilograms	10.0
myriameters	kilometers	10.0
myriawatts	kilowatts	10.0

N

To Convert	Into	Multiply By
nepers	decibels	8.686
Newton	Dynes	1×10^5

O

To Convert	Into	Multiply By
OHM (International)	OHM (absolute)	1.0005
ohms	megohms	10^{-6}
ohms	microhms	10^6
ounces	drams	16.0
ounces	grains	437.5
ounces	grams	28.349527
ounces	pounds	0.0625
ounces	ounces (troy)	0.9115
ounces	tons (long)	2.790×10^{-5}
ounces	tons (metric)	2.835×10^{-5}
ounces (fluid)	cu inches	1.805
ounces (fluid)	liters	0.02957
ounces (troy)	grains	480.0
ounces (troy)	grams	31.103481
ounces (troy)	ounces (avdp.)	1.09714
ounces (troy)	pennyweights (troy)	20.0
ounces (troy)	pounds (troy)	0.08333
Ounce/sq. inch	Dynes/sq. cm.	4309
ounces/sq in.	pounds/sq in.	0.0625

P

To Convert	Into	Multiply By
Parsec	Miles	19×10^{12}
Parsec	Kilometers	3.084×10^{13}
parts/million	grains/U.S. gal	0.0584
parts/million	grains/Imp. gal	0.07016
parts/million	pounds/million gal	8.345
Pecks (British)	cubic inches	554.6
Pecks (British)	liters	9.091901
Pecks (U.S.)	bushels	0.25
Pecks (U.S.)	cubic inches	537.605
Pecks (U.S.)	litters	8.809582
Pecks (U.S.)	quarts (dry)	8
pennyweights (troy)	grains	24.0
pennyweights (troy)	ounces (troy)	0.05
pennyweights (troy)	grams	1.55517
pennyweights (troy)	pounds (troy)	4.1667×10^{-3}
pints (dry)	cu inches	33.60
pints (liq.)	cu cms.	473.2
pints (liq.)	cu feet	0.1671
pints (liq.)	cu inches	28.87
pints (liq.)	cu meters	4.732×10^{-4}
pints (liq.)	cu yards	6.189×10^{-4}
pints (liq.)	gallons	0.125
pints (liq.)	liters	0.4732
pints (liq.)	quarts (liq.)	0.5
Planck's quantum	Erg—second	6.624×10^{-27}
Poise	Gram/cm. sec.	1.00
Pounds (avoirdupois)	ounces (troy)	14.5833
poundals	dynes	13,826.
poundals	grams	14.10
poundals	joules/cm	1.383×10^{-3}
poundals	joules/meter (newtons)	0.1383
poundals	kilograms	0.01410
poundals	pounds	0.03108
pounds	drams	256.
pounds	dynes	44.4823×10^4
pounds	grains	7,000.
pounds	grams	453.5924
pounds	joules/cm	0.04448
pounds	joules/meter (newtons)	4.448

To Convert	Into	Multiply By
pounds	kilograms	0.4536
pounds	ounces	16.0
pounds	ounces (troy)	14.5833
pounds	poundals	32.17
pounds	pounds (troy)	1.21528
pounds	tons (short)	0.0005
pounds (troy)	grains	5,760.
pounds (troy)	grams	373.24177
pounds (troy)	ounces (avdp.)	13.1657
pounds (troy)	ounces (troy)	12.0
pounds (troy)	pennyweights (troy)	240.0
pounds (troy)	pounds (avdp.)	0.822857
pounds (troy)	tons (long)	3.6735×10^{-4}
pounds (troy)	tons (metric)	3.7324×10^{-4}
pounds (troy)	tons (short)	4.1143×10^{-4}
pounds of water	cu feet	0.01602
pounds of water	cu inches	27.68
pounds of water	gallons	0.1198
pounds of water/min	cu ft/sec	2.670×10^{-4}
pound-feet	cm-dynes	1.356×10^{7}
pound-feet	cm-grams	13,825.
pound-feet	meter-kgs	0.1383
pounds/cu ft	grams/cu cm	0.01602
pounds/cu ft	kgs/cu meter	16.02
pounds/cu ft	pounds/cu in.	5.787×10^{-4}
pounds/cu ft	pounds/mil-foot	5.456×10^{-9}
pounds/cu in.	gms/cu cm	27.68
pounds/cu in.	kgs/cu meter	2.768×10^{4}
pounds/cu in.	pounds/cu ft	1,728.
pounds/cu in.	pounds/mil-foot	9.425×10^{-6}
pounds/ft	kgs/meter	1.488
pounds/in.	gms/cm	178.6
pounds/mil-foot	gms/cu cm	2.306×10^{6}
pounds/sq ft	atmospheres	4.725×10^{-4}
pounds/sq ft	feet of water	0.01602
pounds/sq ft	inches of mercury	0.01414
pounds/sq ft	kgs/sq meter	4.882
pounds/sq ft	pounds/sq in.	6.944×10^{-3}
pounds/sq in.	atmospheres	0.06804
pounds/sq in.	feet of water	2.307
poundsTsq in.	inches of mercury	2.036
pounds/sq in.	kgs/sq meter	703.1
pounds/sq in.	pounds/sq ft	144.0

Q

To Convert	Into	Multiply By
quadrants (angle)	degrees	90.0
quadrants (angle)	minutes	5,400.0
quadrants (angle)	radians	1.571
quadrants (angle)	seconds	3.24×10^{5}
quarts (dry)	cu inches	67.20
quarts (liq.)	cu cms	946.4
quarts (liq.)	cu feet	0.03342
quarts (liq.)	cu inches	57.75
quarts (liq.)	cu meters	9.464×10^{-4}
quarts (liq.)	cu yards	1.238×10^{-3}
quarts (liq.)	gallons	0.25
quarts (liq.)	liters	0.9463

R

To Convert	Into	Multiply By
radians	degrees	57.30
radians	minutes	3,438.
radians	quadrants	0.6366
radians	seconds	2.063×10^{5}

To Convert	Into	Multiply By
radians/sec	degrees/sec	57.30
radians/sec	revolutions/min	9.549
radians/sec	revolutions/sec	0.1592
radians/sec/sec	revs/min/min	573.0
radians/sec/sec	revs/min/sec	9.549
radians/sec/sec	revs/sec/sec	0.1592
revolutions	degrees	360.0
revolutions	quadrants	4.0
revolutions	radians	6.283
revolutions/min	degrees/sec	6.0
revolutions/min	radians/sec	0.1047
revolutions/min	revs/sec	0.01667
revolutions/min/min	radians/sec/sec	1.745×10^{-3}
revolutions/min/min	revs/min/sec	0.01667
revolutions/min/min	revs/sec/sec	2.778×10^{-4}
revolutions/sec	degrees/sec	360.0
revolutions/sec	radians/sec	6.283
revolutions/sec	revs/min	60.0
revolutions/sec/sec	radians/sec/sec	6.283
revolutions/sec/sec	revs/min/min	3,600.0
revolutions/sec/sec	revs/min/sec	60.0
Rod	Chain (Gunters)	.25
Rod	Meters	5.029
Rod (Surveyors' meas.)	yards	5.5
rods	feet	16.5

S

To Convert	Into	Multiply By
Scruples	grains	20
seconds (angle)	degrees	2.778×10^{-4}
seconds (angle)	minutes	0.01667
seconds (angle)	quadrants	3.087×10^{-6}
seconds (angle)	radians	4.848×10^{-6}
Slug	Kilogram	14.59
Slug	Pounds	32.17
Sphere	Steradians	12.57
square centimeters	circular mils	1.973×10^{5}
square centimeters	sq ft	1.076×10^{-3}
square centimeters	sq inches	0.1550
square centimeters	sq meters	0.0001
square centimeters	sq miles	3.861×10^{-11}
square centimeters	sq millimeters	100.0
square centimeters	sq yards	1.196×10^{-4}
square feet	acres	2.296×10^{-5}
square feet	circular mils	1.833×10^{8}
square feet	sq cms	929.0
square feet	sq inches	144.0
square feet	sq meters	0.09290
square feet	sq miles	3.587×10^{-8}
square feet	sq millimeters	9.290×10^{4}
square feet	sq yards	0.1111
square inches	circular mils	1.273×10^{6}
square inches	sq cms	6.452
square inches	sq feet	6.944×10^{-3}
square inches	sq millimeters	645.2
square inches	sq mils	10^{6}
square inches	sq yards	7.716×10^{-4}
square kilometers	acres	247.1
square kilometers	sq cms	10^{10}
square kilometers	sq ft	10.76×10^{6}
square kilometers	sq inches	1.550×10^{9}
square kilometers	sq meters	10^{6}
square kilometers	sq miles	0.3861

CONVERSION FACTORS (Continued)

To Convert	Into	Multiply By
square kilometers	sq yards	1.196×10^6
square meters	acres	2.471×10^{-4}
square meters	sq cms	10^4
square meters	sq feet	10.76
square meters	sq inches	1,550.
square meters	sq miles	3.861×10^{-7}
square meters	sq millimeters	10^6
square meters	sq yards	1.196
square miles	acres	640.0
square miles	sq feet	27.88×10^6
square miles	sq kms	2.590
square miles	sq meters	2.590×10^6
square miles	sq yards	3.098×10^6
square millimeters	circular mils	1,973.
square millimeters	sq cms	0.01
square millimeters	sq feet	1.076×10^{-5}
square millimeters	sq inches	1.550×10^{-3}
square mils	circular mils	1.273
square mils	sq cms	6.452×10^{-6}
square mils	sq inches	10^{-6}
square yards	acres	2.066×10^{-4}
square yards	sq cms	8,361.
square yards	sq feet	9.0
square yards	sq inches	1,296.
square yards	sq meters	0.8361
square yards	sq miles	3.228×10^{-7}
square yards	sq millimeters	8.361×10^5

T

To Convert	Into	Multiply By
temperature ($^\circ$C) +273	absolute temperature ($^\circ$C)	1.0
temperature ($^\circ$C) + 17.78	temperature ($^\circ$F)	1.8
temperature ($^\circ$F) +460	absolute temperature ($^\circ$F)	1.0
temperature ($^\circ$F) −32	temperature ($^\circ$C)	5/9
tons (long)	kilograms	1,016.
tons (long)	pounds	2,240.
tons (long)	tons (short)	1.120
tons (metric)	kilograms	1,000.
tons (metric)	pounds	2,205.
tons (short)	kilograms	907.1848
tons (short)	ounces	32,000.
tons (short)	ounces (troy)	29,166.66
tons (short)	pounds	2,000.
tons (short)	pounds (troy)	2,430.56
tons (short)	tons (long)	0.89287
tons (short)	tons (metric)	0.9078
tons (short/sq ft	kgs/sq meter	9,765.

To Convert	Into	Multiply By
tons (short)/sq ft	pounds/sq in.	2,000.
tons of water/24 hrs	pounds of water/hr	83.333
tons of water/24 hrs	gallons/min	0.16643
tons of water/24 hrs	cu ft/hr	1.3349

V

To Convert	Into	Multiply By
Volt/inch	Volt/cm.	.39370
Volt (absolute)	Statvolts	.003336

W

To Convert	Into	Multiply By
watts	Btu/hr	3.4129
watts	Btu/min	0.05688
watts	ergs/sec	10^7
watts	foot-lbs/min	44.27
watts	foot-lbs/sec	0.7378
watts	horsepower	1.341×10^{-3}
watts	horsepower (metric)	1.360×10^{-3}
watts	kg-calories/min	0.01433
watts	kilowatts	0.001
Watts (Abs.)	B.T.U. (mean)/min.	0.056884
Watts (Abs.)	joules/sec.	1
watt-hours	Btu	3.413
watt-hours	ergs	3.60×10^{10}
watt-hours	foot-pounds	2,656.
watt-hours	gram-calories	859.85
watt-hours	horsepower-hrs	1.341×10^{-3}
watt-hours	kilogram-calories	0.8605
watt-hours	kilogram-meters	367.2
watt-hours	kilowatt-hrs	0.001
Watt (International)	Watt (absolute)	1.0002
webers	maxwells	10^8
webers	kilolines	10^5
webers/sq in.	gausses	1.550×10^7
webers/sq in.	lines/sq in.	10^8
webers/sq in.	webers/sq cm	0.1550
webers/sq in.	webers/sq meter	1,550.
webers/sq meter	gausses	10^4
webers/sq meter	lines/sq in.	6.452×10^4
webers/sq meter	webers/sq cm	10^{-4}
webers/sq meter	webers/sq in.	6.452×10^{-4}

Y

To Convert	Into	Multiply By
yards	centimeters	91.44
yards	kilometers	9.144×10^{-4}
yards	meters	0.9144
yards	miles (naut.)	4.934×10^{-4}
yards	miles (stat.)	5.682×10^{-4}
yards	millimeters	914.4

SECTION 6

PHYSICAL DATA

LASER (EYE HAZARD)
NOMOGRAM

This nomogram is used to estimate the safe range at which an object may be illuminated directly. It incorporates a scale for the introduction of loss factors including losses in the eye, optical surfaces external to the laser mirror, and optical losses.

FOR EXAMPLE:
Assume system losses of 50%, a pupil diameter of 4 mm, a laser output of 0.05 J, and a laser beamwidth of 1 mrad. Connect loss factor and pupil size to turning scale ①, from that point to laser output of 0.05 J to turning scale ②, then through safety threshold point to turning scale ③, and finally through laser beamwidth ④ to distance line. In this case the safe range is approximately 4.0 km or 2.6 statute miles.

NOTE: "Safe" threshold levels are a subject of some controversy and the figures specified in the nomogram should be interpreted in the light of most recent information.

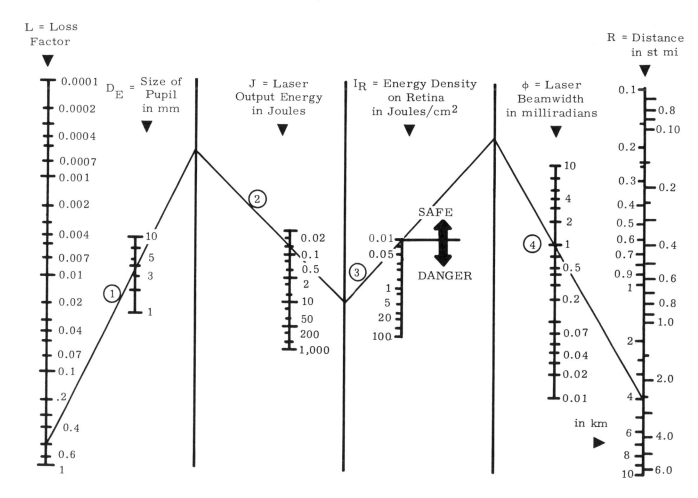

LASER RADIATION
NOMOGRAM

This nomogram relates laser radiation terms, which may be given as photon energy, wave number, frequency, or wavelength. Any of these terms can be converted to the others by a horizontal line across the nomogram.

FOR EXAMPLE:

1. Light at a wavelength of 0.5 μ can also be described as having (1) A wavelength of 5000 Å, (2) a frequency of 600 THz or 6×10^{14} Hz, (3) a wavenumber of 20,000 cm^{-1}, and (4) a photon energy of 2.48 eV.
2. Electrons when falling through 4 V will radiate at 3100 Å.
3. Light at 200 THz will produce conduction in semiconductors with band-gaps up to 0.83 V.

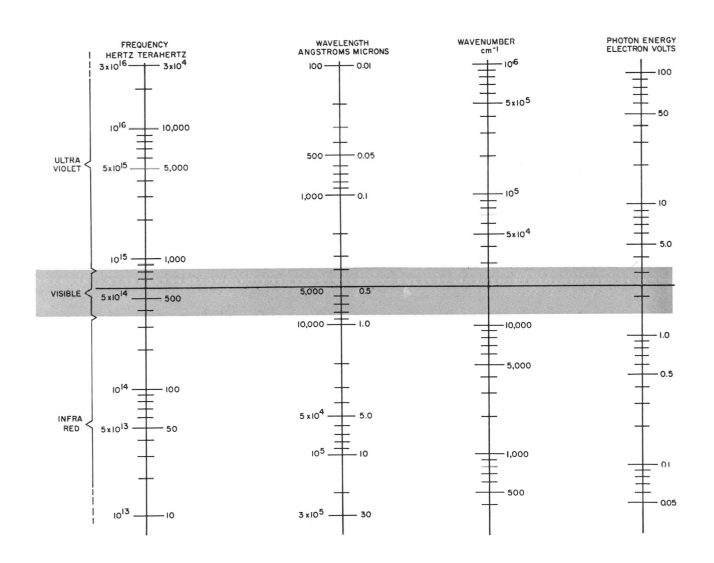

GENERALIZED RADIOACTIVITY DECAY CURVE

Knowing the isotope half-life, its original activity at some particular time, it is an easy matter, using the chart, to determine the residual activity at some subsequent time.

FOR EXAMPLE:

A sample of radioactive iodine—131 has an activity of 10 μC, find the remaining strength 20 days later.

ANSWER:

From an appropriate source determine the half-life of the isotope. For radioactive iodine—131, the half-life is 8.1 days.

Calculate how many "half-lives" there are corresponding to the time interval in question, that is, divide the time interval by the half-life: in this case 20/8.1 = 2.47.

Enter this value on the horizontal axis of the chart and read the "fraction remaining" on the vertical axis as shown by the broken lines. In the case under consideration the value is 0.177.

Multiply this value by the original activity thus giving a final value of 1.77 μC.

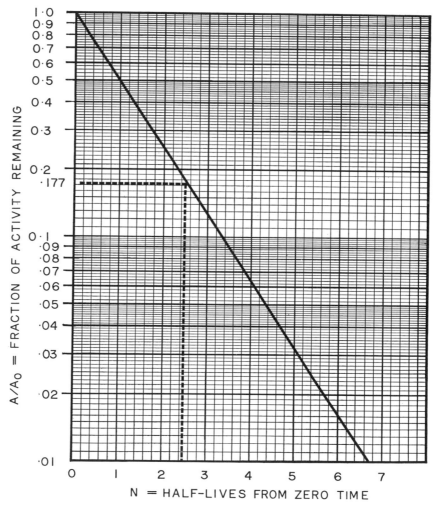

(From *Electronics and Communications*, August 1962.)

SPECTRAL CHARACTERISTICS OF PHOTORECEPTORS AND LIGHT SOURCES

This figure shows spectral sensitivity of various photoreceptors. Response of cadmium sulfide cells is similar to that of the human eye, but other commonly used receptors perform best at wavelengths invisible to the eye.

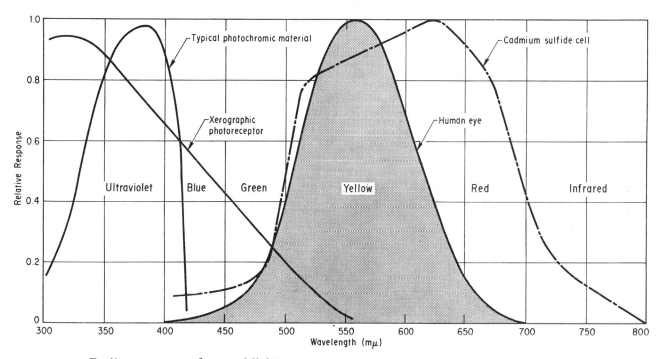

Radiant output of several light sources.

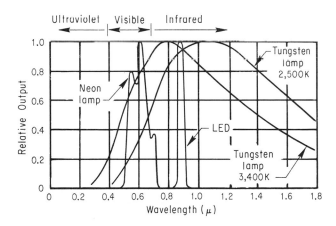

This nomogram solves the light intensity equation:

$$\text{foot-candles} = \frac{\text{candlepower}}{(\text{distance in feet})^2}$$

which assumes a point source (distance greater than five times maximum lamp dimension).

Most lamps are classified according to wattage, and the following approximate relations apply:

1. The shorter the rated life of the lamp, the higher the efficiency (cp/watt) and the higher the color temperature of the light.
2. For standard 120-V inside-frosted incandescent lamps rated for 1000 hr, the following hold true:
   a. Efficiency increases with increasing wattage.
   b. A 25-W lamp is approximately 19 cp, a 60-W lamp about 60 cp, and a 150-W lamp is near 200 cp.
   c. Color temperature increases with increasing wattage (150-W lamp is near 2900 K).
   d. When lamps are operated at constant voltage, light output falls with time, rapidly during the first 50 hr and more slowly thereafter.
   e. When lamps are operated at constant current, light output rises with time, slowly at first, then accelerating to catastrophic failure.

FOR EXAMPLE:
A 6-cp lamp will produce a light intensity of 100 fc, at a distance of 2.94 in. (0.245 ft) from the lamp filament. The same lamp will provide 1 fc at 29.4 in. and 0.01 fc at 294 in.

PHOTOMETRY NOMOGRAM
(Continued)

Several Useful Definitions

A *foot-candle* is the illumination produced when the light from one candle falls normally on a surface at a distance of one foot.

A *lux* (commonly used in Europe) is the illumination produced when the light from one candle falls normally on a surface at a distance of one meter.

A point source emitting light uniformly in all directions radiates 4π lumens/candle.

A *lambert* is the brightness of a perfectly diffusing surface emitting or reflecting one lumen per square centimeter.

A *foot lambert* equals $1/\pi$ candles/ft^2.

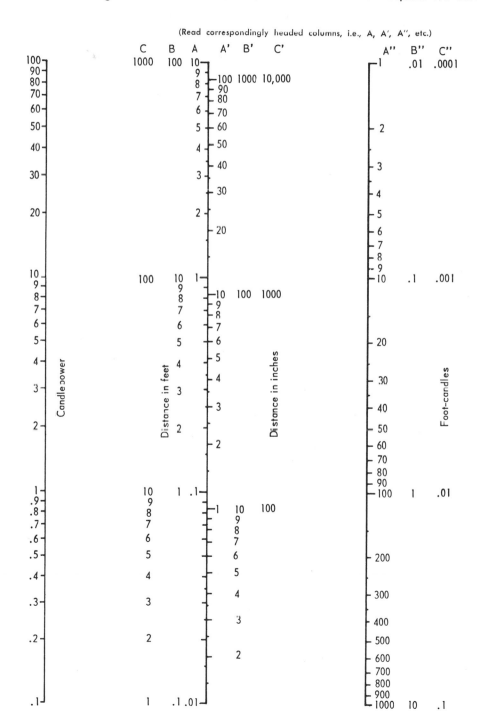

(Read correspondingly headed columns, i.e., A, A', A'', etc.)

MINIMUM DETAIL THAT THE
HUMAN EYE CAN RESOLVE

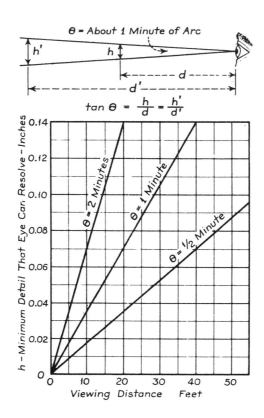

SUGGESTED VALUES OF
ILLUMINANCE

Auditorium	10 fc
Lecture room—library	30 fc
Classroom	30 fc
Drafting room	30 fc
Low-contrast work inspection	250 fc
Hospital operating room	500–1000 fc

ILLUMINATION UNITS
CONVERSION NOMOGRAM

This nomogram relates candles/square foot, foot-candles, lumens/square foot, lamberts, foot-lamberts, lumens/square centimeter, candles/square centimeter, candles/square inch, and lux, and it is based on the following relationships:

foot-lamberts = lumens/square foot = foot-candles = 10.764 lux

lamberts = lumens/square centimeter = 295.72 candles/square foot = 929.03 lumens/square foot

lux = lumens/square centimeter and candles/square centimeter = 3.14159 lambert

A line from any known value through the index point intersects all other scales at corresponding values.

FOR EXAMPLE:

$$4 \, L = 8.2 \, \text{cd/in.}^2$$
$$= 4400 \, \text{fc}$$
$$= 3715 \, \text{lm/ft}^2$$
$$= 1183 \, \text{cd/ft}^2$$

Note that the ranges can be extended by multiplying all scales by the same power of 10.

ILLUMINATION UNITS
CONVERSION NOMOGRAM
(Continued)

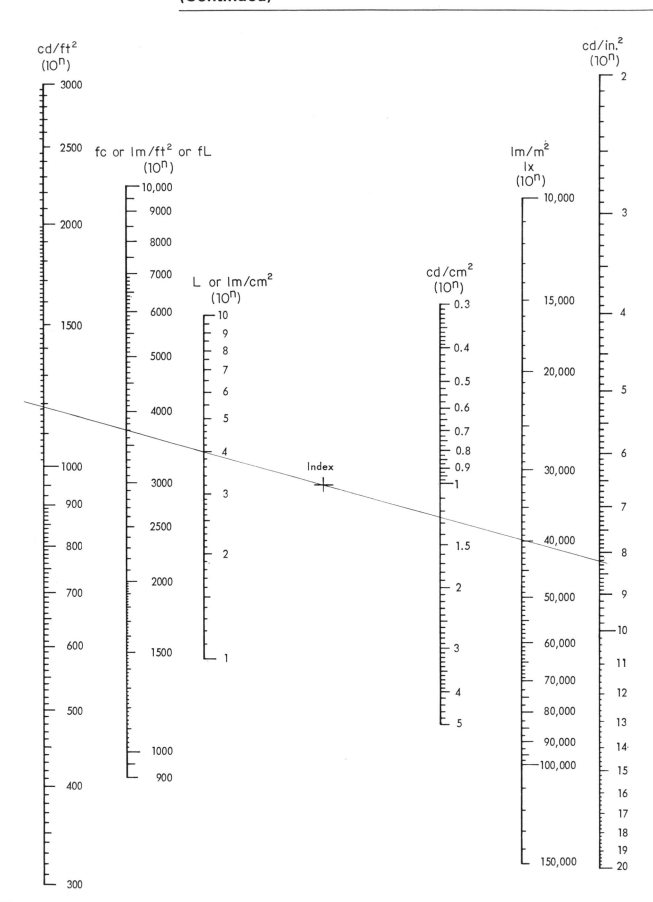

ILLUMINATION POWER
CONVERSION NOMOGRAM

This nomogram relates international lumens, watts, and candlepower. Select the known value. A line from that point through the index point intersects other scales at corresponding values.

FOR EXAMPLE:

5 lm = 0.0074 W

50 lm = 3.98 cp

Note that the ranges can be extended by multiplying all scales by the same power of 10.

The nomogram is based on the following:

1 cp = 12.566 lm
1 lm = 0.001496 W

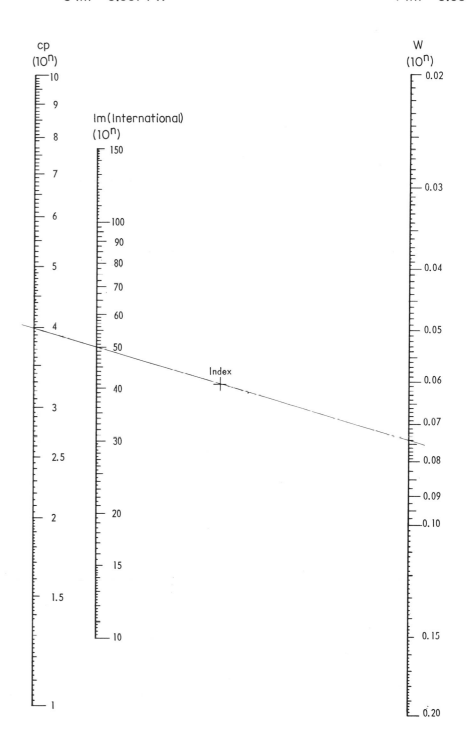

TABULATION OF SOUND INTENSITY LEVELS

This tabulation extends from the barely audible to the unbearable and/or damaging sound intensity levels. The various levels are given in terms of sound pressure in dynes per square centimeter, sound intensity (at the eardrum) in watts per square centimeter, and intensity level in decibels above 10^{-16} W/cm^2 and related to familiar sound situations.

FOR EXAMPLE:
A faint to moderate sound such as can be found in an average residence is equal to a sound pressure of 0.024 dyn/cm^2, which produces a sound intensity at the eardrum of 10^{-12} W/cm^2 (1 pW/cm^2) and is equal to an intensity level of 40 dB above 10^{-16} W/cm^2.

Table of Sound Intensity Levels

Description or Effect	Sound Pressure (dyn/cm²)	Sound Intensity at Eardrum (W/cm²)	Intensity Level (dB above 10^{-16} W/cm²)	Familiar Sources of Sound (number in parentheses shows distance from source)
Impairs hearing		10^{-1}	150	
Pain	2040	10^{-2}	140	jet engine largest air raid siren (100 ft)
Threshold of pain		10^{-3}	130	level of painful sound
Threshold of discomfort	204	10^{-4}	120	pneumatic hammer (5 ft) airplane 1600 rpm (18 ft from propeller) automobile horn
Deafening		10^{-5}	110	engine room of submarine (at full speed) bass drum (maximum)
Discomfort begins	20.4	10^{-6}	100	boiler factory loud bus horn thunder clap subway (express passing a local station) can manufacturing plant
Very loud		10^{-7}	90	very loud musical peaks noisiest spot at Niagera Falls
	2.04	10^{-8}	80	loudest orchestral music noisy factory heavy street traffic { loud speech police whistle very loud radio
Loud		10^{-9}	70	{ average factory average orchestral volume busy street noisy restaurant average conversation (3 ft)
	0.204	10^{-10}	60	quiet typewriter average (quiet) office hotel lobby quiet residential street
Moderate		10^{-11}	50	soft violin solo church quiet automobile
	0.0204	10^{-12}	40	average residence lowest orchestral volume
Faint		10^{-13}	30	quiet suburban garden average whisper
	0.00204	10^{-14}	20	very quiet residence faint whisper (5 ft)
Very faint		10^{-15}	10	{ ordinary breathing (1 ft) outdoor minimum (rustle of leaves) anechoic room
Threshold of hearing	0.000204	10^{-16}	0	normal threshold of hearing reference level

EQUAL LOUDNESS CURVES OF THE AVERAGE HUMAN EAR

(20- to 29-year old subjects)

The curves show that the frequency response characteristic of the human ear varies with the loudness of the sound. At low sound levels the ear is relatively insensitive to the lower frequencies, which must be at least 60 dB to be heard. Higher sound levels are heard nearly equally well at the high and low frequencies. Therefore, for listening at low volume levels, the low frequencies must be boosted considerably to produce the effect of equal loudness and to avoid an apparent lack of low frequency tones. The ear is most sensitive to sounds in the 2000 to 4000 Hz range.

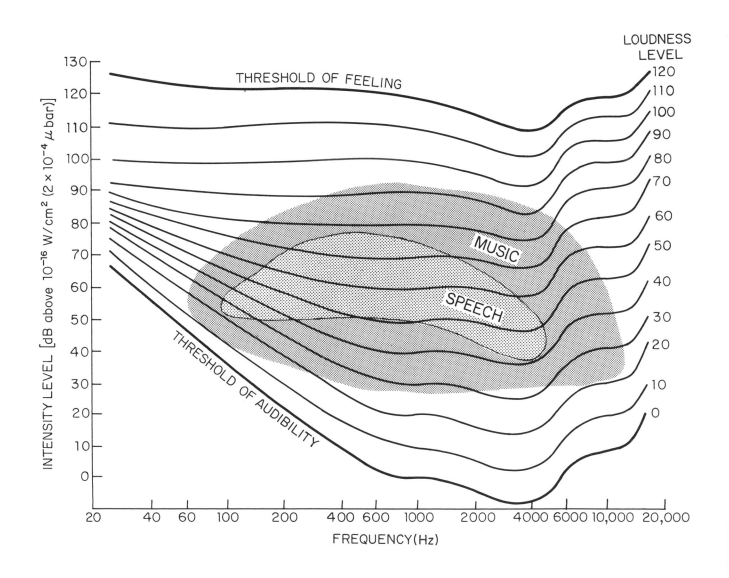

REVERBERATION TIME

These graphs determine the optimum recommended reverberation time as a function of room volume and usage. The optimum times for speech rooms, motion picture theaters, and school auditoriums are given by a single line, whereas the optimum time for music is a broad band. Furthermore, the optimum reverberation time is not the same for all kinds of music. For example, slow organ and choral music require more reverberation than does a brilliant allegro composition played on woodwinds or a harpsichord.

The first chart is used to find the optimum reverberation time for frequencies above 512 Hz. For lower frequencies that value must be multiplied by the appropriate factor in the second graph. For small rooms the lower part of the shaded portion (closer to 1.0 should be used.)

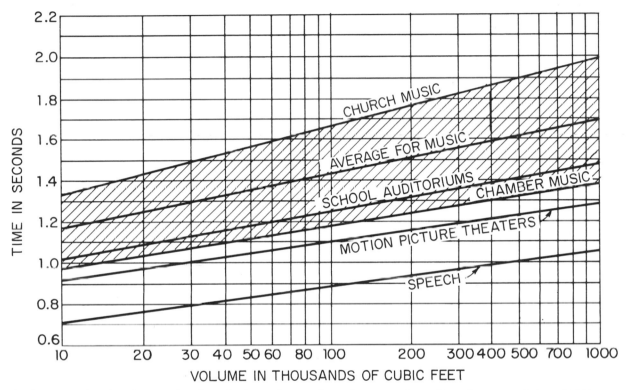

Optimum reverberation time as a function of volume of rooms for various types of sound for a frequency of about 512 Hz.

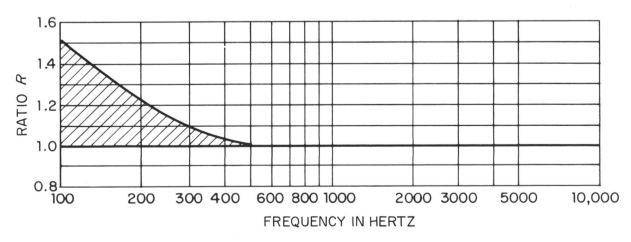

Ratio of the reverberation time for various frequencies as a function of the reverberation for 512 Hz.

PHYSIOLOGICAL EFFECTS OF ELECTRIC CURRENT ON THE HUMAN BODY

The chart shows the physiological effect of various current densities on the human body. Voltage is not the prime consideration, though it takes voltage to produce the current flow. The amount of shock current depends on the body resistance between the points of contact and the skin condition, (that is, moist or dry). For example, the internal resistance between the ears is only 100 ohms (less the skin resistance), while from hand to foot it is close to 500 ohms. Skin resistance may vary from about 1000 ohms for wet skin to over ½ Mohm for dry skin, and is even lower for ac.

The chart shows that shock becomes more severe as current rises. At values as low as 20 mA breathing becomes labored, and as the current approaches 100 mA, ventricular fibrillation of the heart occurs. Above 200 mA, the muscular contractions are so severe that the heart is forcibly clamped during the shock. This clamping protects the heart from going into ventricular fibrillation and the victim's chances for survival are good if the victim is given immediate attention. Resuscitation, consisting of artificial respiration, will usually revive the victim.

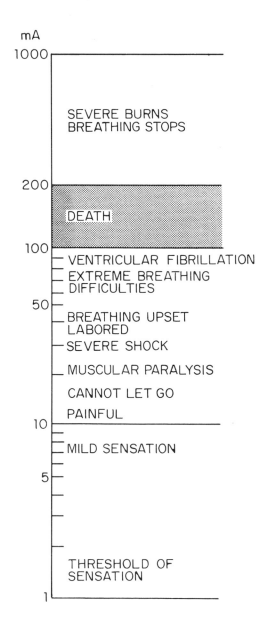

CHARACTERISTICS OF MINIATURE INCANDESCENT LAMPS

This graph relates light output, current, and life of incandescent lamps with rated (design) voltage. The curves show that the light output varies directly as the applied voltage raised to the 3.4th power, while life is inversely proportional to applied voltage raised to the 12th power.

FOR EXAMPLE:

At 110% of rated voltage, the current will increase by 5%, light output increases by 40%, and life will be reduced to nearly 35% of that at design voltage.

At 80% of rated voltage, current decreases by 10%, light output drops by more than 50%, but lamp life is increased to 18 times normal.

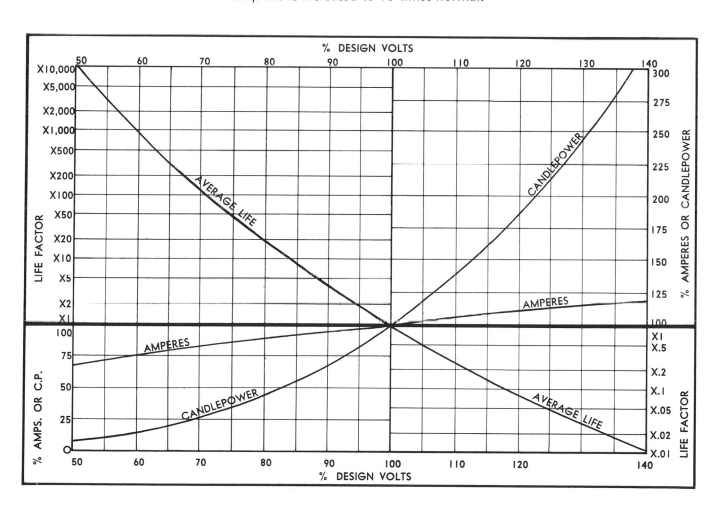

COLOR CODES FOR ELECTRONIC COMPONENTS

STAND-ARD COLORS	NUMERICAL VALUES[1]				FILM RESISTORS			CERAMIC CAPACITORS							CHASSIS WIRING	TWT WIRED LEADS	
	Num Fig.	Decimal Multiplier		Value Tol. (%)	Sig. Fig.	Mult.	Tol. ± %	Sig. Fig.	Multiplier	Tolerance		Temp. Coeff. ppm/°C	Sig. Fig.	Extended Range Temp. Coeff. Multiplier		Tracer	TWT Element
		Power of 10	Mult.							Over 10 pf (± %)	10 pf or less (± pf)						
BLACK	0	10^0	10^0	±20	0	10^0	—	0	1	20	2	0	0.0	−1	Grounds, grounded elements, and returns	none	Grounds, grounded elements
BROWN	1	10^1	10^1	± 1	1	10^1	1	1	10	1	0.1	−30		−10	Heaters or filaments, off-gnd.	none	Heaters or fil. off gnd.
RED	2	10^2	10^3	± 2	2	10^2	2	2	100	2		−75	1.0	−100	Pwr. supply B+	none	Collector
ORANGE	3	10^3	10^3	± 3	3	10^3	—	3	1,000	2.5		−150	1.5	−1,000	Screen grids	none green blue gray	Helix 1 Helix 2 Helix 3 Helix 4
YELLOW	4	10^4	10^4	GMV[2]	4	10^4	—	4	10,000			−220	2.2	−10,000	Cathodes	none	Cathode, also heater-cathode lead if common
GREEN	5	10^5	10^5	± 5 (optional coding)	5	10^5	0.5	5		5	0.5	−330	3.3	+1	Control grids	none black	Grid 1 Grid 5
BLUE	6	10^6	10^6	± 6	6	10^6	0.25	6				−470	4.7	+10	Plates	none black	Grid 2 Grid 6
VIOLET	7	10^7	10^7	±12½	7	10^7	0.10	7				−750	7.5	+100	Not used	—	—
GRAY	8	10^{-2} (alternate)	0.01 (optional coding)	±30	8		0.05	8	0.01		0.25	+30		+1,000	ac pwr. lines	none black	Grid 3 Grid 7
WHITE	9	10^{-1} (alternate)	0.1 (optional coding)	±10 (optional coding)	9		—	9	0.1	10	1				Above or below gnd. returns, AVC, etc.	none black	Grid 4 Grid 8
SILVER	—	10^{-2} (preferred)	0.01	±10	—	0.01	10	—									
GOLD	—	10^{-1} (preferred)	0.1	± 5	—	0.1	5	—									
NO COLOR	—	—	—	±20	—	—	—	—									
ILLUS.	A, B				C			D									

[1] For components such as resistors, capacitors, and wires. Also, for identification of terminals and circuit functions.
[2] GMV = −0 +100% tolerance or Guaranteed Minimum Value.
[3] If heater-cathode elements are internally connected, but have ext. leads, the body color gives the major element and the tracer is the internally connected element.

[4] Orange body identifies elements used principally for modulation purposes or beam focusing.
[5] Elements used to control beam current, beam noise, etc.
[1-8] Elements are numbered according to their relative position from cathode—lowest

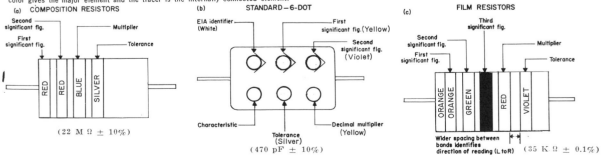

(a) COMPOSITION RESISTORS
(22 M Ω ± 10%)

(b) STANDARD—6-DOT
(470 pF ± 10%)

(c) FILM RESISTORS
(35 K Ω ± 0.1%)

KLYSTRON WIRED LEADS		CROSSED FIELD DEVICES			STEREO PICK-UP LEADS			SEMI-CONDUCTOR DEVICES[8] (Diodes & Rectifiers)		TRANSFORMERS				STANDARD COLORS
Tracer	Tube Element	Magnetron	VTM[6]	BWO[7] (M-Type)	3 Wire	4 Wire	5 Wire	Number	Suf.	A-F	I-F[h]	Power	Center-Tap	
none	Body, or other grounded elements	Body or other grounded elements			Return or grid		Return or gnd.	0	Not Applicable	Grid return (applies whether the secondary is plain or center-tapped)	Grid (or diode) return.	Primary leads if tapped: Common-black. Tap-black and yellow stripe. Finish-black and red stripe.		BLACK
none[a]	Heater	Heaters or filament off-gnd.						1	A	Plate (start) lead on center-tapped primaries. (Blue may be used if polarity isn't important)		Filament winding #2	Brown and yellow stripe.	BROWN
none	Collector (if isolated)	Anode	Anode	Delay Line	Right high	Right high	Right high	2	B	"B +" lead (applies whether the primary is plain or center-tapped)	"B +" lead	H-V plate winding	Red and yellow stripe.	RED
none[a]; green; blue; gray; black; white	Reflector, phase modulation element or electrostatic focusing element #1; element #2; element #3; element #4; element #5; element #6	—	—	Sole				3	C					ORANGE
none[a]	Cathode, also heater cathode lead if common	Cathode or common heater cathode						4	D	Grid (start) lead on center-tapped.		Rectifier filament winding	Yellow and blue stripe.	YELLOW
none[a]; black	Grid 1; Grid 5	—	Injector	Grid		Right low	Right low	5	E	Grid (finish) lead to secondary.	Grid (or diode) lead	Filament winding #1	Green and yellow stripe.	GREEN
none; black	Grid 2; Grid 6	—	—'	Accelerator		Left low	Left low	6	F	Plate (finish) lead of primary	Plate lead			BLUE
—	—	—	—	—				7	G					VIOLET
none; black	Grid 3; Grid 7	—	—	—				8	H			Filament winding #3 (slate color)	Slate and yellow stripe.	GRAY
none; black	Grid 4; Grid 8	—	Cold Cathode	—	Left high	Left high	Left high	9	J					WHITE
														SILVER
														GOLD
														NO COLOR
								E, F, G						ILLUS.

being closest. Where two elements are equidistant, lower voltage element will have the lower number.
[6] Voltage Tunable Magnetron.
[7] Backward Wave Oscillator.

[8] Prefix identification consisting of a number symbol and the letter "N" shall not be indicated in the coding.
[h] If secondary is center-tapped, the second diode plate lead is green-and-black striped and black is used for the center-tapped lead.

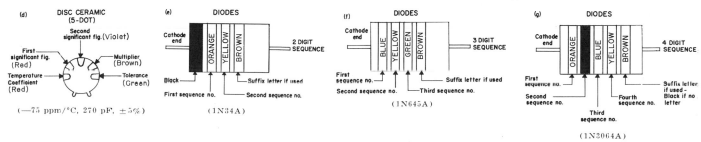

(d) DISC CERAMIC (5-DOT)
First significant fig. (Red)
Second significant fig. (Violet)
Multiplier (Brown)
Temperature Coefficient (Red)
Tolerance (Green)
(—75 ppm/°C, 270 pF, ±5%)

(e) DIODES
Cathode end — ORANGE YELLOW BROWN — 2 DIGIT SEQUENCE
Black
First sequence no.
Second sequence no.
Suffix letter if used
(1N34A)

(f) DIODES
Cathode end — BLUE YELLOW GREEN BROWN — 3 DIGIT SEQUENCE
First sequence no.
Second sequence no.
Third sequence no.
Suffix letter if used
(1N645A)

(g) DIODES
Cathode end — ORANGE BLUE YELLOW BROWN — 4 DIGIT SEQUENCE
First sequence no.
Second sequence no.
Third sequence no.
Fourth sequence no.
Suffix letter if used – Black if no letter
(1N3064A)

CATHODE-RAY TUBE PHOSPHOR CHARACTERISTICS

JEDEC Type	Color Fluorescence	Phosphorescence	Spectral Range A°	Persistence	Application
P1	Yellow-Green	Yellow Green	4900-5800	Medium	Oscillography
P2	Yellow-Green	Yellow-Green	4400-6100	Medium	Oscillography
P3	Yellow-Orange	Yellow-Orange	5040-7000	Medium	No longer in general use
P4	White	White	4100-6900	Medium short	Television
P5	Blue	Blue	3500-5600	Medium short	Photographic
P6	White	White	4160-6950	Short	No longer in general use
P7	White	Yellow-Green	3900-6500	One, medium short; One, long	Radar and oscillography
P10	Dark trace: color depends upon absorption characteristics and type of illumination		4000-5500	Very long	Radar
P11	Blue	Blue	4000-5500	Medium short	Oscillographic recording
P12	Orange	Orange	5450-6800	Long	Radar
P13	Red-Orange	Red-Orange		Medium	No longer in general use
P14	Purple-Blue	Yellow-Orange	3900-7100	One, medium short, One, medium	Radar
P15	Green	Green	3700-6050	Visible, short; Ultraviolet, very short	Flying spot scanning systems; photographic
P16	Blue-Purple and near UV	Blue-Purple and near UV	3450-4450	Very short	Flying spot scanning systems; photographic
P17	Yellow-White to Blue-White	Yellow	3800-6400	One, short; One, long	Radar
P18	White	White	3260-7040	Medium to medium short	Television
P19	Orange	Orange	5450-6750	Long	Radar
P20	Yellow-Green	Yellow-Green	4850-6700	Medium to medium short	Radar
P21	Red-Orange	Red-Orange	5540-6500	Medium	Radar
P22	Tri-color		4000-7200	Medium short	Color Television
P23	White	White	4100-7200	Medium to medium short	Television
P24	Green	Green	4300-6300	Short	Flying spot scanning systems
P25	Orange	Orange	5300-7100	Medium	Radar
P26	Orange	Orange	5450-6650	Very long	Radar
P27	Red-Orange	Red-Orange	5820-7200	Medium	Color television monitor service
P28	Yellow-Green	Yellow-Green	4650-6350	Long	Radar
P29	Two-color phosphor screen composed of a linear array of alternate strips of P2 and P25 phosphors				Radar
P31	Green	Green	4150-6000	Medium short	Oscillography
P32	Purple-Blue	Yellow-Green	3800-6550	Long	Radar
P33	Orange	Orange	5450-6850	Very long	Radar
P34	Blue-Green	Yellow-Green	3900-6800	Very long	Radar and oscillography
P35	Blue-White	Blue-White	4350-6480	Medium short	Photographic

Important operating parameters are listed for various crystal cuts. The impedance of a crystal is close to zero at the resonant frequency (f_s) and rises to a peak at the antiresonant frequency (f_a). The practical parallel resonant operating frequency ranges between f_s and f_a and may include these two limiting values. The operating frequency is expressed as

$$f_p = f_s \sqrt{1 + \frac{C_1}{C_0}}$$

The steep slope of the curve and the corresponding large differential between the impedances at f_s and f_p indicate that the Q of the crystal is high. Also, the frequency separation between f_s and f_p is determined by the capacitance ratio C_0/C_1. For example, the 45° cut is a favorite choice in crystal filters because of its low C_0/C_1 ratio. Thus a larger filter bandwidth is achieved with fewer crystals.

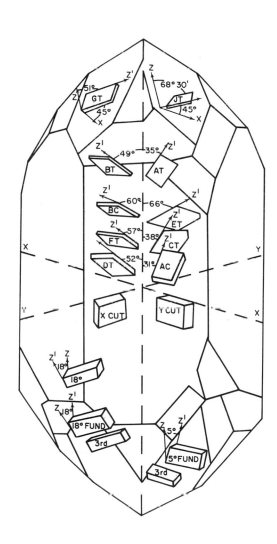

The orientation of the better known crystal cuts shows the difference among the types.

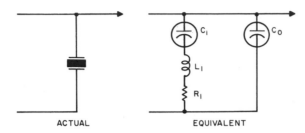

ACTUAL   EQUIVALENT

Equivalent circuit of a crystal includes the capacitances contributed by the wire leads and the holder in C_0. ratio of C_0 and C_1 indicates the frequency separation between the resonant and antiresonant frequencies of the crystal.

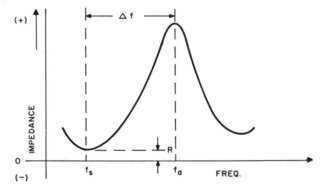

The impedance of a crystal is near zero at the series resonant frequency, f_s, and reaches its peak at the antiresonant frequency, f_a. Steep slope between these two frequencies indicates a high Q.

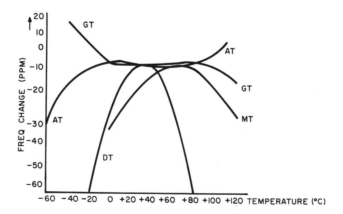

Temperature characteristics of four popular crystal cuts show the extremely stable behavior of the GT cut. Its frequency change is about 1 part per million over a 100° C range.

Cut	Designation	Mode of vibration	Frequency range in kHz	C_0/C_1	Max. drive level	Remarks
Duplex 5°X	J	Length,	0.800-10	190-250	0.20	Used in frequency and oscillator applications. Zero-temperature coefficient occurs at approximately room temperature; therefore the crystal is limited to oven operation and to rigid temperature-control conditions.
XY	Custom-made	Length, width	3-50	600-900	0.1	Suited for oven-control applications, especially in its optimum frequency range.
NT	N	Length,	4-150	800-1500	0.1	Preferred in low-frequency oscillators and filters. It operates over large temperature ranges. Stability of ±5 ppm can be obtained over ±5°C, if oven-controlled in the frequency range. Rugged, if properly mounted. Can obtain frequency stability within ±0.0025% over the normal room-temperature range, without temperature control.
+5°X	H	Flexure	5-140	225	0.1	A relatively large frequency deviation over temperature range restricts filter applications to controlled environments. Low temperature coefficient and large ratio of stored mechanical energy to electrical energy are the characteristic features. Used in wideband filters, below the range of practical size E plates, and in transistor oscillators, where LC circuits are not stable enough, or where there is a space problem. Disadvantages: Fabrication difficulties. The crystal must be made in the form of a long, thin bar to fit in a special holder, to avoid jumping between modes.
BT	D	Thickness	1-75	–	–	Thicker crystal possible at higher frequencies. Disadvantages: Too thick for low frequency. Also, difficult to fabricate and has zero-temperature coefficient over only a very small temperature range. Not as active as the AT.
−18-1/2°X	F	Extensional	50-250	200	–	Used principally in filters where low temperature coefficient is sacrificed for freedom from certain spurious responses. Suitable for multi-electrodes.
+5°X	E	Extensional	50-250	130-160	2.0	Mostly applicable in low-frequency filters, because of low C_0/C_1 and good temperature coefficient.
DT	D	Face shear	80-500	450	2.0	Suitable for oven and non-oven applications. Its low capacity ratio permits many useful filter applications. Used as calibrator crystal and time base for frequency counters. Also used in FM and TV transmitters. Disadvantage: Does not perform well over 500 kHz.

Cut	Designation	Mode of vibration	Frequency range in kHz	C_0/C_1	Max. drive level	Remarks
MT	M	Extensional	50-250	250	2.0	Its low temperature coefficient makes it useful for oscillator control and for filters where low C_0/C_1 ratio is required along with low inductance and good temperature coefficient. However, this crystal is seldom used, because more compact units have replaced it.
GT	G	Extensional	85-400	375	0.1	Has the greatest stability yet attained within a cut. Does not vary more than 1 part per million over a range of 100°C. Offers a low temperature coefficient over a wide frequency range, by coupling any desired mode with another of nearly equal amplitude at a frequency equal to 0.86 times its natural frequency. Used in frequency standards and when stability without temperature control or low impedance is essential. Disadvantages: Most expensive of all types, because of painstaking labor required to obtain exact orientation in dimension.
CT	C	Face shear	300-1100	350-400	2.0	Provides a zero temperature coefficient in the shear mode for low frequencies. Widely used in low-frequency oscillators and filters and does not require constant temperature control over normal operating conditions. Useful in filters because of low C_0/C_1 ratio. Popular in oscillators because of its low series resistance, especially above 400 kHz. Disadvantages: Large face dimensions make it difficult to fabricate for the very low frequencies.
X	Custom-made	Extensional	350-20,000	–	–	Mechanically stable and an economic type of cut. Disadvantages: Large temperature coefficient, with the tendency to jump from one mode to another.
SL	Custom-made	Face shear, coupled to flexure	300-800	450	–	Electrical characteristics similar to DT, but it is larger, has better Q and uniformity of characteristics above 300 kHz. Its various characteristics make it desirable for some filter applications.
Y	Y	Thickness, shear	500-20,000	–	–	Most active. Ratio of stored mechanical to electrical energy is large. Is strong mechanically. Disadvantages: Large temperature coefficient and poor frequency spectrum.
AT	A	Thickness	550-20,000 fundamental 10,000-60,000 (3rd overtone) 100,000 (5th overtone)	10-100,000	1.0-8.0	Excellent temperature and frequency characteristics. Its overtones are used in cases where the frequency should not change with oscillator reactance variations. Designs provide suitable capabilities for satisfying 70-80% of all crystal requirements. Preferred for high-frequency oscillator-control wherever wide variation of temperature is encountered. Because of small size, it can be readily mounted to meet stringent vibration specifications. Disadvantage: Difficult to fabricate for optimum operation without coupling between modes.

The AN nomenclature designation is assigned to:
1. Complete sets of equipment and major components of military design.
2. Groups of articles of commercial or military design which are grouped for a military purpose.
3. Major articles of military design which are not part of, or used with, a set.
4. Commercial articles where nomenclature facilitates identification and/or procedures.

As applied to complete sets, the nomenclature consists of the two letters *AN* followed by a slash and three indicator letters which indicate installation, type of equipment, and purpose. The number that may follow the letters indicates model number, and a subsequent letter refers to modification.

FOR EXAMPLE:

AN/APN—10B   airborne–radar–navigational   aid 10th model–second modification

As applied to components, the AN nomenclature consists of one or two designator letters substituted for AN.

FOR EXAMPLE:

An indicator model 42 for use with APQ-13 is designated as ID-42/APQ-13. Modifications are indicated by letters, for example, ID-42B/APQ-13

MILITARY NOMENCLATURE
SYSTEM (Continued)

Component Indicator Letters

AB—Support, antenna
AM—Amplifier
AS—Antenna assembly
AT—Antenna
BA—Battery, primary type
BB—Battery, secondary type
BZ—Signal device, audible
C—Control article
CA—Commutator assembly, sonar
CB—Capacitor bank
CG—Cable and transmission line, r.f.
CK—Crystal kit
CM—Comparator
CN—Compensator
CP—Computer
CR—Crystal
CU—Coupling device
CV—Converter (electronic)
CW—Cover
CX—Cord
CY—Case
DA—Antenna, dummy
DT—Detecting head
DY—Dynamotor
E—Hoist assembly
F—Filter
FN—Furniture
FR—Frequency measuring device
G—Generator
GO—Goniometer
GP—Ground rod
H—Head, hand, and chest set
HC—Crystal holder
HD—Air conditioning apparatus
ID—Indicating device
IL—Insulator
IM—Intensity measuring device
IP—Indicator, cathode-ray tube
J—Junction device
KY—Keying device
LC—Tool, line construction
LS—Loudspeaker
M—Microphone
MD—Modulator
ME—Meter, portable
MK—Maintenance kit or equipment
ML—Meterological device
MT—Mounting
MX—Miscellaneous

O—Oscillator
OA—Operating assembly
OS—Oscilloscope, test
PD—Prime driver
PF—Fitting, pole
PG—Pigeon article
PH—Photographic article
PP—Power supply
PT—Plotting equipment
PU—Power equipment
R—Radio and radar receiver
RD—Recorder and reproducer
RE—Relay assembly
RF—Radio frequency component
RG—Cable and transmission line, bulk r.f.
RL—Reel assembly
RP—Rope and twine
RR—Reflector
RT—Receiver and transmitter
S—Shelter
SA—Switching device
SB—Switchboard
SG—Generator, signal
SM—Simulator
SN—Synchronizer
ST—Strap
T—Radio and radar transmitter
TA—Telephone apparatus
TD—Timing device
TF—Transformer
TG—Positioning device
TH—Telegraph apparatus
TK—Tool kit or equipment
TL—Tool
TN—Tuning unit
TS—Test equipment
TT—Teletypewriter and facsimile apparatus
TV—Tester, tube
U—Connector, audio and power
UG—Connector, r.f.
V—Vehicle
VS—Signaling equipment, visual
WD—Cable, two-conductor
WF—Cable, two-conductor
WM—Cable, multiple-conductor
WS—Cable, single-conductor
WT—Cable, three-conductor
ZM—Impedance measuring device

MILITARY NOMENCLATURE
SYSTEM (Continued)

Set or Equipment Indicator Letters

1st letter — Designed Installation Classes	2d letter — Type of Equipment	3d letter — Purpose	Model No.	Modification letter	Miscellaneous Identification
A Airborne (installed and operated in aircraft).	A Invisible light, heat radiation.	A Auxiliary assemblies (not complete operating sets used with, or part of, two or more sets or sets series).	1 2 3 4 etc.	A B C D etc.	X } Changes in voltage, phase, or frequency. Y } Z } T Training. (V) Variable grouping.
B Underwater mobile, submarine.	B Pigeon.	B Bombing.			
C Air transportable (inactivated, do not use).	C Carrier.	C Communications (receiving and transmitting).			
D Pilotless Carrier.	D Radiac.	D Direction finder, reconnaissance, and/or surveillance.			
	E Nupac.	E Ejection and/or release.			
F Fixed.	F Photographic.				
G Ground, general ground use (include two or more ground-type installations).	G Telegraph or teletype.	G Fire-control or searchlight directing.			
		H Recording and/or reproducing (graphic meteorological and sound).			
	I Interphone and public address.				
	J Electromechanical or inertial wire covered.				
K Amphibious.	K Telemetering.	K Computing.			
	L Countermeasures.	L Searchlight control (Inactivated, use G).			
M Ground, mobile (installed as operating unit in a vehicle which has no function other than transporting the equipment).	M Meteorological.	M Maintenance and test assemblies (Including tools).			
	N Sound in air.	N Navigational aids (including altimeters, beacons, compasses, racons, depth sounding, approach and landing).			
P Pack or portable (animal or man).	P Radar.	P Reproducing (inactivated, do not use).			
	Q Sonar and underwater sound.	Q Special, or combination of purposes.			
	R Radio.	R Receiving, passive detecting.			
S Water surface craft.	S Special types, magnetic, etc., or combinations of types.	S Detecting and/or range and bearing, search.			
T Ground, transportable.	T Telephone (wire).	T Transmitting.			
U General utility (includes two or more general installation classes, airborne, shipboard, and ground).					
V Ground, vehicular (installed in vehicle designed for functions other than carrying electronic equipment, etc., such as tanks).	V Visual and visible light.				
W Water surface and underwater.	W Armament (peculiar to armament, not otherwise covered).	W Automatic flight or remote control.			
	X Facsimile or television.	X Identification and recognition			
	Y Data processing.				

This nomogram solves for the magnetic field strength, surrounding a power line, as a function of current in the line and the distance from it. Electronic equipment is susceptible to magnetic field interference, and this nomogram helps in determining the magnitude of the problem. For convenience the distance scale is calibrated in inches and centimeters.

FOR EXAMPLE:

The magnetic field strength at a point 5 cm from a line that carries 100 A is 4.2 G.

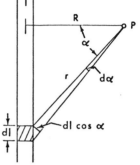

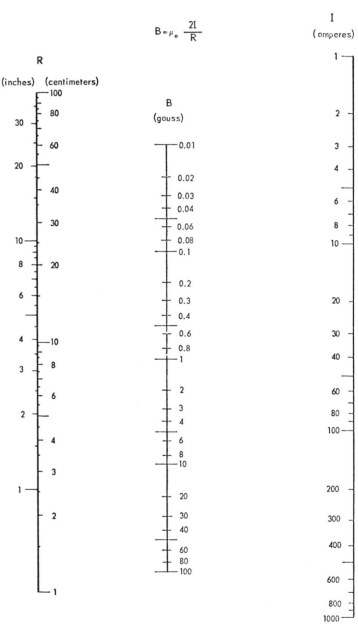

$$B = \mu_0 \frac{2I}{R}$$

Derivation of the Field-Strength Equation

The field at point P resulting from the current in segment dl is given by

$$dB = \mu_0 \frac{I}{r^2} \cos \alpha \, dl$$

If dl is small, then

$$dl \cos \alpha = r \, d\alpha$$
$$r = R/\cos \alpha$$

and

$$\therefore dB = \mu_0 \frac{I}{R} \cos \alpha \, d\alpha$$

If the line is very long with respect to R,

$$B = \int_{-\pi/2}^{\pi/2} \mu_0 \frac{I}{R} \cos \alpha \, d\alpha = \mu_0 \frac{2I}{R}$$

If B is in gauss, I in amperes, and R in centimeters, μ_0 is equal to 0.1.

INTERNATIONAL TIME MAP

This map shows the number of hours to add or subtract from Eastern Standard Time to determine the time anywhere on eart

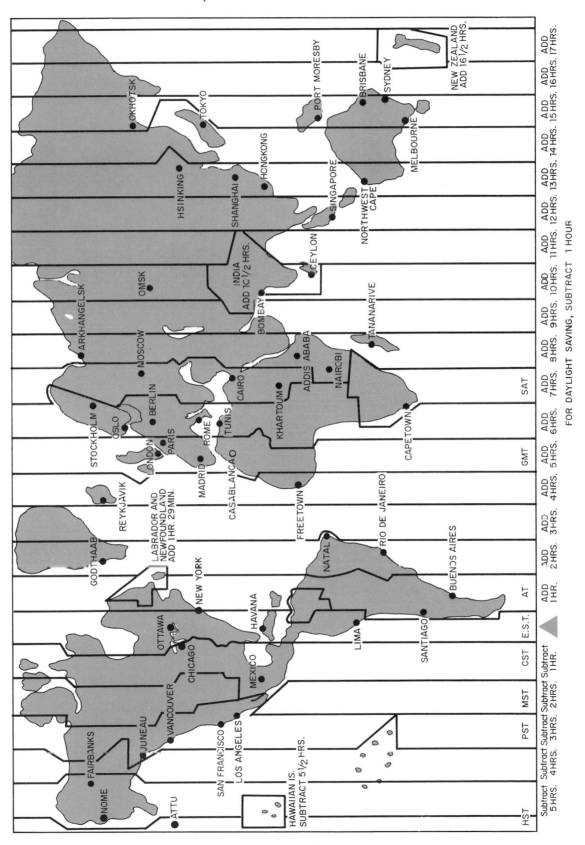

HIGH ALTITUDE CHART

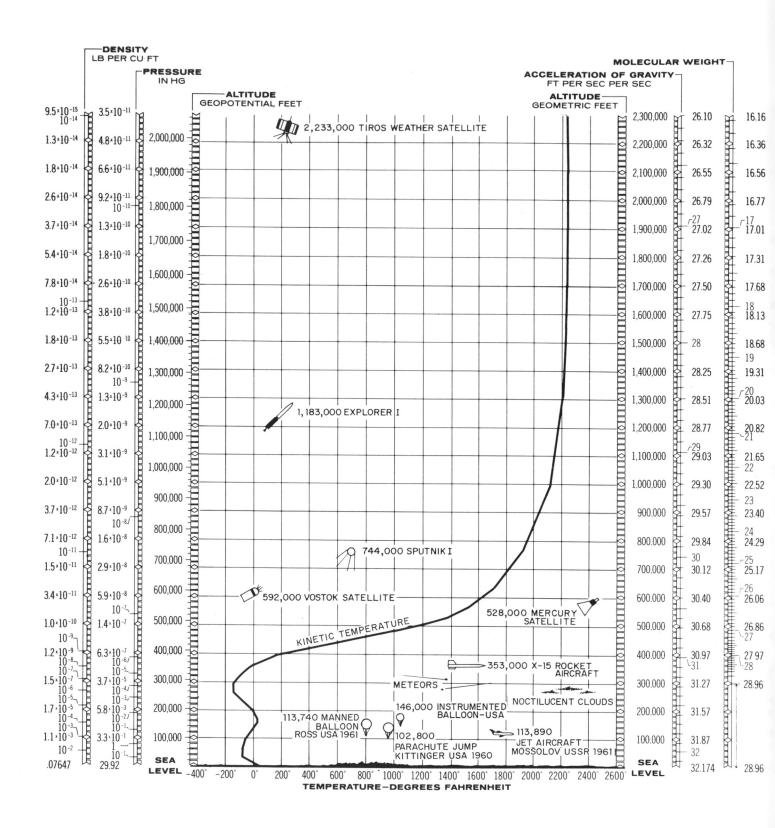

DENSITY LB PER CU FT

PRESSURE IN HG

ALTITUDE GEOPOTENTIAL FEET

ACCELERATION OF GRAVITY FT PER SEC PER SEC

MOLECULAR WEIGHT

ALTITUDE GEOMETRIC FEET

2,233,000 TIROS WEATHER SATELLITE

1,183,000 EXPLORER I

744,000 SPUTNIK I

592,000 VOSTOK SATELLITE

528,000 MERCURY SATELLITE

KINETIC TEMPERATURE

353,000 X-15 ROCKET AIRCRAFT

METEORS

NOCTILUCENT CLOUDS

146,000 INSTRUMENTED BALLOON–USA

113,740 MANNED BALLOON ROSS USA 1961

102,800 PARACHUTE JUMP KITTINGER USA 1960

113,890 JET AIRCRAFT MOSSOLOV USSR 1961

SEA LEVEL

TEMPERATURE—DEGREES FAHRENHEIT

272

ATMOSPHERE CHART

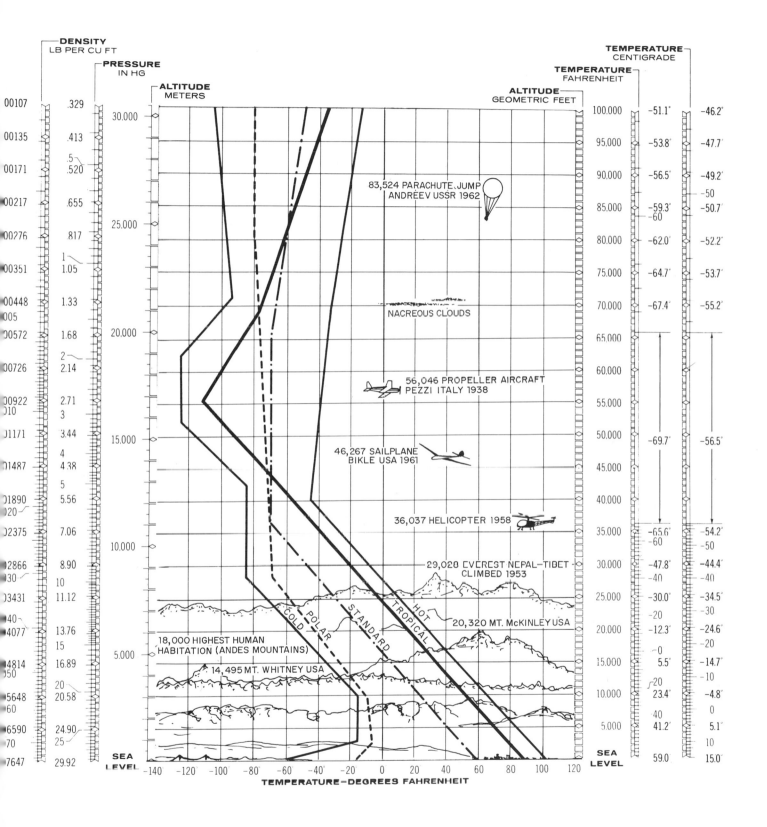

THE TRIBOELECTRIC (OR ELECTROSTATIC) SERIES

The table below is so arranged that any material becomes positively charged (that is, it gives up electrons) when rubbed with any material lower on the list. The farther apart the materials are on the list, the higher the charge will be. Surface conditions and variations in the characteristics of some materials may alter some positions slightly.

Positive polarity (+)

Asbestos
Rabbit's fur
Glass
Mica
Nylon
Wool
Cat's fur
Silk
Paper
Cotton
Wood
Lucite
Sealing wax
Amber
Polystyrene
Polyethylene
Rubber balloon
Sulphur
Celluloid
Hard rubber
Vinylite
Saran wrap

Negative polarity (−)

FOR EXAMPLE:
A rubber balloon rubbed with nylon will produce a negative charge on the balloon and leave the nylon positively charged.

CORROSION

Galvanic corrosion occurs when two dissimilar metals are in contact, in a liquid capable of carrying an electric current. Under these conditions the least noble metal (the anode) corrodes, while the more noble metal (the cathode) is not attacked.

In general, galvanic corrosion may be avoided by uniformity in the types of metals used. If uniformity is not practical, then metals should be used that are as close as possible to each other in the galvanic table, which lists metals in order of increasing nobility.

Stainless steel is "active" when chemicals present do not allow the formation of an oxide film on the surface of the metal. The treatment of stainless steel in a passivating solution accelerates the formation of the oxide film, thus making it "passive" and thereby increasing its resistance to galvanic corrosion.

Table 1. Listings of base-to-noble metal sequence, activity series, and galvanic series. Base metals at the top of the list function as the anode when used with metals lower in the series (more noble), and are subject to corrosion. The activity series, with hydrogen gas as the arbitrary reference, indicate the relative inertness of reactivity of metals. The reactive elements are above hydrogen while the inert elements are below. The galvanic series, the most used series in considering the electronics of corrosion, indicate voltage readings recorded between the indicated metal and a silver/silver-chloride reference electrode while immersed in a relatively unpolluted sea-water electrolyte.

BASE-NOBLE METAL SEQUENCE	ACTIVITY SERIES	GALVANIC SERIES	
BASE ▼	Magnesium ▼	Material	Voltage
Magnesium	Aluminum	Magnesium	1.5
Zinc	Zinc	Zinc	1.03
Aluminum	Chromium	Aluminum	0.75*
Cadmium	Iron	Cast iron & carbon steel	0.61
Steel or Iron	Cadmium	Stainless steel	0.55*
Chromium-iron (active)	Nickel	Bronze	0.4
Lead-tin solders	Tin	Yellow brass	0.36
Lead	Lead	Copper	0.36
Nickel (active)	Hydrogen	Red brass	0.33
Brasses	Copper	Admiralty brass	0.29
Copper	Silver	Copper-nickel	0.27*
Bronzes	Palladium	Nickel	0.2
Copper-nickel alloys	Platinum ▼	Monel	0.075
Nickel-copper alloys	Gold		
Silver solder			
Nickel (passive)			
Chromium-iron (passive)			
Silver			
Graphite			
Gold			
Platinum ▼			
NOBLE			

*Represents an "average" reading taken of varying alloys of each of the respective metals.

THERMOPLASTICS FOR ELECTRICAL APPLICATIONS

Material and Major Application Considerations	Common Available Forms	Representative Tradenames and Suppliers
Acetals Good electrical properties at most frequencies, which are little changed in humid environments to 125° C. Outstanding mechanical strength, stiffness, toughness, and dimensional stability.	Extrusions, injection moldings, stock shapes.	Delrin (DuPont); Celcon (Celanese Corp.)
Acrylics Excellent resistance to arcing and electrical tracking. Excellent clarity and resistance to outdoor weathering.	Castings, extrusions, injection moldings, thermoformed parts, stock shapes, film, fiber.	Lucite (DuPont); Plexiglas (Rohm and Haas Co.)
Cellulosics Good electrical properties and toughness. Used more for general-purpose applications than for ultimate in any electrical requirement. Several types available.	Blow moldings, extrusions, injection moldings, thermoformed parts, film, fiber, stock shapes.	Tenite (Eastman Chemical Co.); Ethocel-EC (Dow Chemical Co.); Forticel-CAP (Celanese Corp.)
Chlorinated Polyethers Good electrically, but most outstanding properties are corrosion resistance and physical and thermal stability.	Extrusions, injection moldings, stock shapes, film.	Penton (Hercules Powder Co.)
Fluorocarbons **TFE:** Electrically one of the most outstanding thermoplastic materials. Very low electrical losses; very high electrical resistivity. Useful from −300° to over 500°F. Excellent high frequency dielectric. Has excellent combination of mechanical and electrical properties but is relatively weak in cold-flow properties. Nearly inert chemically, as are most fluorocarbons. Very low coefficient of friction. Nonflammable.	Compression moldings, stock shapes, film.	Teflon TFE (DuPont); Halon TFE (Allied Chemical Corp.)
FEP: Similar to TFE, except useful temperature limited to about 400°F. Easier to mold than TFE.	Extrusions, injection moldings, laminates, film.	Teflon FEP (DuPont)
CTFE: Excellent electrical properties and relatively good mechanical properties. Stiffer than TFE and FEP, but does have some cold flow. Useful to about 400°F.	Extrusions, isostatic moldings, injection moldings, film, stock shapes.	Kel-F (3M Co.); Plaskon CTFE (Allied Chemical Corp.)
PVF_2: One of the easiest of the fluorocarbons to process. Stiffer and more resistant to cold flow than TFE. Good electrically. Useful to about 300°F. Major electrical application is wire jacketing.	Extrusions, injection moldings, laminates, film.	Kynar (Pennsalt Chemicals Corp.)
Nylons **Conventional:** Good general-purpose electrical properties. Easily processed. Good mechanical strength and abrasion resistance and low coefficient of friction. Commonly used types of nylon are nylon 6, nylon 6/6 and nylon 6/10. Some have limited use in electrical applications because of moisture-absorption properties. Nylon 6/10 is best here.	Extrusions, injection moldings, laminates, rotational moldings, stock shapes, film, fiber.	Zytel (DuPont); Plaskon (Allied Chemical Co.); Bakelite (Union Carbide Corp.)
High-Temperature: Has excellent combination of thermal endurance (to 200°C) and electrical properties. Exhibits relatively low dielectric constant, high volume resistivity, and good dielectric strength. Has high tensile strength and wear resistance.	Fiber, sheet, tape, paper, fabric.	Nomex (DuPont)
Polysulfones Good combination of thermal endurance (to over 300°F) and dielectric properties. Relatively low dielectric constant and dissipation factor, and high volume resistivity. Electrical properties are maintained at 90% of initial values after one year at 300°F. Good dimensional stability and high creep resistance. Flame resistant, and good chemical resistance.	Extrusions, injection moldings, thermoformed parts, stock shapes, film, sheet.	Polysulfone (Union Carbide Corp.)

THERMOPLASTICS FOR
ELECTRICAL APPLICATIONS
(Continued)

Material and Major Application Considerations	Common Available Forms	Representative Tradenames and Suppliers
Parylenes Excellent low-loss dielectric properties and good dimensional stability. Low permeability to gases and moisture. Produced as a film on a substrate, from a vapor phase. Used primarily as thin films in capacitors and dielectric coatings.	Film coatings.	Parylene (Union Carbide Corp.)
Polycarbonates Relatively low electrical losses and high volume resistivity. Loss properties are stable to about 150°C. Excellent dimensional stability, low water absorption, low creep, and outstanding impact resistance.	Extrusions, injection moldings, thermoformed parts, stock shapes, film.	Lexan (G. E. Co.); Merlon (Mobay Chemical Co.)
Polyesters Outstanding dielectric strength and tear strength. Widely used for machine-applied tape insulation. Has high volume resistivity and low moisture absorption.	Films and tapes.	Mylar (DuPont); Scotchpar (3M Co.); Celanar (Celanese Corp.)
Polyethylenes, Polypropylenes, Polyallomers Excellent electrical properties, especially low electrical losses. Tough and chemically resistant, but weak to varying degrees in creep and thermal resistance. Thermal stability generally increases with density classes of polyethylene. Polypropylenes are generally similar to polyethylenes, but offer about 50°F higher heat resistance. Polyallomers are electrically similar to polyethylene and polypropylene but have better stress-crack resistance and surface hardness. Crosslinked polyethylenes provide improved thermal endurance.	Blow moldings, extrusions, injection molding, thermoformed parts, stock shapes, film, fiber, foam.	Alathon Polyethylene (DuPont); Petrothene Polyethylene (USI Chemical Co.); Grex H. D. Polyethylene (Allied Chemical Corp.), Hi-Fax H. D. Polyethylene, Pro-Fax Polypropylene (Hercules Powder Co.); Tenite Polyethylene, Polypropylene, and Polyallomer (Eastman Chemical Co.)
Polyimides and Polyamide-imides Among the highest-temperature thermoplastics available, having useful operating temperatures to about 700°F or higher. Excellent electrical properties, good rigidity, and excellent thermal stability.	Films, coatings, molded and machined parts, resin solutions.	Vespel parts and shapes, Kapton film, and Pyre-M.L. resin (DuPont); AI (Amoco); Skybond (Monsanto Co.)
Polyphenylene Oxides (PPO) Excellent electrical properties, especially loss properties to above 350°F, and over a wide frequency range. Good mechanical strength and toughness. A lower-cost grade, Noryl, has similar properties to PPO, but with a 75° to 100°F reduction in heat resistance.	Extrusions, injection moldings, thermoformed parts, stock shapes, film.	PPO and Noryl (G. E. Co.)
Polystyrenes **General-Purpose:** Excellent electrical properties, especially loss properties. Conventional polystyrene is temperature-limited, but high-temperature modifications such as Rexolite or Polypenco crosslinked polystyrene are widely used, especially for high-frequency applications.	Blow moldings, extrusions, injection moldings, rotational moldings, thermoformed parts, foam.	Styron (Dow Chemical Co.); Lustrex (Monsanto Co.); Rexolite (American Enka Corp.); Polypenco Q-200.5 (Polymer Corp.)
ABS: Good general electrical properties but not outstanding for any specific electric application. Extremely tough, with high impact resistance. Can be formulated over a wide range of hardness and toughness properties. Special grades available for plated surfaces.	Extrusions, injection moldings, thermoformed parts, laminates, stock shapes, foam.	Marbon Cycolac (Borg-Warner Corp.); Lustran (Monsanto Co.); Abson (Goodrich Chemical Co.)
Vinyls Good low-cost, general-purpose thermoplastic materials, but electrical properties are not outstanding. Properties are greatly influenced by plasticizers. Many variations available, including flexible and rigid types. Flexible vinyls, especially PVC, are widely used for wire insulation.	Blow moldings, extrusions, injection moldings, rotational moldings, film, sheet.	Diamond PVC (Diamond Alkali Co.); Pliovic (Goodyear Chemical Co.); Saran (Dow Chemical Co.)

THERMOSETTING PLASTICS FOR ELECTRICAL APPLICATIONS

Material and Major Application Considerations	Common Available Forms	Representative Tradenames and Suppliers
Alkyds Excellent dielectric strength, arc resistance, and dry insulation resistance. Low dielectric constant and dissipation factor. Good dimensional stability. Easily molded.	Compression and transfer moldings.	Plaskon (Allied Chemical Corp.); Glaskyd (American Cyanamid Co.)
Aminos (Melamine and Urea) Good general electrical properties, but not outstanding except for glass-filled melamines whose hardness and arc resistance make them useful for molded connectors.	Compression and transfer moldings, extrusions, laminates.	Plaskon (Allied Chemical Corp.); Resimene (Monsanto Co.); Cymel melamine, Beetle urea (American Cyanamid Co.)
Diallyl Phthalates (Allylics) Unsurpassed among thermosets in retention of electrical properties in high-humidity environments. Also, they have among the highest volume and surface resistivities in thermosets. Low dissipation factor and heat resistance to 400°F or higher. Excellent dimensional stability. Easily molded.	Compression, injection, and transfer moldings; extrusions; laminates.	Dapon (FMC Corp.); Diall (Allied Chemical Corp.)
Epoxies Good electrical properties, low shrinkage, excellent dimensional stability, and good to excellent adhesion. Easy to compound, using nonpressure processes, for a variety of end properties. Useful over a wide range of environments.	Castings; compression, injection, and transfer moldings; extrusions; laminates; matched-die moldings; filament windings; foam.	Epon (Shell Chemical Co.); Epi-Rez (Jones-Dabney Co.); D.E.R. (Dow Chemical Co.); Araldite (Ciba Products Co.); ERL (Union Carbide Corp.); Scotchcast (3M Co.)
Phenolics Good general electrical properties, leading to wide use for general-purpose molded parts. Not outstanding in any specific electric property, but some formulations have excellent thermal stability above 300°F.	Castings; compression, injection, and transfer moldings; extrusions; laminates; matched-die moldings; stock shapes; foam.	Bakelite (Union Carbide Corp.); Durez (Hooker Chemical Corp.)
Polyesters Very low dissipation factor. Low-cost and extremely easy to compound using nonpressure processes. Like epoxies, they can be formulated for either room temperature or elevated temperature use. Not equivalent to epoxies in environmental resistance.	Compression, injection, and transfer moldings; extrusions; laminates; matched-die moldings; filament windings; stock shapes.	Selectron (Pittsburgh Plate Glass Co.); Laminac (American Cyanamid Co.); Paraplex (Rohm & Haas Co.)
Silicones (rigid) Excellent electrical properties, especially low dielectric constant and dissipation factor, which change little to 400°F.	Castings, compression and transfer moldings, laminates.	DC Resins (Dow Corning Corp.)

SIGNIFICANCE OF PROPERTIES OF ELECTRICAL INSULATING MATERIALS

Property and Definition	Significance of Values
Dielectric Strength All insulating materials fail at some level of applied voltage for a given set of operating conditions. The dielectric strength is the voltage an insulating material can withstand before dielectric breakdown occurs. Dielectric strength is normally expressed in voltage gradient terms, such as volts per mil. In testing for dielectric strength, two methods of applying the voltage (gradual or by steps) are used. Type of voltage, temperature, and any pre-conditioning of the test part must be noted. Also, thickness of the piece being tested must be recorded because the voltage per mil at which breakdown occurs varies with thickness of test piece. Normally, breakdown occurs at a much higher volt-per-mil value in very thin test pieces (a few mils thick) than In thicker sections (⅛ in. thick, for example).	The higher the value, the better the insulator. Dielectric strength of a material (per mil of thickness) usually increases considerably with decrease in insulation thickness. Materials suppliers can provides curves of dielectric strength vs thickness for their insulating materials.
Resistance and Resistivity Resistance of an insulating material, like that of a conductor, is the resistance offered by the conducting path to passage of electrical current. Resistance is expressed in ohms. Insulating materials are very poor conductors, offering high resistance. For insulating materials, the term *volume resistivity* is more commonly applied. Volume resistivity is the electrical resistance between opposite faces of a unit cube for a given material and at a given temperature. The relationship between resistance and resistivity is expressed by the equation $\rho = RA/l$ where ρ = volume resistivity in ohm-cm, R = resistance in ohms between faces; A = area of the faces, and l = distance between faces of the piece on which measurement is made. This is not resistance per unit volume, which would be ohm/cm³—although this term is sometimes erroneously used. Other terms are sometimes used to describe a specific application or condition. One such term is *surface resistivity*, which is the resistance between two opposite edges of a surface film 1 cm square. Since the length and width of the path are the same, the centimeter terms cancel. Thus, units of surface resistivity are actually ohms. However, to avoid confusion with usual resistance values, surface resistivity is normally given in ohms/sq. Another broadly used term is *insulation resistance*, which, again, is a measurement of ohmic resistance for a given condition, rather than a standardized resistivity test. For both surface resistivity and insulation resistance, standardized comparative tests are normally used. Such tests can provide data such as effects of humidity on a given insulating material configuration.	The higher the value, the better for a good insulating material. The resistance value for a given material depends upon a number of factors. It varies inversely with temperature, and is affected by humidity, moisture content of the test part, level of the applied voltage, and time during which the voltage is applied. When tests are made on a piece that has been subjected to moist or humid conditions, it is important that measurements be made at controlled time intervals during or after the test condition has been applied, since dry-out and resistance increase occur rapidly. Comparing or interpreting data is difficult unless the test period is controlled and defined.
Dielectric Constant The dielectric constant of an insulating material is the ratio of the capacitance of a capacitor containing that particular material to the capacitance of the same electrode system with air replacing the insulation as the dielectric medium. The dielectric constant is also sometimes defined as the property of an insulation which determines the electrostatic energy stored within the solid material. The dielectric constant of most commercial insulating materials varies from about 2 to 10, air having the value 1.	Low values are best for high-frequency or power applications, to minimize electrical power losses. Higher values are best for capacitance applications. For most insulating materials, dielectric constant increases with temperature, especially above a critical temperature region which is unique for each material. Dielectric constant values are also affected (usually to a lesser degree) by frequency. This variation is also unique for each material.
Power Factor and Dissipation Factor Power factor is the ratio of the power dissipated (watts) in an insulating material to the product of the effective voltage and current (volt-ampere input) and is a measure of the relative dielectric loss in the insulation when the system acts as a capacitor. Power factor is nondimensional and is a commonly used measure of insulation quality. It is of particular interest at high levels of frequency and power in such applications as microwave equipment, transformers, and other inductive devices. Dissipation factor is the tangent of the dielectric loss angle. Hence, the term *tan delta* (tangent of the angle) is also sometimes used. For the low values ordinarily encountered in insulation, dissipation factor is practically the equivalent of power factor, and the terms are used interchangeably.	Low values are favorable, indicating a more efficient system, with lower power losses.
Arc Resistance Arc resistance is a measure of an electrical breakdown condition along an insulating surface, caused by the formation of a conductive path on the surface. It is a common ASTM measurement, especially used with plastic materials because of the variations among plastics in the extent to which a surface breakdown occurs. Arc resistance is measured as the time, in seconds, required for breakdown along the surface of the material being measured. Surface breakdown (arcing or electrical tracking along the surface) is also affected by surface cleanliness and dryness.	The higher the value, the better. Higher values indicate greater resistance to breakdown along the surface due to arcing or tracking conditions.

To convert from Fahrenheit to Celsius*—locate temperature (°F) in center column and read °C in left column.

To convert from Celsius* to Fahrenheit—locate temperature (°C) in center column and read °F in right column.

−459.4 To −70

C		F
−273	−459.4	
−268	−450	
−262	−440	
−257	−430	
−251	−420	
−246	−410	
−240	−400	
−234	−390	
−229	−380	
−223	−370	
−218	−360	
−212	−350	
−207	−340	
−201	−330	
−196	−320	
−190	−310	
−184	−300	
−179	−290	
−173	−280	
−169	−273	−459.4
−168	−270	−454
−162	−260	−436
−157	−250	−418
−151	−240	−400
−146	−230	−382
−140	−220	−364
−134	−210	−346
−129	−200	−328
−123	−190	−310
−118	−180	−292
−112	−170	−274
−107	−160	−256
−101	−150	−238
−95.6	−140	−220
−90.0	−130	−202
−84.4	−120	−184
−78.9	−110	−166
−73.3	−100	−148
−72.6	−99	−146.2
−72.2	−98	−144.4
−71.7	−97	−142.6
−71.1	−96	−140.8
−70.6	−95	−139.0
−70.0	−94	−137.2
−69.5	−93	−135.4
−68.9	−92	−133.6
−68.4	−91	−131.8
−67.8	−90	−130.0
−67.2	−89	−128.2
−66.6	−88	−126.4
−66.1	−87	−124.6
−65.5	−86	−122.8
−65.0	−85	−121.0
−64.4	−84	−119.2
−63.9	−83	−117.4
−63.3	−82	−115.6
−62.8	−81	−113.8
−62.2	−80	−112.0
−61.7	−79	−110.2
−61.1	−78	−108.4
−60.6	−77	−106.6
−60.0	−76	−104.8
−59.5	−75	−103.0
−58.9	−74	−101.2
−58.4	−73	−99.4
−57.8	−72	−97.6
−57.3	−71	−95.8
−56.7	−70	−94.0

−69 To 0

C		F
−56.1	−69	−92.2
−55.5	−68	−90.4
−55.0	−67	−88.6
−54.4	−66	−86.8
−53.9	−65	−85.0
−53.3	−64	−83.2
−52.8	−63	−81.4
−52.2	−62	−79.6
−51.7	−61	−77.8
−51.1	−60	−76.0
−50.6	−59	−74.2
−50.0	−58	−72.4
−49.5	−57	−70.6
−48.9	−56	−68.8
−48.4	−55	−67.0
−47.8	−54	−65.2
−47.3	−53	−63.4
−46.7	−52	−61.6
−46.2	−51	−59.8
−45.6	−50	−58.0
−45.0	−49	−56.2
−44.4	−48	−54.4
−43.9	−47	−52.6
−43.3	−46	−50.8
−42.8	−45	−49.0
−42.2	−44	−47.2
−41.7	−43	−45.4
−41.1	−42	−43.6
−40.6	−41	−41.8
−40.0	−40	−40.0
−39.4	−39	−38.2
−38.8	−38	−36.4
−38.3	−37	−34.6
−37.8	−36	−32.8
−37.2	−35	−31.0
−36.6	−34	−29.2
−36.1	−33	−27.4
−35.5	−32	−25.6
−35.0	−31	−23.8
−34.4	−30	−22.0
−33.9	−29	−20.2
−33.3	−28	−18.4
−32.8	−27	−16.6
−32.2	−26	−14.8
−31.7	−25	−13.0
−31.1	−24	−11.2
−30.6	−23	−9.4
−30.0	−22	−7.6
−29.5	−21	−5.8
−28.9	−20	−4.0
−28.3	−19	−2.2
−27.7	−18	−0.4
−27.2	−17	1.4
−26.6	−16	3.2
−26.1	−15	5.0
−25.5	−14	6.8
−25.0	−13	8.6
−24.4	−12	10.4
−23.9	−11	12.2
−23.3	−10	14.0
−22.8	−9	15.8
−22.2	−8	17.6
−21.7	−7	19.4
−21.1	−6	21.2
−20.6	−5	23.0
−20.0	−4	24.8
−19.5	−3	26.6
−18.9	−2	28.4
−18.4	−1	30.2
−17.8	0	32.0

1 To 69

C		F
−17.2	1	33.8
−16.7	2	35.6
−16.1	3	37.4
−15.6	4	39.2
−15.0	5	41.0
−14.4	6	42.8
−13.9	7	44.6
−13.3	8	46.4
−12.8	9	48.2
−12.2	10	50.0
−11.7	11	51.8
−11.1	12	53.6
−10.6	13	55.4
−10.0	14	57.2
−9.44	15	59.0
−8.89	16	60.8
−8.33	17	62.6
−7.78	18	64.4
−7.22	19	66.2
−6.67	20	68.0
−6.11	21	69.8
−5.56	22	71.6
−5.00	23	73.4
−4.44	24	75.2
−3.89	25	77.0
−3.33	26	78.8
−2.78	27	80.6
−2.22	28	82.4
−1.67	29	84.2
−1.11	30	86.0
−0.56	31	87.8
0	32	89.6
0.56	33	91.4
1.11	34	93.2
1.67	35	95.0
2.22	36	96.8
2.78	37	98.6
3.33	38	100.4
3.89	39	102.2
4.44	40	104.0
5.00	41	105.8
5.56	42	107.6
6.11	43	109.4
6.67	44	111.2
7.22	45	113.0
7.78	46	114.8
8.33	47	116.6
8.89	48	118.4
9.44	49	120.2
10.0	50	122.0
10.6	51	123.8
11.1	52	125.6
11.7	53	127.4
12.2	54	129.2
12.8	55	131.0
13.3	56	132.8
13.9	57	134.6
14.4	58	136.4
15.0	59	138.2
15.6	60	140.0
16.1	61	141.8
16.7	62	143.6
17.2	63	145.4
17.8	64	147.2
18.3	65	149.0
18.9	66	150.8
19.4	67	152.6
20.0	68	154.4
20.6	69	156.2

70 To 139

C		F
21.1	70	158.0
21.7	71	159.8
22.2	72	161.6
22.8	73	163.4
23.3	74	165.2
23.9	75	167.0
24.4	76	168.8
25.0	77	170.6
25.6	78	172.4
26.1	79	174.2
26.7	80	176.0
27.2	81	177.8
27.8	82	179.6
28.3	83	181.4
28.9	84	183.2
29.4	85	185.0
30.0	86	186.8
30.6	87	188.6
31.1	88	190.4
31.7	89	192.2
32.2	90	194.0
32.8	91	195.8
33.3	92	197.6
33.9	93	199.4
34.4	94	201.2
35.0	95	203.0
35.6	96	204.8
36.1	97	206.6
36.7	98	208.4
37.2	99	210.2
37.8	100	212.0
38.3	101	213.8
38.9	102	215.6
39.4	103	217.4
40.0	104	219.2
40.6	105	221.0
41.1	106	222.8
41.7	107	224.6
42.2	108	226.4
42.8	109	228.2
43.3	110	230.0
43.9	111	231.8
44.4	112	233.6
45.0	113	235.4
45.6	114	237.2
46.1	115	239.0
46.7	116	240.8
47.2	117	242.6
47.8	118	244.4
48.3	119	246.2
48.9	120	248.0
49.4	121	249.8
50.0	122	251.6
50.6	123	253.4
51.1	124	255.2
51.7	125	257.0
52.2	126	258.8
52.8	127	260.6
53.3	128	262.4
53.9	129	264.2
54.4	130	266.0
55.0	131	267.8
55.6	132	269.6
56.1	133	271.4
56.7	134	273.2
57.2	135	275.0
57.8	136	276.8
58.3	137	278.6
58.9	138	280.4
59.4	139	282.2

140 To 290

C		F
60.0	140	284.0
60.6	141	285.8
61.1	142	287.6
61.7	143	289.4
62.2	144	291.2
62.8	145	293.0
63.3	146	294.8
63.9	147	296.6
64.4	148	298.4
65.0	149	300.2
65.6	150	302.0
66.1	151	303.8
66.7	152	305.6
67.2	153	307.4
67.8	154	309.2
68.3	155	311.0
68.9	156	312.8
69.4	157	314.6
70.0	158	316.4
70.6	159	318.2
71.1	160	320.0
71.7	161	321.8
72.2	162	323.6
72.8	163	325.4
73.3	164	327.2
73.9	165	329.0
74.4	166	330.8
75.0	167	332.6
75.6	168	334.4
76.1	169	336.2
76.7	170	338.0
77.2	171	339.8
77.8	172	341.6
78.3	173	343.4
78.9	174	345.2
79.4	175	347.0
80.0	176	348.8
80.6	177	350.6
81.1	178	352.4
81.7	179	354.2
82.2	180	356.0
82.8	181	357.8
83.3	182	359.6
83.9	183	361.4
84.4	184	363.2
85.0	185	365.0
85.6	186	366.8
86.1	187	368.6
86.7	188	370.4
87.2	189	372.2
87.8	190	374.0
88.3	191	375.8
88.9	192	377.6
89.4	193	379.4
90.0	194	381.2
90.6	195	383.0
91.1	196	384.8
91.7	197	386.6
92.2	198	388.4
92.8	199	390.2
93.3	200	392
98.9	210	410
100	212	413
104	220	428
110	230	446
116	240	464
121	250	482
127	260	500
132	270	518
138	280	536
143	290	554

300 To 1000

C		F
149	300	572
154	310	590
160	320	608
166	330	626
171	340	644
177	350	662
182	360	680
188	370	698
193	380	716
199	390	734
204	400	752
210	410	770
216	420	788
221	430	806
227	440	824
232	450	842
238	460	860
243	470	878
249	480	896
254	490	914
260	500	932
266	510	950
271	520	968
277	530	986
282	540	1004
288	550	1022
293	560	1040
299	570	1058
304	580	1076
310	590	1094
316	600	1112
321	610	1130
327	620	1148
332	630	1166
338	640	1184
343	650	1202
349	660	1220
354	670	1238
360	680	1256
366	690	1274
371	700	1292
377	710	1310
382	720	1328
388	730	1346
393	740	1364
399	750	1382
404	760	1400
410	770	1418
416	780	1436
421	790	1454
427	800	1472
432	810	1490
438	820	1508
443	830	1526
449	840	1544
454	850	1562
460	860	1580
466	870	1598
471	880	1616
477	890	1634
482	900	1652
488	910	1670
493	920	1688
499	930	1706
504	940	1724
510	950	1742
516	960	1760
521	970	1778
527	980	1796
532	990	1814

TEMPERATURE CONVERSION
TABLES AND FORMULAS
(Continued)

1000 to 1490		1500 to 1990		2000 to 2490		2500 to 3000	
C	F	C	F	C	F	C	F
538	1000 1832	816	1500 2732	1093	2000 3632	1371	2500 4532
543	1010 1850	821	1510 2750	1099	2010 3650	1377	2510 4550
549	1020 1868	827	1520 2768	1104	2020 3668	1382	2520 4568
554	1030 1886	832	1530 2786	1110	2030 3686	1388	2530 4586
560	1040 1904	838	1540 2804	1116	2040 3704	1393	2540 4604
566	1050 1922	843	1550 2822	1121	2050 3722	1399	2550 4622
571	1060 1940	849	1560 2840	1127	2060 3740	1404	2560 4640
577	1070 1958	854	1570 2858	1132	2070 3758	1410	2570 4658
582	1080 1976	860	1580 2876	1138	2080 3776	1416	2580 4676
588	1090 1994	866	1590 2894	1143	2090 3794	1421	2590 4694
593	1100 2012	871	1600 2912	1149	2100 3812	1427	2600 4712
599	1110 2030	877	1610 2930	1154	2110 3830	1432	2610 4730
604	1120 2048	882	1620 2948	1160	2120 3848	1438	2620 4748
610	1130 2066	888	1630 2966	1166	2130 3866	1443	2630 4765
616	1140 2084	893	1640 2984	1171	2140 3884	1449	2640 4784
621	1150 2102	899	1650 3002	1177	2150 3902	1454	2650 4802
627	1160 2120	904	1660 3020	1182	2160 3920	1460	2660 4820
632	1170 2138	910	1670 3038	1188	2170 3938	1466	2670 4838
638	1180 2156	916	1680 3056	1193	2180 3956	1471	2680 4856
643	1190 2174	921	1690 3074	1199	2190 3974	1477	2690 4874
649	1200 2192	927	1700 3092	1204	2200 3992	1482	2700 4892
654	1210 2210	932	1710 3110	1210	2210 4010	1488	2710 4910
660	1220 2228	938	1720 3128	1216	2220 4028	1493	2720 4928
666	1230 2246	943	1730 3146	1221	2230 4046	1499	2730 4945
671	1240 2264	949	1740 3164	1227	2240 4064	1504	2740 4964
677	1250 2282	954	1750 3182	1232	2250 4082	1510	2750 4982
682	1260 2300	960	1760 3200	1238	2260 4100	1516	2760 5000
688	1270 2318	966	1770 3218	1243	2270 4118	1521	2770 5018
693	1280 2336	971	1780 3236	1249	2280 4136	1527	2780 5036
699	1290 2354	977	1790 3254	1254	2290 4154	1532	2790 5054
704	1300 2372	982	1800 3272	1260	2300 4172	1538	2800 5072
710	1310 2390	988	1810 3290	1266	2310 4190	1543	2810 5090
716	1320 2408	993	1820 3308	1271	2320 4208	1549	2820 5108
721	1330 2426	999	1830 3326	1277	2330 4226	1554	2830 5126
727	1340 2444	1004	1840 3344	1282	2340 4244	1560	2840 5144
732	1350 2462	1010	1850 3362	1288	2350 4262	1566	2850 5162
738	1360 2480	1016	1860 3380	1293	2360 4280	1571	2860 5180
743	1370 2498	1021	1870 3398	1299	2370 4298	1577	2870 5198
749	1380 2516	1027	1880 3416	1304	2380 4316	1582	2880 5216
754	1390 2534	1032	1890 3434	1310	2390 4334	1588	2890 5234
760	1400 2552	1038	1900 3452	1316	2400 4352	1593	2900 5252
766	1410 2570	1043	1910 3470	1321	2410 4370	1599	2910 5270
771	1420 2588	1049	1920 3488	1327	2420 4388	1604	2920 5288
777	1430 2606	1054	1930 3506	1332	2430 4406	1610	2930 5306
782	1440 2624	1060	1940 3524	1338	2440 4424	1616	2940 5324
788	1450 2642	1066	1950 3542	1343	2450 4442	1621	2950 5342
793	1460 2660	1071	1960 3560	1349	2460 4460	1627	2960 5360
799	1470 2678	1077	1970 3578	1354	2470 4478	1632	2970 5378
804	1480 2696	1082	1980 3596	1360	2480 4496	1638	2980 5396
810	1490 2714	1088	1990 3614	1366	2490 4514	1643	2990 5414
						1649	3000 5432

Interpolation Factors

C		F
0.56	1	1.8
1.11	2	3.6
1.67	3	5.4
2.22	4	7.2
2.78	5	9.0
3.33	6	10.8
3.89	7	12.6
4.44	8	14.4
5.00	9	16.2
5.56	10	18.0

*The term Centigrade was officially changed to Celsius by international agreement in 1948. The Celsius scale uses the triple phase point of water, at 0.01° Centigrade, in place of the ice point as a reference, but for all practical purposes the two terms are interchangeable.

Given	Temperature Conversion				
	Celsius	Fahrenheit	Kelvin	Reaumur	Rankine
Cels.	—	$\left(\frac{9}{5}C\right) + 32$	$C + 273.16$	$\frac{4}{5}C$	$1.8\,(C + 273.16)$
Fahr.	$\frac{5}{9}(F - 32)$	—	$\left[\frac{5}{9}(F - 32)\right] + 273.16$	$\frac{4}{9}(F - 32)$	$F + 459.7$
Kelvin	$K - 273.16$	$\left[\frac{9}{5}(K - 273.16)\right] + 32$	—	$\frac{4}{5}(K - 273.16)$	$K \times 1.8$
Reau.	$Re \times \frac{5}{4}$	$\left(\frac{9}{4}Re\right) + 32$	$\left(\frac{5}{4}Re\right) + 273.16$	—	$\left(\frac{9}{4}Re\right) + 491.7$
Rank.	$\frac{Ra}{1.8} - 273.16$	$Ra - 459.7$	$\frac{Ra}{1.8}$	$\frac{4}{9}(Ra - 491.7)$	—

Five major temperature scales are in use at present. They are: Fahrenheit, Celsius, Kelvin (Absolute), Rankine, and Reaumur. The interrelationship among the scales is shown here.

TEMPERATURE CONVERSION
TABLES AND FORMULAS
(Continued)

°F

		(F) FAHRENHEIT	(C) CELSIUS	(K) KELVIN
10,000	Surface of the SUN	(approx.) 9890°	5476°	5749°
9000				
8000	CARBON boils	7592°	4199°	4472°
7000				
6000	TUNGSTEN melts	6098°	3370°	3643°
	ROCKET exhaust	(approx.) 5500°	3037°	3300°
5000				
4000				
3000	IRON melts	2795°	1535°	1808°
2000	COPPER melts	1981°	1083°	1356°
	ALUMINUM melts	1219°	659°	932°
1000				
200	WATER boils	212°	100°	373.2°
	High U.S. temperature	134°	57°	330°
100	Human body	98.6°	37°	310°
	WATER freezes	32°	0.0°	273.2°
0				
	MERCURY freezes	−38°	−38.9°	234.3°
	Low U.S. temperature	−69.7°	−56.5°	216.7°
−100	DRY ICE	−109.3°	−78.5°	194.7°
	Low WORLD temperature	−126°	−88°	185.2°
−200	ETHYL ALCOHOL freezes	−202°	−130°	143°
	NATURAL GAS liquefies	−258°	−161°	112°
−300	OXYGEN liquefies	−297.4°	−183.0°	90.0°
	NITROGEN liquefies	−320.5°	−195.8°	77.4°
	NITROGEN freezes	−345.8°	−209.9°	63.3°
	OXYGEN freezes	−361.1°	−218.4°	61.8°
−400	HYDROGEN liquefies	−422.9°	−252.7°	20.5°
	HELIUM liquefies	−452°	−269°	4°
	ABSOLUTE ZERO	−459.7°	−273.2°	0.0°

COMPARATIVE TEMPERATURE SCALES

DIFFERENCE BETWEEN WET AND DRY BULB READINGS DEGREES CELSIUS: 760 mm Hg

Percent Relative Humidity For Celsius Temperature Difference Up To 45°

$$F = \frac{9}{5}C + 32$$

$$C = \frac{5}{9}(F - 32)$$

DRY BULB TEMP. °C	1	2	3	4	5	6	7	8	9	10	11	12	13	14	15	16	17	18	19	20	21	22	23	24	25	26	27	28	29	30	31	32	33	34	35	36	37	38	39	40	41	42	43	44	45
0	82	64	47	31	14																																								
1	83	66	50	34	18																																							
2	84	68	52	37	22																																							
3	84	69	54	40	25	12																																						
4	85	70	56	42	29	16																																						
5	86	72	58	45	32	19																																						
6	86	73	60	47	35	23	3																																					
7	87	74	61	49	37	26	7																																					
8	87	75	62	51	40	29	11																																					
9	88	76	64	53	42	31	21																																					
10	88	77	65	55	44	34	24	14	5																																			
11	88	77	66	56	46	36	26	17	8																																			
12	89	78	68	57	48	38	29	20	11	3																																		
13	89	79	69	59	49	40	31	23	14	6																																		
14	90	79	70	60	51	42	33	25	17	9	2																																	
15	90	80	71	61	53	44	36	27	20	12	5																																	
16	90	81	72	63	54	46	38	30	22	15	8	1																																
17	91	81	72	63	55	47	39	32	24	17	10	4																																
18	91	82	73	65	57	49	41	34	27	20	13	6																																
19	91	82	74	66	58	50	43	36	29	22	15	9	3																															
20	91	83	74	66	59	51	44	37	31	24	18	12	6																															
21	91	83	75	67	60	52	45	39	32	26	20	14	8	3																														
22	92	84	76	68	61	54	47	41	34	28	22	16	11	5																														
23	92	84	76	69	62	55	48	42	36	30	24	18	13	8	3																													
24	92	85	77	70	63	56	49	43	37	31	25	20	15	10	5																													
25	93	85	78	71	64	57	51	45	39	33	27	22	17	12	7	3																												
26	93	86	79	72	65	58	52	46	40	34	29	24	19	14	10	5									2																			
27	93	86	79	72	66	59	53	47	41	36	31	26	21	16	11	7	3								4																			
28	93	86	79	73	66	60	54	48	43	37	32	27	23	18	13	9	5					2			6																			
29	93	87	80	74	67	61	54	49	43	38	33	28	24	19	15	11	7	3				4			8																			
30	93	86	80	74	68	62	57	51	46	40	36	31	27	22	17	13	9	5							10					3														
31	93	87	80	75	69	62	57	52	46	41	37	32	28	24	19	16	12	8	3						12					5														
32	94	87	81	75	69	63	58	52	47	42	38	34	30	25	21	17	14	10	5	3					14					7														
33	94	87	81	76	70	64	58	53	48	44	39	35	31	27	24	19	15	12	8	5					15					9		2												
34	94	88	82	76	70	65	59	54	49	45	40	36	32	29	25	21	17	14	11	7				2	17					10		5												
35	94	88	82	77	71	65	60	55	50	46	41	37	34	30	27	23	20	17	13	10				3	18					12		7			3									
36	94	88	82	77	72	66	61	56	51	47	43	39	35	32	28	25	22	19	16	13				5	20					13		8			4									
37	94	88	83	77	72	67	62	57	52	48	44	40	36	33	30	26	23	20	17	15				6	22					15		10			6	2								
38	94	89	83	78	72	67	62	57	53	49	44	41	37	34	31	28	25	22	18	16				7						16		11			7	3								
39	94	89	83	78	73	68	63	58	54	49	45	42	38	35	32	29	26	23	20	17				8						17		13			9	4								
40	94	89	84	78	73	69	64	59	55	50	46	43	39	36	33	30	27	24	21	18				10						19		14			10	5	3	2	2	2	2	3	3		
42	94	89	84	79	74	69	65	60	56	51	47	44	40	37	34	31	28	25	22	20				11						20		15			11	7	4	4	3	3	3	4	5		
44	95	90	84	79	75	70	65	61	57	52	48	45	41	38	35	32	29	26	24	21				12						21		16			12	8	5	5	4	5	5	6	6		
46	95	90	85	80	75	71	66	62	58	53	49	46	42	39	36	33	31	28	25	22				13						22		17			13	9	7	6	6	6	6	7	7		
48	95	90	85	81	76	72	67	63	59	54	50	47	43	40	37	34	31	29	26	24				14						23		18			14	10	8	7	7	7	8	8	9		
50	94	89	83	79	74	70	65	61	57	53	49	55	51	54	51	48	46	44	42	40					25							19				13	9	8	8	8	9	9	10		
52	95	90	83	79	74	71	66	62	58	54	51	56	52	54	52	49	47	45	43	41																									
54	95	90	83	79	75	71	66	63	59	55	52	56	53	55	53	50	48	46	44	42																									
56	95	91	83	80	76	72	67	64	60	57	53	58	54	55	53	51	49	46	45	43																									
58	95	91	83	81	77	72	68	64	61	57	54	58	55	56	54	51	49	48	46	44																									
60	95	91	88	81	78	74	69	65	61	58	55	59	56	57	54	52	50	48	47	45																									
62	95	91	88	82	78	74	69	66	62	59	56	59	57	57	55	53	51	49	47	46																									
64	95	91	88	82	79	76	70	67	63	60	57	60	58	58	55	53	51	50	48	47																									
66	96	92	88	84	80	76	71	67	64	61	58	61	58	59	55	54	52	50	49	48																									
68	96	92	88	84	80	77	72	68	65	61	58	61	58	59	56	54	52	51	49	48																									
70	95	91	87	81	81	78	72	69	65	62	59	62	59	59	56	55	53	51	50	49																									
72	95	91	87	82	81	78	73	69	66	63	60	63	60	59	57	56	53	52	51	49																									
74	95	91	87	84	82	79	73	70	66	64	61	63	60	60	57	56	54	52	51	50																									
76	95	92	88	84	83	80	74	70	67	64	61	63	60	60	58	57	54	53	52	51																									
78	96	92	88	84	83	80	74	71	67	65	62	63	61	61	59	57	55	53	52	51																									
80	96	92	89	83	82	78	72	69	65	62	59	62	59	54	51	49										18									19			13						3
82	96	93	89	83	82	79	73	69	66	62	59	62	59	54	52											19									20									4
84	96	93	89	84	83	79	73	70	66	63	60	63	60	55	52											20									21									6
86	96	93	89	85	83	80	74	70	67	63	60	63	60	55	53											21									22									7
88	96	93	89	85	83	80	74	71	67	64	61	63	61	56	53											22									23									8
90	96	93	89	86	84	81	75	71	68	65	64	62	59	56	54											24									24									9
92	96	93	89	86	84	81	75	71	68	65	64	62	59	57	55											25									25									10
94	96	93	89	86	84	81	75	72	69	66	65	62	60	57	55											26									26									11
96	96	93	89	86	85	82	76	72	69	66	65	62	60	58	56											27									27									12
98	96	93	89	86	85	82	76	73	70	67	66	63	60	58	56											28									28									13
100	96	93	89	86	85	82	77	74	71	68	66	63	61	59	56																														

RELATIVE HUMIDITY TABLES
(Continued)

To determine relative humidity from wet and dry bulb temperature readings, subtract the wet-bulb temperature from the dry-bulb temperature and find the number representing this difference in the top row. Follow that column vertically to find the relative humidity at the intersection of the horizontal column representing the dry-bulb reading. Tables are given for Celsius and Fahrenheit readings at sea level.

FOR EXAMPLE:
A dry-bulb reading of 88°F and a wet-bulb reading of 80°F (difference 8°F) indicates a relative humidity of 70%.

DIFFERENCE IN DEGREES FAHRENHEIT BETWEEN WET AND DRY BULB THERMOMETERS.

DRY BULB Thermometer Readings	1	2	3	4	5	6	7	8	9	10	11	12	13	14	15	16	17	18	19	20
100	96	93	89	86	83	80	77	73	70	68	65	62	59	56	54	51	49	46	44	41
98	96	93	89	86	83	79	76	73	70	67	64	61	58	56	53	50	48	45	43	40
96	96	93	89	86	82	79	76	73	69	66	63	61	58	55	52	50	47	44	42	39
94	96	93	89	85	82	79	75	72	69	66	63	60	57	54	51	49	46	43	41	38
92	96	92	89	85	82	78	75	72	68	65	62	59	56	53	50	48	45	42	40	37
90	96	92	89	85	81	78	74	71	68	65	61	58	55	52	49	47	44	41	39	36
88	96	92	88	85	81	77	74	70	67	64	61	57	54	51	48	46	43	40	37	35
86	96	92	88	84	81	77	73	70	66	63	60	57	53	50	47	44	42	39	36	33
84	96	92	88	84	80	76	73	69	66	62	59	56	52	49	46	43	40	37	35	32
82	96	92	88	84	80	76	72	69	65	61	58	55	51	48	45	42	39	36	33	30
80	96	91	87	83	79	75	72	68	64	61	57	54	50	47	44	41	38	35	32	29
79	96	91	87	83	79	75	71	68	64	60	57	53	50	46	43	40	37	34	31	28
78	96	91	87	83	79	75	71	67	63	60	56	53	49	46	43	39	36	33	30	27
77	96	91	87	83	79	74	71	67	63	59	56	52	48	45	42	39	35	32	29	26
76	96	91	87	82	78	74	70	66	62	59	55	51	48	44	41	38	34	31	28	25
75	96	91	86	82	78	74	70	66	62	58	54	51	47	44	40	37	34	30	27	24
74	95	91	86	82	78	74	69	65	61	58	54	50	47	43	39	36	33	29	26	23
73	95	91	86	82	78	73	69	65	61	57	53	50	46	42	39	35	32	29	25	22
72	95	91	86	82	77	73	69	65	61	57	53	49	45	42	38	34	31	28	24	21
71	95	90	86	81	77	72	68	64	60	56	52	48	45	41	37	33	30	27	23	20
70	95	90	86	81	77	72	68	64	59	55	51	48	44	40	36	33	29	25	22	19
69	95	90	85	81	76	72	67	63	59	55	51	47	43	39	35	32	28	24	21	18
68	95	90	85	80	76	71	67	62	58	54	50	46	42	38	34	31	27	23	20	16
67	95	90	85	80	75	71	66	62	58	53	49	45	41	37	33	30	26	22	19	15
66	95	90	85	80	75	71	66	61	57	53	48	44	40	36	32	29	25	21	17	14
65	95	90	85	80	75	66	59	54	49	44	39	35	31	27	24	20	16	12		
64	95	90	84	79	74	70	65	60	56	51	47	43	38	34	30	26	22	18	15	11
63	95	89	84	79	74	69	64	60	55	50	46	41	37	33	29	25	21	17	13	10
62	94	89	84	79	74	69	64	59	54	50	45	41	36	32	28	24	20	16	12	8
61	94	89	84	78	73	68	63	58	53	48	44	40	35	31	27	22	18	14	10	7
60	94	89	83	78	73	68	63	58	53	48	43	39	34	30	26	21	17	13	9	5
59	94	89	83	78	72	67	62	57	52	47	42	38	33	29	24	20	16	11	7	3
58	94	88	83	77	72	66	61	56	51	46	41	37	32	27	23	18	14	10	6	1
57	94	88	82	77	71	66	61	55	50	45	40	35	31	26	22	17	13	8	4	
56	94	88	82	76	71	65	60	55	50	44	39	34	30	25	20	16	11	7	2	
55	94	88	82	76	70	65	59	54	49	43	38	33	28	23	19	14	9	5	0	
54	94	88	82	76	70	64	59	53	48	42	37	32	27	22	17	12	8	3		
53	94	87	81	75	69	63	58	52	47	41	36	31	26	20	16	10	6	1		
52	94	87	81	75	69	63	57	52	46	40	35	29	24	19	14	9	4			
51	94	87	81	75	68	62	56	50	45	39	34	28	23	17	12	7				
50	93	87	80	74	67	61	55	49	43	38	32	27	21	16	10	5	0			
49	93	86	80	73	67	61	54	48	42	36	31	25	19	14	9	3				
48	93	86	79	73	66	60	54	47	41	35	29	23	18	12	6	1				
47	93	86	79	72	66	59	52	46	40	34	28	22	16	10	5					
46	93	86	79	72	65	58	52	45	39	32	26	20	14	8	2					
45	93	86	78	71	64	57	51	44	38	31	25	18	12	6						
44	93	85	78	71	63	56	49	43	36	30	23	16	10	4						
43	92	85	77	70	63	55	48	42	35	28	21	14	8	1						
42	92	85	77	70	62	55	47	41	34	26	19	13	6							
41	92	84	76	69	61	54	46	39	31	24	17	10	3							
40	92	83	75	68	60	52	45	37	29	22	15	7	0							
39	92	83	75	67	59	51	43	35	27	20	12	5								
38	91	83	75	66	58	50	42	33	25	19	10	2								
37	91	83	74	65	57	48	40	31	23	15	7									
36	91	82	73	64	56	46	39	29	21	13	5									
35	91	81	72	63	54	45	36	27	19	10	2									
34	90	81	71	62	52	43	34	25	16	8										
33	90	80	70	60	51	41	32	23	14	5										
32	89	79	69	59	49	39	30	20	11	2										
31	89	78	68	58	47	37	28	18	8											
30	89	78	67	56	46	36	26	16	6											
29	88	77	66	55	44	34	23	13	3											
28	88	76	65	54	43	32	21	10												
27	88	76	64	52	41	29	18	7												
26	87	75	63	51	39	27	16	4												
25	87	74	62	49	37	25	13	1												
24	87	73	60	47	35	22	10													
23	86	72	59	46	33	20	7													
22	86	71	58	44	31	17	4													
21	85	71	56	42	28	15	1													
20	85	70	55	40	26	12														

DRY BULB Thermometer Readings	1	2	3	4	5	6	7	8	9	10	11	12	13	14	15	16	17	18	19	20	21	22	23	24	25	26	27	28	29	30
200	98	96	94	92	90	88	86	84	82	80	79	77	75	74	72	70	69	67	66	64	63	61	60	58	57	55	54	53	52	51
190	98	96	94	92	90	88	85	84	82	80	78	76	75	73	71	69	68	66	65	63	62	60	58	57	56	54	53	51	50	49
188	98	96	94	92	90	87	85	84	82	80	78	76	74	73	71	69	68	66	64	63	62	61	59	58	57	55	54	52	51	50
186	98	96	94	92	90	87	85	83	82	80	78	76	74	72	71	69	67	66	64	62	61	59	58	56	55	53	52	51	49	48
184	98	96	94	92	89	87	85	83	82	79	77	76	74	72	70	69	67	65	64	62	61	59	57	56	55	53	52	50	49	48
182	98	96	94	91	89	87	85	83	81	79	77	75	73	72	70	68	67	65	63	62	60	59	57	56	54	53	51	50	49	48
180	98	96	94	91	89	87	85	83	81	79	77	75	73	72	70	68	67	65	63	62	60	58	57	55	54	52	51	50	48	47
178	98	96	94	91	89	87	85	83	81	79	77	75	73	72	70	68	66	64	63	61	60	58	56	55	53	52	50	49	48	46
176	98	96	94	91	89	87	85	83	81	79	77	75	73	71	70	68	66	64	63	61	60	58	56	55	53	51	50	49	47	46
174	98	96	93	91	89	87	84	83	81	78	76	75	73	71	69	67	66	64	62	61	59	57	56	54	53	51	50	49	47	46
172	98	95	93	91	89	86	84	82	81	78	76	74	73	71	69	67	66	64	62	60	59	57	55	54	53	51	50	48	47	46
170	98	95	93	91	89	86	84	82	80	78	76	74	72	70	69	67	65	63	62	60	59	57	55	54	53	51	51	49	48	47
168	98	95	93	91	88	86	84	82	80	78	76	74	72	70	68	67	65	63	61	60	58	57	55	53	52	50	49	47	46	45
166	98	95	93	91	88	86	84	82	80	78	76	74	72	70	68	66	65	63	61	59	58	56	54	52	51	50	48	47	46	44
164	98	95	93	91	88	86	84	82	80	78	75	73	71	69	68	66	64	62	61	59	58	56	54	52	51	49	48	47	45	44
162	98	95	93	90	88	86	84	82	80	77	75	73	71	69	68	66	64	62	60	59	57	55	53	52	50	49	47	46	45	44
160	98	95	93	90	88	86	83	81	79	77	75	73	71	69	67	65	64	62	60	58	57	55	53	52	50	49	47	46	45	43
158	98	95	93	90	88	86	83	81	79	77	75	73	71	69	67	65	63	61	60	58	56	55	53	51	50	48	47	45	44	43
156	98	95	93	90	88	85	83	81	79	77	74	72	70	68	66	65	63	61	59	57	56	54	53	51	49	48	46	45	44	42
154	98	95	93	90	88	85	83	81	79	77	74	72	70	68	66	65	63	61	59	57	56	54	52	50	49	47	46	44	43	42
152	98	95	93	90	88	85	83	81	79	76	74	72	70	68	66	64	62	60	59	57	55	53	52	50	48	47	46	44	42	41
150	98	95	92	90	87	85	82	80	78	76	74	72	70	68	66	64	62	60	58	57	55	53	51	49	48	46	45	43	42	41
148	97	95	92	90	87	85	82	80	78	76	74	71	69	67	65	63	61	60	58	56	54	53	51	49	48	46	45	43	42	40
146	97	95	92	90	87	85	82	80	78	75	73	71	69	67	65	63	61	59	57	56	54	52	50	49	47	45	44	43	41	40
144	97	95	92	89	87	84	82	80	78	75	73	71	69	67	65	63	61	59	57	55	53	51	50	48	47	45	44	42	41	40
142	97	94	92	89	87	84	82	80	77	75	73	70	68	66	64	62	60	58	57	55	53	51	49	48	46	44	43	42	40	39
140	97	94	92	89	87	84	82	79	77	75	72	70	68	66	64	62	60	58	56	54	52	50	49	47	45	44	42	41	40	38
138	97	94	92	89	87	84	82	79	77	75	72	70	68	66	64	62	60	58	56	54	52	50	48	47	45	43	42	40	39	37
136	97	94	92	89	86	84	81	79	76	74	72	69	67	65	63	61	59	57	55	53	51	50	48	46	45	43	41	39	38	37
134	97	94	92	89	86	84	81	79	76	74	72	69	67	65	63	61	59	57	55	53	51	49	48	46	44	43	41	39	37	36
132	97	94	92	89	86	83	81	78	76	73	71	69	67	64	62	60	58	56	54	52	50	48	47	45	43	41	40	38	37	35
130	97	94	91	89	86	83	81	78	76	73	71	69	67	64	62	60	58	56	54	52	50	48	47	45	43	41	40	38	37	35
128	97	94	91	89	86	83	81	78	76	73	70	68	65	63	61	59	57	55	53	51	49	47	45	44	42	41	39	37	36	34
126	97	94	91	88	86	83	80	78	75	73	70	68	66	64	61	59	57	55	53	51	49	47	45	44	42	40	39	38	36	34
124	97	94	91	88	85	83	80	77	75	72	70	68	65	63	61	59	57	54	52	50	48	46	44	42	41	39	38	36	34	33
122	97	94	91	88	85	82	80	77	74	72	69	67	65	63	60	58	56	54	52	50	48	46	44	43	41	39	38	36	34	32
120	97	94	91	88	85	82	80	77	74	72	69	67	65	62	60	58	55	53	51	49	47	45	43	42	40	38	36	34	33	31
118	97	94	91	88	85	82	79	77	74	71	69	67	64	62	60	58	56	54	52	50	48	46	44	42	40	38	36	34	33	31
116	97	94	91	88	85	82	79	76	74	71	69	66	64	61	59	57	55	53	51	49	47	45	43	41	39	37	35	33	31	29
114	97	94	91	88	85	82	79	76	73	71	68	66	63	61	59	57	55	53	51	49	47	45	43	41	39	37	36	34	32	31
112	97	94	90	87	84	81	79	76	73	70	68	65	63	60	58	56	55	53	51	49	47	44	42	40	38	36	34	33	31	29
110	97	93	90	87	84	81	78	75	73	70	67	65	62	60	57	55	52	50	48	46	44	42	40	38	36	34	33	31	29	27
108	97	93	90	87	84	81	78	75	72	70	67	64	62	59	57	54	52	49	47	45	43	41	39	37	35	33	31	29	27	25
106	97	93	90	87	84	81	78	75	72	69	66	64	61	59	56	54	51	49	47	45	43	41	39	37	35	33	31	29	26	24
104	97	93	90	87	83	80	77	74	71	69	66	63	60	58	55	53	50	48	46	43	41	39	37	35	33	31	29	26	25	23
102	96	93	90	86	83	80	77	74	71	68	65	62	60	57	55	52	49	47	45	42	40	38	36	34	32	30	28	26	24	22
100	96	93	89	86	83	80	77	73	70	68	65	62	59	56	54	51	49	46	44	41	39	37	35	33	30	28	26	24	22	21

TEMPERATURE-HUMIDITY INDEX

The United States Weather Bureau developed the formula for temperature—humidity index. It is based on temperature and relative humidity.

$$THI = 15 + 0.4(T_{dry\ bulb} + T_{wet\ bulb})$$

where temperatures are in degrees Fahrenheit. It has been determined that when the THI reaches 72, some people are uncomfortable; when it reaches 76 most everyone is uncomfortable.

Actually it is the combination of both high temperature and high humidity which causes discomfort. Lowering either one will increase comfort. On the other hand, lower temperature plus low humidity can cause discomfort on the cool side. Thus, in the wintertime, when the humidity in heated buildings is low, a higher temperature is needed for comfort than is required during other seasons when the humidity is higher.

FOR EXAMPLE:
At a dry-bulb temperature of 75°F and a relative humidity of 60%, the THI is 71.

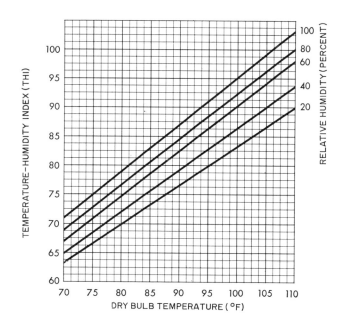

COLOR SCALE OF TEMPERATURE

Commonly used terms to describe the color of heat are related to the approximate range of temperature.

Incipient red heat	500– 550
Dark red heat	650– 750
Bright red heat	800– 900
Orange-red heat	900–1000
Yellow heat	1050–1150
Incipient white heat	1250–1350
White heat	Above 1450

STANDARD ANNEALED COPPER WIRE TABLE

AWG B & S Gauge	Diameter in Mils	Cross Section		Ohms/ 1000 Ft at 20°C (68°F)	Lb/ 1000 Ft	Ft/Lb	Ft/Ohm at 20°C (68°F)	Ohms/Lb at 20°C (68°F)	Lb/Ohm at 20°C (= 68°F)
		Circular Mils	Square Inches						
0000	460.0	211,600	0.1662	0.04901	640.5	1.561	20,400	0.00007652	13,070
000	409.6	167,800	0.1318	0.06180	507.9	1.968	16,180	0.0001217	8,219
00	364.8	133,100	0.1045	0.07793	402.8	2.482	12,830	0.0001935	5,169
0	324.9	105,500	0.08289	0.09827	319.5	3.130	10,180	0.0003076	3,251
1	289.3	83,690	0.06573	0.1239	253.3	3.947	8,070	0.0004891	2,044
2	257.6	66,370	0.05213	0.1563	200.9	4.977	6,400	0.0007778	1,286
3	229.4	52,640	0.04134	0.1970	159.3	6.276	5,075	0.001237	808.6
4	204.3	41,740	0.03278	0.2485	126.4	7.914	4,025	0.001966	508.5
5	181.9	33,100	0.02600	0.3133	100.2	9.980	3,192	0.003127	319.8
6	162.0	26,250	0.02062	0.3951	79.46	12.58	2,531	0.004972	201.1
7	144.3	20,820	0.01635	0.4982	63.02	15.87	2,007	0.007905	126.5
8	128.5	16,510	0.01297	0.6282	49.98	20.01	1,592	0.01257	79.55
9	114.4	13,090	0.01028	0.7921	39.63	25.23	1,262	0.01999	50.03
10	101.9	10,380	0.008155	0.9989	31.43	31.82	1,001	0.03178	31.47
11	90.74	8,234	0.006467	1.260	24.92	40.12	794	0.05053	19.79
12	80.81	6,530	0.005129	1.588	19.77	50.59	629	0.08035	12.45
13	71.96	5,178	0.004067	2.003	15.68	63.80	499.3	0.1278	7.827
14	64.08	4,107	0.003225	2.525	12.43	80.44	396.0	0.2032	4.922
15	57.07	3,257	0.002558	3.184	9.858	101.4	314.0	0.3230	3.096
16	50.82	2,583	0.002028	4.016	7.818	127.9	249.0	0.5136	1.947
17	45.26	2,048	0.001609	5.064	6.200	161.3	197.5	0.8167	1.224
18	40.30	1,624	0.001276	6.385	4.917	203.4	156.6	1.299	0.7700
19	35.89	1,288	0.001012	8.051	3.899	256.5	124.2	2.065	.4843
20	31.96	1,022	0.0008023	10.15	3.092	323.4	98.50	3.283	.3046
21	28.46	810.1	0.0006363	12.80	2.452	407.8	78.11	5.221	.1915
22	25.35	642.4	0.0005046	16.14	1.945	514.2	61.95	8.301	.1205
23	22.57	509.5	0.0004002	20.36	1.542	648.4	49.13	13.20	.07576
24	20.10	404.0	0.0003173	25.67	1.223	817.7	38.96	20.99	.04765
25	17.90	320.4	0.0002517	32.37	0.9699	1,031.0	30.90	33.37	.02997
26	15.94	254.1	0.0001996	40.81	0.7692	1,300	24.50	53.06	.01885
27	14.20	201.5	0.0001583	51.47	0.6100	1,639	19.43	84.37	.01185
28	12.64	159.8	0.0001255	64.90	0.4837	2,067	15.41	134.2	.007454
29	11.26	126.7	0.00009953	81.83	0.3836	2,607	12.22	213.3	.004688
30	10.03	100.5	0.00007894	103.2	0.3042	3,287	9.691	339.2	.002948
31	8.928	79.70	0.00006260	130.1	0.2413	4,145	7.685	539.3	.001854
32	7.950	63.21	0.00004964	164.1	0.1913	5,227	6.095	857.6	.001166
33	7.080	50.13	0.00003937	206.9	0.1517	6,591	4.833	1,364	.0007333
34	6.305	39.75	0.00003122	260.9	0.1203	8,310	3.833	2,168	.0004612
35	5.615	31.52	0.00002476	329.0	0.09542	10,480	3.040	3,448	.0002901
36	5.000	25.00	0.00001964	414.8	0.07568	13,210	2.411	5,482	.0001824
37	4.453	19.83	0.00001557	523.1	0.06001	16,660	1.912	8,717	.0001147
38	3.965	15.72	0.00001235	659.6	0.04759	21,010	1.516	13,860	.00007215
39	3.531	12.47	0.000009793	831.8	0.03774	26,500	1.202	22,040	.00004538
40	3.145	9.888	0.000007766	1049.0	0.02993	33,410	0.9534	35,040	.00002854

Temperature coefficient of resistance: The resistance of a conductor at temperature t in degrees Celsius is given by

$$R_i = R_{20} [1 + a_{20} (t - 20)]$$

where R_{20} is the resistance at 20°C and a_{20} is the temperature coefficient of resistance at 20°C. For copper, $a_{20} = 0.00393$. That is, the resistance of a copper conductor increases approximately 0.4% per degree celsius rise in temperature.

PROPERTIES OF COMMON WIRE
AND CABLE INSULATIONS

Insulation Material	Breakdown Voltage	R. F. Losses	Operating Temp. (°C)	Weather Resistance	Flex- ibility	Suggested Use
Standard PVC	High	Medium	−20 to +80	Good	Fair	General purpose
Premium PVC	High	Medium	−55 to +105	Good	Fair	General purpose
Polyethylene	High	Low	−60 to +80	Good	Good	R. f. cables
Natural rubber	High	High	−40 to +70	Poor	Good	Light duty
Neoprene	Low	High	−30 to +90	Good	Good	Rough service
Waxed cotton	Low	High		Poor	Good	Experimenting
Teflon	High	Low	−70 to +260	Good	Fair	High temperature

TEMPERATURE CLASSIFICATION
OF INSULATING MATERIALS

Temperature Classifications
Definitions of Insulating Materials (IEEE)

Class	Definition	
0	Materials or combinations of materials such as cotton, silk, and paper without impregnation. Other materials or combinations of materials may be included in this class if by experience or accepted tests they can be shown to be capable of operation at	90C
A	Materials or combinations of materials such as cotton, silk, and paper when suitably impregnated or coated or when immersed in a dielectric liquid such as oil. Other materials or combinations of materials may be included in this class if by experience or accepted tests they can be shown to be capable of operation at	105C
B	Materials or combinations of materials such as mica, glass fiber, asbestos, etc., with suitable bonding substances. Other materials or combinations of materials, not necessarily inorganic, may be included in this class if by experience or accepted tests they can be shown to be capable of operation at	130C
F	Materials or combinations of materials such as mica, glass fiber, asbestos, etc., with suitable bonding substances. Other materials or combinations of materials, not necessarily inorganic, may be included in this class if by experience or accepted tests they can be shown to be capable of operation at	155C
H	Materials or combinations of materials such as silicone elastomer, mica, glass fiber, asbestos, etc., with suitable bonding substances such as appropriate silicone resins. Other materials or combinations of materials may be included in this class if by experience or accepted tests they can be shown to be capable of operation at	180C
220C	Materials or combinations of materials which by experience or accepted tests can be shown to be capable of operation at	220C
Over 220C (class C)	Insulation that consists entirely of mica, porcelain, glass, quartz, and similar inorganic materials. Other materials or combinations of materials may be included in this class if by experience or accepted tests they can be shown to be capable of operation at temperatures over	220C

NOTES:

1. Insulation is considered to be "impregnated" when a suitable substance provides a bond between components of the structure and also a degree of filling and surface coverage sufficient to give adequate performance under the extremes of temperature, surface contamination (moisture, dirt, etc.), and mechanical stress expected in service. The impregnant must not flow or deteriorate enough at operating temperature so as to seriously affect performance in service.

2. The electrical and mechanical properties of the insulation must not be impaired by the prolonged application of the limiting insulation temperature permitted for the specific insulation class. The word "impaired" is here used in the sense of causing any change which could disqualify the insulating material for continuously performing its intended function whether creepage spacing, mechanical support, or dielectric barrier action.

3. In the above definitions the words "accepted tests" are intended to refer to recognized Test Procedures established for the thermal evaluation of materials by themselves or in simple combinations. Experience or test data, used in classifying insulating materials, are distinct from the experience or test data derived for the use of materials in complete insulation systems. The thermal endurance of complete systems may be determined by Test Procedures specified by the responsible Technical Committees. A material that is classified as suitable for a given temperature may be found suitable for a different temperature, either higher or lower, by an insulation system Test Procedure. For example, it has been found that some materials suitable for operation at one temperature in air may be suitable for a higher temperature when used in a system operated in an inert gas atmosphere.

4. It is important to recognize that other characteristics, in addition to thermal endurance, such as mechanical strength, moisture resistance and corona endurance, are required in varying degrees in different applications for the successful use of insulating materials.

VOLTAGE-CURRENT-WIRE SIZE NOMOGRAM

This nomogram can be used to determine:
1. The minimum wire size for any given load current and voltage drop;
2. the mV drop/foot for any given wire size and load current;
3. the maximum recommended* current for any given size wire.

*Based on an arbitrary minimum 500 circular mils per ampere. High-temperature class insulation will safely allow higher currents.

FOR EXAMPLE:
1. With a permissible voltage drop of 5 mV/ft, the minimum wire size in a 3-A circuit is #12 AWG.
2. At 300 mA the voltage drop across #22 AWG wire is 4.5 mV/ft.
3. The maximum recommended current for #18 AWG wire is 3.5 A. (This is found by connecting point A on the IR drop scale with the wire gage scale, and reading the intersect point on the Current scale).

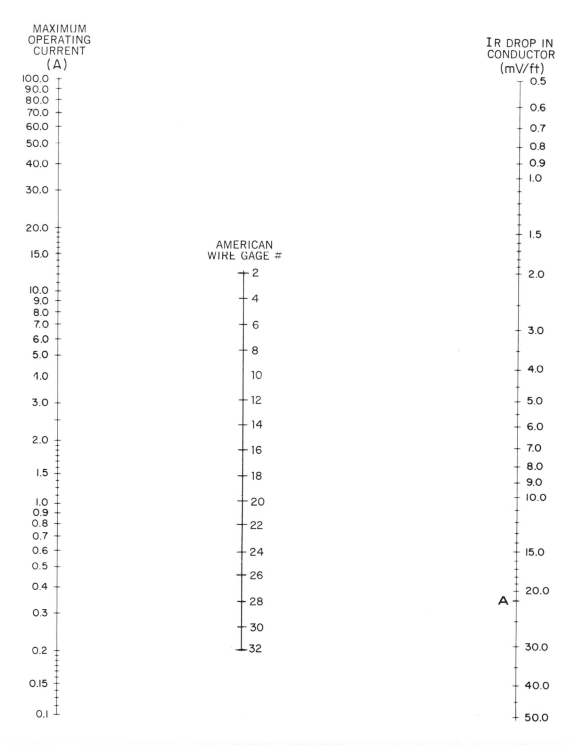

FUSING CURRENTS OF WIRES

This table gives the fusing currents in amperes for five commonly used types of wires. The current I in amperes at which a wire will melt can be calculated from $I = Kd^{3/2}$ where d is the wire diameter in inches and K is a constant that depends on the metal concerned. A wide variety of factors influence the rate of heat loss, and these figures must be considered approximations.

AWG B & S Gauge	d (in.)	Copper K = 10,244	Aluminum K = 7585	German Silver K = 5230	Iron K = 3148	Tin K = 1642
40	0.0031	1.77	1.31	0.90	0.54	0.28
38	0.0039	2.50	1.85	1.27	0.77	0.40
36	0.0050	3.62	2.68	1.85	1.11	0.58
34	0.0063	5.12	3.79	2.61	1.57	0.82
32	0.0079	7.19	5.32	3.67	2.21	1.15
30	0.0100	10.2	7.58	5.23	3.15	1.64
28	0.0126	14.4	10.7	7.39	4.45	2.32
26	0.0159	20.5	15.2	10.5	6.31	3.29
24	0.0201	29.2	21.6	14.9	8.97	4.68
22	0.0253	41.2	30.5	21.0	12.7	6.61
20	0.0319	58.4	43.2	29.8	17.9	9.36
19	0.0359	69.7	51.6	35.5	21.4	11.2
18	0.0403	82.9	61.4	42.3	25.5	13.3
17	0.0452	98.4	72.9	50.2	30.2	15.8
16	0.0508	117	86.8	59.9	36.0	18.8
15	0.0571	140	103	71.4	43.0	22.4
14	0.0641	166	123	84.9	51.1	26.6
13	0.0719	197	146	101	60.7	31.7
12	0.0808	235	174	120	72.3	37.7
11	0.0907	280	207	143	86.0	44.9
10	0.1019	333	247	170	102	53.4
9	0.1144	396	293	202	122	63.5
8	0.1285	472	349	241	145	75.6
7	0.1443	561	416	287	173	90.0
6	0.1620	668	495	341	205	107

This chart shows the maximum length of line that can be used between an amplifier and speaker(s) that would assure that the power loss does not exceed 15% in low-impedance circuits, and 5% in high-impedance circuits.

When several speaker lines are brought separately to an amplifier, calculations must be made for each line independently.

FOR EXAMPLE:
Four 16-ohm speakers are connected in parallel to the 4-ohm tap for perfect impedance match. Line losses are calculated for each line on the basis of the 16-ohm impedance rather than the combined 4-ohm impedance.

Maximum Length of Line for 15% Power Loss—Low Impedance Lines

Wire Size (B and S)	Load Impedance		
	4 ohms	8 ohms	16 ohms
14	125 ft	250 ft	450 ft
16	75 ft	150 ft	300 ft
18	50 ft	100 ft	200 ft
20	25 ft	50 ft	100 ft

Maximum Length of Line for 5% Power Loss—High Impedance Lines

Wire Size (B and S)	Load Impedance		
	100 ohms	250 ohms	500 ohms
14	1000 ft	2500 ft	5000 ft
16	750 ft	1500 ft	3000 ft
18	400 ft	1000 ft	2000 ft
20	250 ft	750 ft	1500 ft

SPARK-GAP BREAKDOWN VOLTAGES

The curves are for a voltage that is continuous or at a frequency low enough to permit complete de-ionization between cycles, between needle points, or clean, smooth, spherical surfaces (electrodes ungrounded) in dust-free clean air. Temperature is 25°C and pressure is 760 mm (29.9 in.) of mercury. Peak kilovolts shown in the graph should be multiplied by the factors given in the table for other atmospheric conditions.

An approximate rule for uniform fields at all frequencies up to at least 300 MHz is that the voltage breakdown gradient of air is 30 peak kV/cm or 75 peak kV/in. at sea level (760 mm of mercury) and normal temperature (25°C). The breakdown voltage is approximately equal to pressure and inversely proportional to absolute (° Kelvin) temperature.

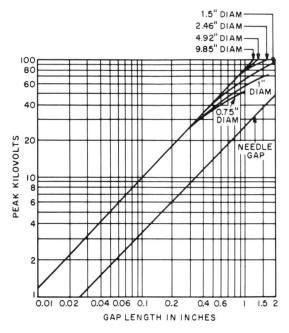

Spark-gap breakdown voltages.

Table of Multiplying Factors

Pressure				Temperature (°C)			
(in. Hg)	(mm Hg)	-40	-20	0	20	40	60
5	127	0.26	0.24	0.23	0.21	0.20	0.19
10	254	0.47	0.44	0.42	0.39	0.37	0.34
15	381	0.68	0.64	0.60	0.56	0.53	0.50
20	508	0.87	0.82	0.77	0.72	0.68	0.64
25	635	1.07	0.99	0.93	0.87	0.82	0.77
30	762	1.25	1.17	1.10	1.03	0.97	0.91
35	889	1.43	1.34	1.26	1.19	1.12	1.05
40	1016	1.61	1.51	1.42	1.33	1.25	1.17
45	1143	1.79	1.68	1.58	1.49	1.40	1.31
50	1270	1.96	1.84	1.73	1.63	1.53	1.44
55	1397	2.13	2.01	1.89	1.78	1.67	1.57
60	1524	2.30	2.17	2.04	1.92	1.80	1.69

CONVERSION TABLE FOR BASIC PHYSICAL UNITS

	CGS-ESU	Multiply by	to get CGS-EMU	Multiply by	to get Rationalized MKS
1. Length	Centimeter	1	Centimeter	10^{-2}	Meter
2. Mass	Gram	1	Gram	10^{-3}	Kilogram
3. Force	Dyne	1	Dyne	10^{-5}	Newton, Dyne-five
4. Energy, Work	Erg	1	Erg	10^{-7}	Joule
5. Power	Erg/second	1	Erg/second	10^{-7}	Watt
6. Electric Charge	Statcoulomb	3.335×10^{-11}	Abcoulomb	10	Coulomb
7. Linear Charge Density	Statcoulomb/cm.	3.335×10^{-11}	Abcoulomb/cm.	10^3	Coulomb/m.
8. Surface Charge Density	Statcoulomb/cm.^{2}	3.335×10^{-11}	Abcoulomb/cm.^{2}	10^5	Coulomb/m.^{2}
9. Volume Charge Density	Statcoulomb/cm.^{3}	3.335×10^{-11}	Abcoulomb/cm.^{3}	10^7	Coulomb/m.^{3}
10. Electric Flux	Statcoulomb	3.335×10^{-11}	Abcoulomb	10	Coulomb
11. Displacement, Electric Flux Density	Statcoulomb/cm.^{2}	3.335×10^{-11}	Abcoulomb/cm.^{2}	10^5	Coulomb/m.^{2}
12. Polarization	Statcoulomb/cm.^{2}	3.335×10^{-11}	Abcoulomb/cm.^{2}	10^5	Coulomb/m.^{2}
13. Electric Dipole Moment	Statcoulomb-cm.	3.335×10^{-11}	Abcoulomb-cm.	10^{-1}	Coulomb-m.
14. Potential	Statvolt	2.998×10^{10}	Abvolt	10^{-8}	Volt
15. Electric Field Intensity	Statvolt/cm.	2.998×10^{10}	Abvolt/cm.	10^{-6}	Volt/m.
16. Current	Statampere	3.335×10^{-11}	Abampere	10	Ampere
17. Surface Current Density	Statampere/cm.	3.335×10^{-11}	Abampere/cm.	10^3	Ampere/m.
18. Volume Current Density	Statampere/cm.^{2}	3.335×10^{-11}	Abampere/cm.^{2}	10^5	Ampere/m.^{2}
19. Resistance	Statohm	8.988×10^{20}	Abohm	10^{-9}	Ohm
20. Resistivity	Statohm-cm.	8.988×10^{20}	Abohm-cm.	10^{-11}	Ohm-m.
21. Conductance	Statmho	1.113×10^{-21}	Abmho	10^9	Mho
22. Conductivity	Statmho/cm.	1.113×10^{-21}	Abmho/cm.	10^{11}	Mho/m.
23. Capacity	Statfarad, Cm.	1.113×10^{-21}	Abfarad	10^9	Farad
24. Elastance	Statdaraf	8.988×10^{20}	Abdaraf	10^{-9}	Daraf
25. Dielectric Constant, Permittivity	—	1.113×10^{-21}	—	$.7958 \times 10^{10}$	Farad/m.
26. Inductance	Stathenry	8.988×10^{20}	Abhenry (Centimeter)	10^{-9}	Henry
27. Permeability	—	8.988×10^{20}	Gauss/Oersted	1.257×10^{-6}	Henry/m.
28. Reluctivity	—	1.113×10^{-21}	Oersted/Gauss	10^7	—
29. Magnetic Charge	—	2.998×10^{10}	Unit Pole	1.257×10^{-7}	Weber
30. Magnetic Flux	—	2.998×10^{10}	Maxwell (Line)	10^{-8}	Weber
31. Magnetic Flux Density, Magnetic Induction	—	2.998×10^{10}	Gauss, lines/cm.^{2}	10^{-4}	Weber/m.^{2}
32. Magnetization	—	2.998×10^{10}	Pole/cm.^{2}	1.257×10^{-3}	Weber/m.^{2}
33. Magnetic Dipole Moment	—	2.998×10^{10}	Pole-cm.	1.257×10^{-9}	Weber-m.
34. Magnetic Field Intensity, Magnetizing Force	—	3.335×10^{-11}	Oersted (Gilbert/cm.) (Gauss)	10^3 / $.7958 \times 10^2$	Praoersted / Ampere-turn/m.
35. Magnetomotive Force	—	3.335×10^{-11}	Gilbert	10 / .7958	Pragilbert / Ampere-turn
36. Reluctance	—	1.113×10^{-21}	Gilbert/Maxwell (Oersted)	10^9 / $.7958 \times 10^8$	Pragilbert/Weber / Ampere-turn/Weber
37. Permeance	—	8.988×10^{20}	Maxwell/Gilbert	10^{-9} / 1.257×10^{-8}	— / Weber/Ampere-turn

Practical System: *Incomplete system similar to MKS, but using centimeters and grams.*
For all Systems: *Temperature is in °C. Time is in seconds.*
For MKS System: *Space Permittivity 8.854×10^{-12} F/m. Space permeability 1.257×10^{-6} H/m.*
Older or obsolete names are shown in parentheses.
To convert CGS-ESU to Rationalized MKS, multiply by both factors.

TERMINAL AND CONTROL MARKINGS ON FOREIGN EQUIPMENT

RADIO-PHONO

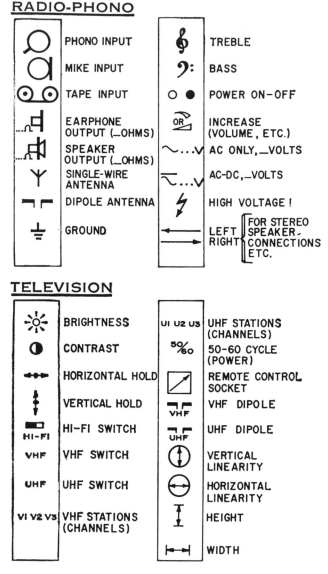

	PHONO INPUT
	MIKE INPUT
	TAPE INPUT
	EARPHONE OUTPUT (_OHMS)
	SPEAKER OUTPUT (_OHMS)
	SINGLE-WIRE ANTENNA
	DIPOLE ANTENNA
	GROUND
	TREBLE
	BASS
	POWER ON-OFF
	INCREASE (VOLUME, ETC.)
	AC ONLY,_VOLTS
	AC-DC,_VOLTS
	HIGH VOLTAGE !
	LEFT RIGHT ⎰FOR STEREO SPEAKER-CONNECTIONS ETC.

TELEVISION

	BRIGHTNESS
	CONTRAST
	HORIZONTAL HOLD
	VERTICAL HOLD
HI-FI	HI-FI SWITCH
VHF	VHF SWITCH
UHF	UHF SWITCH
V1 V2 V3	VHF STATIONS (CHANNELS)
U1 U2 U3	UHF STATIONS (CHANNELS)
50/60	50-60 CYCLE (POWER)
	REMOTE CONTROL SOCKET
VHF	VHF DIPOLE
UHF	UHF DIPOLE
	VERTICAL LINEARITY
	HORIZONTAL LINEARITY
	HEIGHT
	WIDTH

(Reprinted from *Radio Electronics*, copyright Gernsback Publications Inc., September 1964.)

TORQUE-POWER-SPEED NOMOGRAM

This nomogram relates power, torque, and speed.

FOR EXAMPLE:
200 oz-in. at 500 rpm is 0.1 hp, which equals approximately 75 W. The nomogram is based on the formula:

Horsepower = 9.92 × torque × speed × 10^{-7}

where torque is in ounce-inches and speed in revolutions per minute.

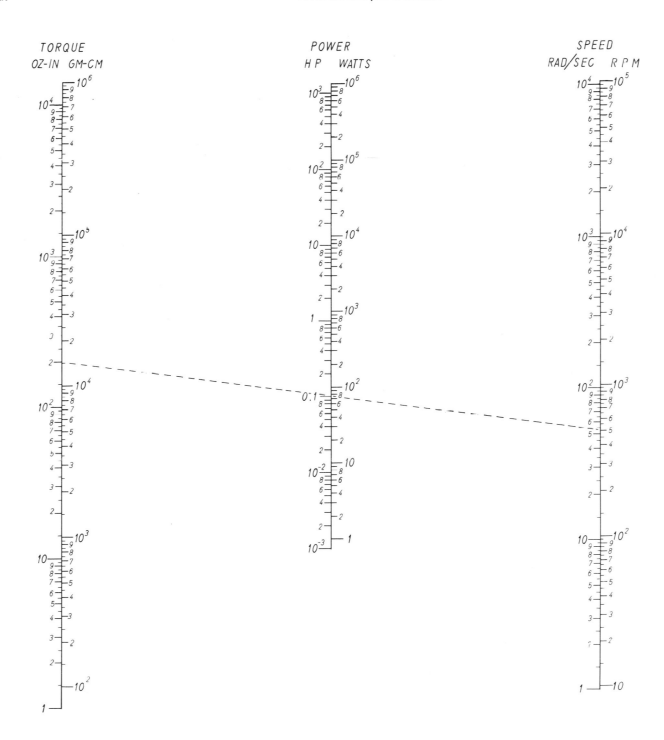

APPROXIMATE FULL-LOAD CURRENT FOR CONTINUOUS-DUTY MOTORS

Direct-Current Motors[a]
(Amperes at Full Load)

HP	115 V	230 V	550 V
1/2	4.6	2.3	
3/4	6.6	3.3	1.4
1	8.6	4.3	1.8
$1\frac{1}{2}$	12.6	6.3	2.6
2	16.4	8.2	3.4
3	24	12	5.0
5	40	20	8.3
$7\frac{1}{2}$	58	29	12.0
10	76	38	16.0
15	112	56	23.0
20	148	74	31
25	184	92	38
30	220	110	46
40	292	146	61
50	360	180	75
60	430	215	90
75	536	268	111
100		355	148
125		443	184
150		534	220
200		712	295

Single-Phase, Alternating-Current Motors[b]
(Amperes at Full Load)

HP	115 V	230 V	440 V
1/6	3.2	1.6	
1/4	4.6	2.3	
1/2	7.4	3.7	
3/4	10.2	5.1	
1	13	6.5	
$1\frac{1}{2}$	18.4	9.2	
2	24	12	
3	34	17	
5	56	28	
$7\frac{1}{2}$	80	40	21
10	100	50	26

For full-load currents of 208- and 200-V motors, increase corresponding 230-V motor full-load current by 10% and 15%, respectively.

[a]These values for full-load current are average for all speeds.
[b]These values of full-load current are for motors running at speeds usual for belted motors and motors with normal torque characteristics. Motors built for especially low speeds or high torques may require more running current, in which case the name plate current rating should be used.

NOMOGRAM RELATING AMPLITUDE, FREQUENCY, AND ACCELERATION OF A BODY WITH SIMPLE HARMONIC MOTION

This nomogram is based on the formula

$$g = 0.10225 \, (d) \, (f)^2$$

where

g = acceleration in g-units

f = frequency of vibration in cps

d = amplitude of vibration (peak displacement each side of resting point) in inches

FOR EXAMPLE:
A vibrating body with a displacement of 0.01 in. each side of center at 200 Hz, has an acceleration of 40 g's.

Note: To find the acceleration in a rotating body resulting from centrifugal force, substitute radius of rotation for amplitude (d), and revolutions per second for vibrations per second (f). g = 32 ft/sec/sec in the MKS system of units.

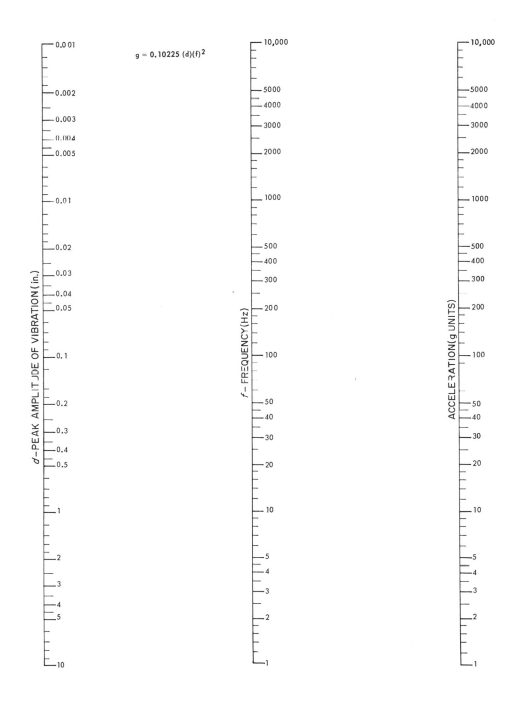

$g = 0.10225 \, (d)(f)^2$

SHOCK DECELERATION
NOMOGRAM

This nomogram relates deceleration (G load), stopping distance, and drop height as an aid to designers and engineers who must deal with problems of shock caused by violent or sudden deceleration.

The equation used to plot the nomograph is $\log G = \log g + \log H - \log D$. Relating deceleration (G load), stopping distance, and drop height, it is based on the following relationships:

$H = gt^2/2$

$D = GT'^2/2$

$V_t = gt$

$V_i = Gt'$

where:

H = free-fall distance

g = acceleration due to free fall

t = free-fall time

D = stopping or deflection distance

G = G load due to impact shock

t' = deceleration time

V_t = terminal velocity due to free fall at instant of impact

V_i = initial deceleration velocity at instant of impact

Since at the moment of impact the terminal velocity (V_t) caused by acceleration is equal to the initial velocity (V_i), it follows that:

$$gt = Gt'$$

Combining the equations:

$$H/D = \frac{gt^2/2}{Gt'^2/2} = gt(t)/Gt'(t')$$

Since $gt = Gt'$, $H/D = t/t'$. Also, since $G/g = t/t'$, $H/D = G/g$. Transposing, $G = g(H/D)$ or $\log G =$ $\log g + \log H - \log D$. This equation is based on a constant or uniformly decelerating force. For linear deceleration the equation for load distance relationship is: $G = 2gH/D$.

Neither formula includes the stopping distance as part of the distance traveled because its effect is negligible for small values of stopping distance (D).

FOR EXAMPLE:

1. Find the G load on a shock-mounted case that endures a 30-in. drop height with a maximum mount deflection of 0.4 in. Assume a rigid case and uniform deceleration in the mount.

ANSWER
Intersect impact shock (G) scale with a line connecting the 30-in. drop height with 0.4 in. on the absorber deflection scale. Read answer off impact shock scale. In this example, it is 73G.

2. Find the impact shock on a piece of equipment that is dropped 20 in. on expanded rubber foam gasket. The foam is compressed a total of 0.1 in. and is assumed to have a linear deceleration characteristic.

ANSWER
Intersect the impact shock (G) scale with a line connecting the 20-in. drop height with 0.1 in. on the absorber deflection scale. Since peak impact shock (G) load due to linear deceleration is approximately twice as severe as that due to uniform deceleration, the value of 200G obtained is multiplied by 2 for linear deflection. Answer is 400G.

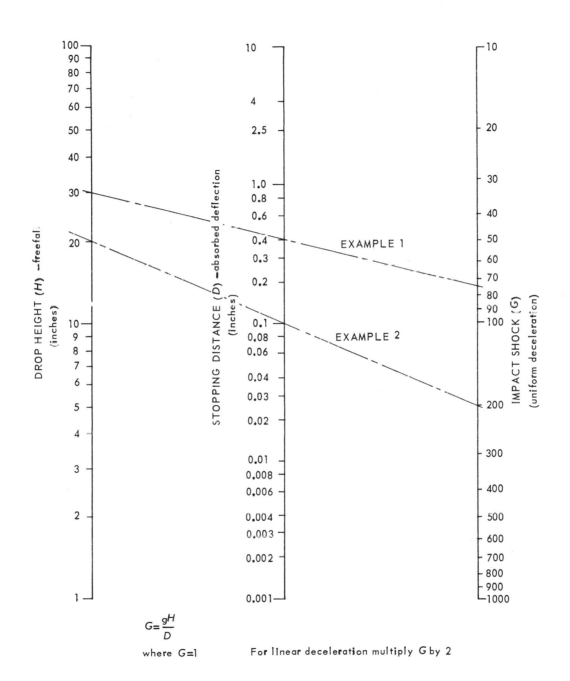

$$G = \frac{gH}{D}$$

where $G=1$ For linear deceleration multiply G by 2

AIR-COOLING NOMOGRAM

For a given power dissipation and air density, this nomogram solves for the air flow (cubic feet per minute) that is required to keep the temperature rise of an equipment at a specified value. At sea level (760 mm Hg), 0°C, and an air density of 0.079 lb/ft³, the temperature rise is approximately equal to 3000 P/Q, where P is power dissipation in kilowatts and Q is the air flow in cubic feet per minute.

To use the nomogram first determine the ambient temperature and altitude at which the equipment must operate and note from the graph the appli-

cable air density for these conditions. On the nomogram align the permissible temperature rise with the equipment's power dissipation and note the intersect point on the turning scale. Align this point with the applicable air density and read required air flow in cubic feet per minute on scale B.

FOR EXAMPLE:

To operate an equipment with a power consumption of 500 W at sea level, an ambient temperature of 20°C, and a permissible heat rise of 15°C, requires an air flow of 50 ft³/min.

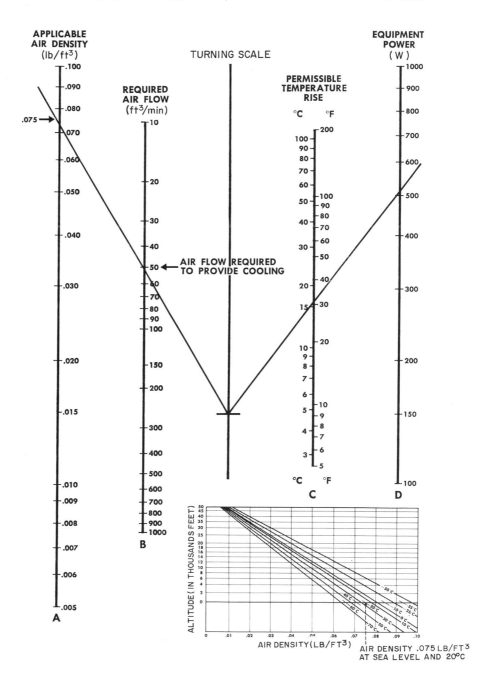

SPECIFIC GRAVITIES AND WEIGHTS

Substance	Specific Gravity	Weight (lb/ft^3)	Substance	Specific Gravity	Weight (lb/ft^3)	Substance	Specific Gravity	Weight (lb/ft^3)
Metals, Alloys, Ores			**Timber, U. S. Seasoned (Continued**			**Earth, Etc., Excavated**		
Aluminum, cast-hammered	2.55-2.75	165	Redwood, California	0.42	26	Clay, dry	—	63
Aluminum, bronze	7.7	481	Spruce, white, black	0.40-0.46	27	Clay, damp, plastic	—	110
Brass, cast-rolled	8.4-8.7	534	Walnut, black	0.61	38	Clay and gravel, dry	—	100
Bronze, 7.9 to 14% Sn	7.4-8.9	509	Walnut, white	0.41	26	Earth, dry, loose	—	76
Copper, cast-rolled	8.8-9.0	556	Moisture Contents:			Earth, dry, packed	—	95
Copper ore, pyrites	4.1-4.3	262	Seasoned timber 15 to 20%.			Earth, moist, loose	—	78
Gold, cast-hammered	19.25-19.3	1205	Green timber up to 50%			Earth, moist, packed	—	96
Iron, cast, pig	7.2	450				Earth, mud, flowing	—	108
Iron, wrought	7.6-7.9	485	**Various Liquids**			Earth, mud, packed	—	115
Iron, steel	7.8-7.9	490	Alcohol, 100%	0.79	49	Riprap, limestone	—	80-85
Iron, spiegel-eisen	7.5	468	Acids, muriatic 40%	1.20	75	Riprap, sandstone	—	90
Iron, ferro-silicon	6.7-7.3	437	Acids, nitric 91%	1.50	94	Riprap, shale	—	105
Iron ore, hematite	5.2	325	Acids, sulphuric 87%	1.80	112	Sand, gravel, dry, loose	—	90-105
Iron ore, hematite in bank	—	160-180	Lye, soda 66%	1.70	106	Sand, gravel, dry, packed	—	100-120
Iron ore, hematite loose		130-160	Oils, vegetable	0.91-0.94	58	Sand, gravel, dry, wet	—	118-120
Iron ore, limonite	3.6-4.0	237	Oils, mineral, lubricants	0.90-0.93	57	**Excavation in Water**		
Iron ore, magnetite	4.9-5.2	315	Water, 4°C, max. density	1.0	62.428	Sand or gravel	—	60
Iron slag	2.5-3.0	172	Water, 100°C	0.9584	59.830	Sand or gravel and clay	—	65
Lead	11.37	710	Water, ice	0.88-0.92	56	Clay	—	80
Lead ore, galena	7.3-7.6	465	Water, snow, fresh fallen	.125	8	River mud	—	90
Manganese	7.2-8.0	475	Water, sea water	1.02-1.03	64	Soil	—	70
Manganese ore, pyrolusite	3.7-4.6	259	**Gases, Air – 1**			Stone riprap	—	65
Mercury	13.6	849	Air, 0°C, 760 mm	1.0	.08071	**Minerals**		
Nickel	8.9-9.2	565	Ammonia	0.5920	.0478	Asbestos	2.1-2.8	153
Nickel, monel metal	8.8-9.0	556	Carbon dioxide	1.5291	.1234	Barytes	4.50	281
Platinum, cast-hammered	21.1-21.5	1330	Carbon monoxide	0.9673	.0781	Basalt	2.7-3.2	184
Silver, cast-hammered	10.4-10.6	656	Gas, illuminating	0.35-0.45	.028-.036	Bauxite	2.55	159
Tin, cast-hammered	7.2-7.5	459	Gas, natural	0.47-0.48	.038-.039	Borax	1.7-1.8	109
Tin ore, cassiterite	6.4-7.0	418	Hydrogen	0.0693	.00559	Chalk	1.8-2.6	137
Zinc, cast-rolled	6.9-7.2	440	Nitrogen	0.9714	.0784	Clay, marl	1.8-2.6	137
Zinc ore, blende	3.9-4.2	253	Oxygen	1.1056	.0892	Dolomite	2.9	181
Various Solids			**Ashlar Masonry**			Feldspar, orthoclase	2.5-2.6	159
Cereals, oats, bulk	—	32	Granite, syenite, gneiss	2.3-3.0	165	Gneiss, serpentine	2.4-2.7	159
Cereals, barley, bulk	—	39	Limestone, marble	2.3-2.8	160	Granite, syenite	2.5-3.1	175
Cereals, corn, rye, bulk	—	48	Sandstone, bluestone	2.1-2.4	140	Greenstone, trap	2.8-3.2	187
Cereals, wheat, bulk		48	**Mortar Rubble Masonry**			Gypsum, alabaster	2.3-2.8	159
Hay and Straw, bales	—	20	Granite, syenite, gneiss	2.2-2.8	155	Hornblende	3.0	187
Cotton, Flax, Hemp	1.47-1.50	93	Limestone, marble	2.2-2.6	150	Limestone, marble	2.5-2.8	165
Fats	0.90-0.97	58	Sandstone, bluestone	2.0-2.2	130	Magnesite	3.0	187
Flour, loose	0.40-0.50	28	**Dry Rubble Masonry**			Phosphate rock, apatite	3.2	200
Flour, pressed	0.70-0.80	47	Granite, syenite, gneiss	1.9-2.3	130	Porphyry	2.6-2.9	172
Glass, common	2.40-2.60	156	Limestone, marble	1.9-2.1	125	Pumice, natural	0.37-0.90	40
Glass, plate or crown	2.45-2.72	161	Sandstone, bluestone	1.8-1.9	110	Quartz, flint	2.5-2.8	165
Glass, crystal	2.90-3.00	184	**Brick Masonry**			Sandstone, bluestone	2.2-2.5	147
Leather	0.86-1.02	59	Pressed brick	2.2-2.3	140	Shale, slate	2.7-2.9	175
Paper	0.70-1.15	58	Common brick	1.8-2.0	120	Soapstone, talc	2.6-2.8	169
Potatoes, piled	—	42	Soft brick	1.5-1.7	100	**Stone, Quarried, Piled**		
Rubber, caoutchouc	0.92-0.96	59	**Concrete Masonry**			Basalt, granite, gneiss	—	96
Rubber goods	1.0-2.0	94	Cement, stone, sand	2.2-2.4	144	Limestone, marble, quartz	—	95
Salt, granulated, piled	—	48	Cement, slag, etc.	1.9-2.3	130	Sandstone	—	82
Saltpeter	—	67	Cement, cinder, etc.	1.5-1.7	100	Shale	—	92
Starch	1.53	96	**Various Building Material**			Greenstone, hornblende	—	107
Sulphur	1.93-2.07	125	Ashes, cinders	—	40-45	**Bituminous Substances**		
Wool	1.32	82	Cement, portland, loose	—	90	Asphaltum	1.1-1.5	81
Timber, U. S. Seasoned			Cement, portland, set	2.7-3.2	183	Coal, anthracite	1.4-1.7	97
Ash, white-red	0.62-0.65	40	Lime, gypsum, loose	—	53-64	Coal, bituminous	1.2-1.5	84
Cedar, white-red	0.32-0.38	22	Mortar, set	1.4-1.9	103	Coal, lignite	1.1-1.4	78
Chestnut	0.66	41	Slags, bank slag	—	67-72	Coal, peat, turf, dry	0.65-0.85	47
Cypress	0.48	30	Slags, bank screenings	—	98-117	Coal, charcoal, pine	0.28-0.44	23
Fir, Douglas spruce	0.51	32	Slags, machine slag	—	96	Coal, charcoal, oak	0.47-0.57	33
Fir, eastern	0.40	25	Slags, slag sand	—	49-55	Coal, coke	1.0-1.4	75
Elm, white	0.72	45				Graphite	1.9-2.3	131
Hemlock	0.42-0.52	29				Paraffine	0.87-0.91	56
Hickory	0.74-0.84	49				Petroleum	0.87	54
Locust	0.73	46				Petroleum, refined	0.79-0.82	50
Maple, hard	0.68	43				Petroleum, benzine	0.73-0.75	46
Maple, white	0.53	33				Petroleum, gasoline	0.66-0.69	42
Oak, chestnut	0.86	54				Pitch	1.07-1.15	69
Oak, live	0.95	59				Tar, bituminous	1.20	75
Oak, red, black	0.65	41				**Coal and Coke, Piled**		
Oak, white	0.74	46				Coal, anthracite	—	47-58
Pine, Oregon	0.51	32				Coal, bituminous, lignite	—	40-54
Pine, red	0.48	30				Coal, peat, turf	—	20-26
Pine, white	0.41	26				Coal, charcoal	—	10-14
Pine, yellow, long-leaf	0.70	44				Coal, coke	—	23-32
Pine, yellow, short-leaf	0.61	38						
Poplar	0.48	30						

Note: The specific gravities of solids and liquids refer to water at 4°C, those of gases to air at 0°C and 760 mm pressure. The weights per cubic foot are derived from average specific gravities, except where stated that weights are for bulk, heaped, or loose material, etc.

The term solder alloys covers a broad range of materials with greatest emphasis placed on compositions of tin and lead. The tin lead system of alloys has a general solidus temperature of 361°F. The eutectic composition, the alloy with a single sharp melting point and no plastic range, is 63% tin, 37% lead. This alloy is in widest use in the electronic industry.

The specific tin lead alloy selected is determined by the nature of the joining operation and the degree to which a plastic or "mushy" solder state can be tolerated or is desirable. Tin lead alloys with a tin content from 20% up through and including 97.5% have the same 361°F solidus line. Alloys containing lower percentages of tin have an increased solidus temperature. This is also true of tin antimony, tin silver, and lead silver alloys. The higher solidus line permits operation of the soldered part in higher ambient temperatures. It also permits sequential or piggy-back soldering. Where two soldering connections are to be made in areas very close to each other, the first joint can be made with one of the high-temperature alloys. When the second joint is made with an alloy in the normal tin lead system, the first joint will not be disturbed.

Solder Alloy Chart

Percent Tin	Percent Lead	Percent Silver	Percent Antimony	Temperature at which Solder Becomes Plastic °C	°F	Temperature at which Solder Becomes Liquid °C	°F
0	100					327	621
5	95			300	572	315	599
10	90			267.5	514	300	572
15	85			223	433	290	554
20	80			183	361	280	536
25	75			183	361	267	513
30	70			183	361	255	491
35	65			183	361	245	473
40	60			183	361	235	455
45	55			183	361	223	433
50	50			183	361	212	414
55	45			183	361	200	392
60	40			183	361	189	372
63	37			eutectic alloy[a]		183	361
65	35			183	361	186	367
70	30			183	361	191	376
75	25			183	361	195	383
80	20			183	361	201	394
85	15			183	361	207	404
90	10			183	361	214	417
95	5			183	361	222	432
97.5	2.5			183	361	227	441
100	0					232	450
35	63		2	187	369	237	459
20	78.7	1.3		181	358	276	529
27	70	3		178	352	253	487
	95	5		305	581	360	680

[a]A eutectic alloy is that composition of two or more metals that has one sharp melting point and no plastic range.

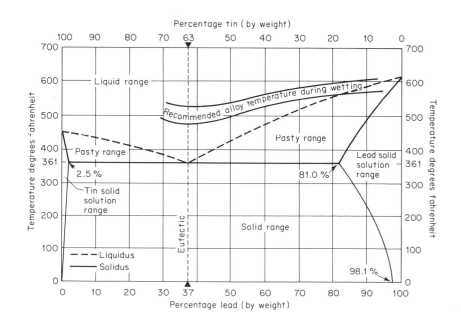

VELOCITY OF SOUND IN SOLIDS, GASES, AND LIQUIDS

Velocity of Sounds in Solids

Medium	Velocity (ft/sec)	Medium	Velocity (ft/sec)
Aluminum	17,192	Magnesium	16,079
Brass	11,221	Nickel	15,615
Cadmium	7,874	Quartz Glass	17,618
Copper	11,745	Silver	8,661
Cork	1,640	Steel	16,569
Iron	16,962	Tin	8,957

Velocity of Sound in Gases at 0°C

Medium	Symbol	Velocity (ft/sec)
Air		1,087
Ammonia	NH_3	1,361
Argon	A	1,046
Carbon Monoxide	CO	1,106
Carbon Dioxide	CO_2	881 (above 100 Hz)
Chlorine	CL	674
Ethylene	C_2H_4	1,040
Helium	He	3,182
Hydrogen	H_2	4,165
Methane	CH_4	1,417
Neon	Ne	1,427
Nitric Oxide	NO	1,066
Nitrous Oxide	N_2O	859
Nitrogen	N_2	1,096
Oxygen	O_2	1,041

Velocity of Sounds in Liquids

Medium	Temperature (°C)	Velocity (ft/sec)
Water (fresh)	17	4,691
Water (sea)	15	4,937
Alcohol (ethyl)	20	3,838
Benzene	20	4,330
Ether (ethyl)	20	3,313
Glycerin	20	6,299
Mercury	20	4,757

DEFINED VALUES AND
PHYSICAL CONSTANTS

A consistent set of physical values has been adapted by the National Bureau of Standards. The values presented below are at least as accurate as any others available, and have the advantage of being self-consistent, thus preventing the necessity of having to make a choice between different answers derived in different ways.

Table 1—Defined Values and Equivalents

Meter	(m)	1 650 763.73 wavelengths *in vacuo* of the unpertubed transition $2p_{10} - 5d_5$ in ^{86}Kr
Kilogram	(kg)	mass of the international kilogram at Sèvres, France
Second	(s)	1/31 556 925.974 7 of the tropical year at 12^h ET 0 January 1900
Degree Kelvin	($^\circ$K)	defined in the thermodynamic scale by assigning 273.16 $^\circ$K to the triple point of water (freezing point, 273.15 $^\circ$K = 0 $^\circ$C)
Unified atomic mass unit	(u)	1/12 the mass of an atom of the ^{12}C nuclide
Mole	(mol)	amount of substance containing the same number of atoms as 12 g of pure ^{12}C
Standard acceleration of free fall	(g_n)	9.806 65 m s^{-2}, 980.665 cm s^{-2}
Normal atmospheric pressure	(atm)	101 325 N m^{-2}, 1 013 250 dyn cm^{-2}
Thermochemical calorie	(cal$_{th}$)	4.1840 J, 4.1840 $\times$ 10^7 erg
International Steam Table calorie	(cal$_{IT}$)	4.1868 J, 4.1868 $\times$ 10^7 erg
Liter	(l)	0.001 000 028 m^3, 1 000.028 cm^3 (recommended by CIPM, 1950)
Inch	(in.)	0.0254 m, 2.54 cm
Pound (avdp.)	(lb)	0.453 592 37 kg, 453,592 37 g

(From *Electronics and Communications,* February 1964.)

Table 2—Physical Constants

Constant	Symbol	Value	Est.[a] Error Limit	Unit Système International (MKSA)		Unit Centimeter-Gram-Second (CGS)	
Speed of light in vacuum	c	2.997925	3	$\times 10^8$	m s^{-1}	$\times 10^{10}$	cm s^{-1}
Elementary charge	e	1.60210	7	10^{-19}	C	10^{-20}	cm$^{1/2}$g$^{1/2}$ [b]
		4.80298	20	—	—	10^{-10}	cm$^{3/2}$g$^{1/2}$s^{-1} [c]
Avogadro constant	N_A	6.02252	28	10^{23}	mol^{-1}	10^{23}	mol^{-1}
Electron rest mass	m_e	9.1091	4	10^{-31}	kg	10^{-28}	g
Proton rest mass	m_p	1.67252	8	10^{-27}	kg	10^{-24}	g
Neutron rest mass	m_n	1.67482	8	10^{-27}	kg	10^{-24}	g
Faraday constant	F	9.64870	16	10^4	C mol^{-1}	10^3	cm$^{1/2}$g$^{1/2}$mol^{-1} [b]
Planck constant	h	6.6256	5	10^{-34}	J s	10^{-27}	erg s
	$\hbar/2\pi$	1.05450	7	10^{-34}	J s	10^{-27}	erg s
Fine structure constant	α	7.29720	10	10^{-3}	—	10^{-3}	
Charge to mass ratio for electron	e/m_e	1.758796	19	10^{11}	C kg^{-1}	10^7	cm$^{1/2}$g$^{1/2}$ [b]
		5.27274	6	—	—	10^{17}	cm$^{3/2}$g$^{-1/2}$s^{-1} [c]
Rydberg constant	R_∞	1.0973731	3	10^7	m^{-1}	10^5	cm^{-1}
Bohr radius	a_0	5.29167	7	10^{-11}	m	10^{-9}	cm
Gyromagnetic ratio of proton	$\gamma\prime$	2.67519	2	10^8	rad s^{-1}T^{-1}	10^4	rad s^{-1}G^{-1} [b]
(uncorrected for diamagnetism, H$_2$O)	γ	2.67512	2	10^8	rad s^{-1}T^{-1}	10^4	rad s^{-1}G^{-1} [b]
Bohr magneton	μ_B	9.2732	6	10^{-24}	J T^{-1}	10^{-21}	erg G^{-1} [b]
Gas constant	R	8.3143	12	10^0	J °K^{-1} mol^{-1}	10^7	erg °K^{-1} mol^{-1}
Normal volume perfect gas	V_0	2.24136	30	10^{-2}	m^3 mol^{-1}	10^4	cm^3 mol^{-1}
Boltzmann constant	k	1.38054	18	10^{-23}	J °K^{-1}	10^{-16}	erg °K^{-1}
First radiation constant ($2\pi hc^2$)	c_1	3.7405	3	10^{-16}	W m^2	10^{-5}	erg cm^2 s^{-1}
Second radiation constant	c_2	1.43879	19	10^{-2}	m °K	10^0	cm °K
Wien displacement constant	b	2.8978	4	10^{-3}	m °K	10^{-1}	cm °K
Stefan Boltzmann constant	σ	5.6697	29	10^{-8}	W m^{-2} °K^{-4}	10^{-5}	erg cm^{-2}s^{-1} °K^{-4}
Gravitational constant	G	6.670	15	10^{-11}	N m^2 kg^{-2}	10^{-8}	dyn cm^2 g^{-2}

[a]Based on 3 std. dev, applied to last digits in preceding column.
[b]Electromagnetic system.
[c]Electrostatic system.
C—coulomb; J—joule; Hz—hertz; W—watt; N—newton; T—tesla; G—gauss.

(From *Electronics and Communications*, February 1964.)

INDEX

311

58685   WITHDRAWAL